Virtual Reality in Geography

Virtual Reality in Geography

Edited by Peter Fisher and David Unwin

London and New York

First published 2002 by Taylor & Francis
11 New Fetter Lane, London EC4P 4EE

Simultaneously published in the USA and Canada
by Taylor & Francis Inc,
29 West 35th Street, New York, NY 10001

Taylor & Francis is an imprint of the Taylor & Francis Group

Typeset in Times by Wearset Ltd, Boldon, Tyne and Wear
Printed and bound in Great Britain by Biddles Ltd, Guildford and King's Lynn

British Library Cataloguing in Publication Data
A catalogue record for this book is available from the British Library

Library of Congress Cataloging in Publication Data
Unwin, D. (David John)
Virtual reality in geography / David Unwin and Peter Fisher.
p. cm.
1. Geography–Computer simulation. 2. Virtual reality.
I. Fisher, Peter. II. Title.
G70.28 .U69 2001
910'.285'6–dc21 2001027965

ISBN 0-7484-0905-X

Contents

Contributors

David Arnold, Information Systems, University of East Anglia, Norwich, NR4 7TJ, United Kingdom. E-mail: dba@sys.uea.ac.uk

Michael Batty, Centre for Advanced Spatial Analysis (CASA), University College London, 1–19 Torrington Place, London, WC1E 6BT, United Kingdom. E-mail: m.batty@ucl.ac.uk

Ken Brodlie, School of Computing, University of Leeds, Leeds, LS2 9JT, United Kingdom. E-mail: kwb@comp. leeds.ac.uk

Iain M. Brown, Environmental Change Institute, University of Oxford, 12–16 St Michael's Street, Oxford, OX1 2DU, United Kingdom. E-mail: iain.brown@ukcip.org.uk

Jo Cheesman, School of Geography, University of Manchester, Mansfield Cooper Building, Oxford Road, Manchester, M13 9PL, United Kingdom. E-mail: j.cheesman@man.ac.uk

Weiso Chen, Macaulay Land Use Research Institute, Craigiebuckler, Aberdeen, AB15 8QH, United Kingdom. E-mail: w.chen@ mluri.sari.ac.uk

Dick Cobb, Schools of Environmental Sciences, University of East Anglia, Norwich, NR4 7TJ, United Kingdom. E-mail: D.Cobb@uea.ac.uk

Martin Dodge, Centre for Advanced Spatial Analysis, University College London, 1–19 Torrington Place, London, WC1E 6BT, United Kingdom. E-mail: m.dodge@ucl.ac.uk.

Paul Dolman, School of Environmental Sciences, University of East Anglia, Norwich, NR4 7TJ, United Kingdom. E-mail: P.Dolman@ uea.ac.uk

Steve Dowers, Department of Geography, University of Edinburgh, Drummond St, Edinburgh, EH8 9XP, Scotland. E-mail: sd@geo.ed.ac.uk

Roger A. Dunham, The Forestry Commission, Northern Research Station, Roslin, Midlothian, EH25 9SY, Scotland. Also Macaulay Land Use

Research Institute, Craigiebuckler, Aberdeen, AB15 8QH, Scotland. E-mail: r.dunham@mluri.sari.ac.uk

Jason Dykes, Department of Information Science, City University, Northampton Square, London, EC1V 0HB, United Kingdom. E-mail: jad7@soi.city.ac.uk

Nuha El-Khalili, University of Petra, Amman, Jordan.

David Fairbairn, Department of Geomatics, University of Newcastle upon Tyne, Newcastle, NE1 7RU, United Kingdom. E-mail: Dave.Fairbairn@ncl.ac.uk

Peter Fisher, Department of Geography, University of Leicester, Leicester, LE1 7RH, United Kingdom. E-mail: pete.fisher@le.ac.uk

A. John W. Gerrard, Department of Geography, University of Birmingham, Birmingham, United Kingdom. E-mail: gerraajw@novell9.bham.ac.uk

Mark Gillings, School of Archaeological Studies, University of Leicester, Leicester, LE1 7RH, United Kingdom. E-mail: mg41@le.ac.uk

Reginald Golledge, Department of Geography, University of California at Santa Barbara, Santa Barbara, CA 93106, USA. E-mail: golledge@geog.ucsb.edu

Mordechay E. Haklay, Centre for Advanced Spatial Analysis, University College London, 1–19 Torrington Place, London, WC1E 6BT, United Kingdom. E-mail: m.haklay@ucl.ac.uk

Francis Harvey, Department of Geography, University of Kentucky, Lexington, KY 40506-0027, USA. E-mail: fharvey@pop.uky.edu

R. Daniel Jacobson, Department of Geography, Florida State University, Tallahassee, FL 32306, USA. E-mail: djacobson@geog.ucsb.edu

Richard Kennaway, Information Systems, University of East Anglia, Norwich, NR4 7TJ, United Kingdom. E-mail: jrk@sys.uea.ac.uk

David B. Kidner, University of Glamorgan, GIS Research Centre, School of Computing, Pontypridd, Wales, CF37 1DL, United Kingdom. E-mail: dbkidner@glam.ac.uk

Rob Kitchin, Department of Geography, National University of Ireland, Maynooth, County Kildare, Ireland. E-mail: Rob.Kitchin@may.ie

Menno-Jan Kraak, Division of Geoinformatics, Cartography and Visualization, ITC, PO Box 6, 7500 AA Enschede, The Netherlands. E-mail: kraak@itc.nl

Andrew Lovett, School of Environmental Sciences, University of East

Anglia, Norwich, NR4 7TJ, United Kingdom. E-mail: A.Lovett@uea.ac.uk

William Mackaness, Department of Geography, University of Edinburgh, Drummond St, Edinburgh, EH8 9XP, Scotland. E-mail: wam@geo.ed.ac.uk

David R. Miller, Macaulay Land Use Research Institute, Craigiebuckler, Aberdeen, AB15 8QH, United Kingdom. E-mail: d.miller@mluri.sari.ac.uk

Kate E. Moore, Department of Geography, University of Leicester, Leicester, LE1 7RH, United Kingdom. E-mail: mek@le.ac.uk

Cliff Ogleby, Department of Geomatics, The University of Melbourne, Parkville, 3052, Australia. E-mail: c.ogleby@eng.unimelb.edu.au

Tim O'Riordan, School of Environmental Sciences, University of East Anglia, Norwich, NR4 7TJ, United Kingdom. E-mail: T.Oriordan@uea.ac.uk

Chris Perkins, School of Geography, University of Manchester, Mansfield Cooper Building, Oxford Road, Manchester, M13 9PL, United Kingdom. E-mail: c.perkins@man.ac.uk

Ross Purves, Department of Geography, University of Edinburgh, Drummond St, Edinburgh, EH8 9XP, Scotland. E-mail: rsp@geo.ed.ac.uk

Jonathan Raper, Department of Information Science, City University, London, EC1V 0HB, United Kingdom. E-mail: jraper@soi.city.ac.uk

Andy Smith, Centre for Advanced Spatial Analysis (CASA), University College London, 1–19 Torrington Place, London, WC1E 6BT, United Kingdom. E-mail: asmith@geog.ucl.ac.uk

Gilla Süennenberg, Schools of Environmental Sciences, University of East Anglia, Norwich, NR4 7TJ, United Kingdom. E-mail: G.Sunnenberg@uea.ac.uk

George Taylor, Department of Geomatics, University of Newcastle upon Tyne, Newcastle, NE1 7RU, United Kingdom. E-mail: George.Taylor@ncl.ac.uk

David Unwin, School of Geography, Birkbeck College, 7–15 Gresse St, London, WIP 1PY, United Kingdom. E-mail: d.unwin@bbk.ac.uk

J. Mark Ware, University of Glamorgan, GIS Research Centre, School of Computing, Pontypridd, Wales, CF37 1DL, United Kingdom. E-mail: jmware@glam.ac.uk

Jo Wood, Department of Information Science, City University, London, EC1V 0HB, United Kingdom. E-mail: jwo@soi.city.ac.uk

Acknowledgements

Thanks are due to Taylor & Francis and Anne Mumford (Advisory Group on Computer Graphics) for having the confidence to help fund the meeting, and to colleagues on the Virtual Field Course project for stimulating interactions over the years. Any errors are, of course, the faults of the editors.

1 Virtual reality in geography

An introduction

Peter Fisher and David Unwin

A question of definition?

This is a book about virtual reality (VR) in geography. We assume that most people reading it will have an understanding of the term 'geography', but unfortunately although, and perhaps because, 'virtual reality' is a very trendy term, neither 'virtual' nor 'reality' is either well defined or strictly appropriate. For example, much more helpful would be the earlier notion of 'alternative reality' or the more recent one of 'virtual worlds'.

In preparing this book we struggled to find a definition of VR that met the aspirations of those involved in editing and writing the major chapters. The most acceptable definition that could be agreed on was that:

> Virtual reality is the ability of the user of a constructed view of a limited digitally-encoded information domain to change their view in three dimensions causing update of the view presented to any viewer, especially the user.

This definition is catholic, but it is one which encapsulates the essence of all visualizations that can legitimately be called virtual reality. More prosaically, it also brings in all the participants in this workshop and others into a unified framework.

We suspect that to many people VR is associated with games parlours, films, etc., involving many and varied but always very specialised computer technology using hardware such as headsets, VR theatres, caves, data-gloves and other haptic feedback devices. All of these are associated with experiences where users are immersed in a seemingly real world that may be entirely artificial or apparently real. This is the high-technology, immersive image of VR, but it seems that:

- there have been widely-reported ergonomic issues from using these immersive environments such as feelings of user disorientation and nausea,
- producers have an enormous ability to deliver the necessary information

for virtual worlds to the public using the simpler technology of the World Wide Web, and
- immersive equipment remains expensive to the average user.

In practice, therefore, and in common with much of the current writing on VR, most of the applications described in this book make use of a 'through-the-window' approach using conventional interaction methods, basic, if top-of-the-range, desktop computers, and industry-standard operating systems.

VR is of enormous commercial importance in the computer games industry and similar technology is often used in training simulations. Currently, many academic disciplines are involved in developing and using virtual worlds and this list includes geography (Brown, 1999; Câmara and Raper, 1999; Martin and Higgs, 1997). The most obvious geographical applications are in traditional cartography, for example, in creating navigable, computer-generated block diagrams, and especially in the same discipline re-invented as scientific visualization (Cartwright *et al.*, 1999; Hearnshaw and Unwin, 1994). Second, because of the importance of the spatial metaphor in those worlds, basic concepts of cartographic visualization of the world are fundamental to our ability to navigate and negotiate almost any applications of VR. Basic geographical concepts have thus much to contribute to the more general world of VR. Third, some geographers are developing concepts that extend VR environments into completely artificial realms such as abstract data realms (Harvey, Chapter 22) and even completely imaginary, but interesting, *AlphaWorlds* (Dodge, Chapter 21).

Why this book now?

Our purposes in convening the expert workshop on which this book is based were three:

- To highlight the fact that geography in general, and cartography in particular, has much to offer those developing VR environments. This is a result of the historical concern in these disciplines with the rendering and communication of spatial information.
- To demonstrate that VR is a technology that overtly requires the construction of a world to be explored. In so doing, it necessitates a statement of how that world is constructed, and therefore makes clear, at least to informed users, the multiple representations which may have been possible for the same subject. In many ways VR relates back to traditional concerns within geography as to the nature of representation, the primacy of particular representations, and the tensions between alternative constructions which we share with colleagues.

- To review current state-of-the-art work in the field of VR applications in the spatial sciences.

These purposes are underscored by an entry in the Benchmarking Statement for Geography in Higher Education prepared for the Quality Assurance Agency for Higher Education in England, Wales and Northern Ireland which states that:

> Geographers should show knowledge and critical understanding of the diversity of forms of representations of the human and physical worlds. Maps are one important form of representation of the world, and geographers should be conversant with their basic cartographic, interpretational and social dimensions. However, geographers should show a similar depth of understanding of other representational forms, including texts, visual images and digital technologies, particularly geographic information systems (GIS).

and, one might add, virtual reality.

To meet these objectives, this book has been structured into four major sections:

- positioning VR in technology and computing,
- explorations of virtual natural environments,
- constructions of virtual cities,
- other virtual worlds which may be imagined, or constructed as windows onto unreal worlds.

How it was put together

Initially each section was built from papers submitted to a workshop organised by the editors in association with the Royal Geographical Society with the Institute of British Geographers Annual Conference held in Leicester in January 1999.

Most papers were presented at that workshop, but have subsequently been extensively revised in the light of discussion at the workshop. Most importantly, authors collaborated in preparing introductions to each section in which they put their individual contributions into a wider context, providing an overview of other relevant work. Reading these introductions on their own will provide a good introduction to the field, but we hope that this will only serve to hone an interest in the remaining chapters, and in the potential of this technology for representations in geography.

References

Brown, I.M. 1999. Developing a virtual reality user-interface (VRUI) for geographic information retrieval on the Internet. *Transactions in GIS*, 3, 3, 207–20.

Câmara, A.S. and Raper, J.F. (eds). 1999. *Spatial Multimedia and Virtual Reality*. London: Taylor and Francis.

Cartwright, W., Peterson, M.P. and Gartner, G. (eds). 1999. *Multimedia Cartography*. Berlin: Springer.

Hearnshaw, H.M. and Unwin, D.J. (eds). 1994. *Visualization in Geographical Information Systems*. London: John Wiley and Sons.

Martin, D. and Higgs, G. 1997. The visualization of socio-economic GIS data using virtual reality tools. *Transactions in GIS*, 1, 4, 255–66.

Part I

Introduction to VR and technology

2 Geography in VR

Context

Ken Brodlie, Jason Dykes, Mark Gillings, Mordechay E. Haklay, Rob Kitchin and Menno-Jan Kraak

VR in geography

Given its ubiquity, researchers could be forgiven for believing that a concise and coherent definition of virtual reality (VR) exists, bolstered by a carefully-charted developmental history, a comprehensive list of the ways in which VR can be most profitably applied, and, perhaps most fundamentally of all, an encompassing critique of the technology.

As noted in our editors' Introduction (Chapter 1), reading the chapters in this book will show quite clearly that such a consensus doesn't exist even in the restricted field of academic geography. The diversity of technologies and approaches employed by the authors of this first section demonstrates this point clearly. It would appear that there are as many 'virtual realities' as there are researchers actively involved with VR. This has not prevented these authors from offering definitions or frameworks derived from a bewilderingly wide range of fields (including computational mathematics, education, cartography and aesthetics), based on technologies, or by using purely pragmatic approaches. The latter are the most commonly encountered and are founded upon the assertion that if you are engaged with a representation to the point where your body is responding involuntarily to it as though it were the real world, then you are probably dealing with VR! In such a scheme the definition of VR is reduced to the creation of representations that are so convincing that were a virtual glass to fall from a virtual hand, the user would involuntarily reach out to catch it (Brodlie and El-Khalili, Chapter 4).

Defining VR

Without wanting to anticipate these more detailed discussions, by way of introduction to this section we offer a very broad definition of the technology that emphasises the common elements in our work. We agree that VR is a form of human–computer interface (HCI). More specifically, in each case the process of using VR, or producing a VR simulation, involves the creation of a construct on the basis of a source reality, in our case the

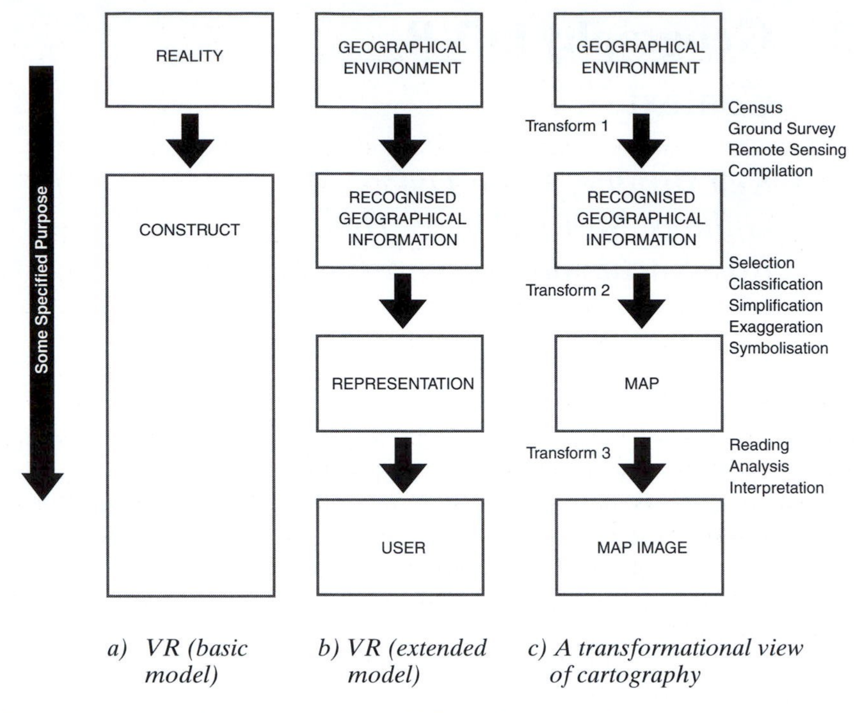

Figure 2.1 Schematic representations of VR and cartography.

a) VR identified as creation of construct from reality; b) An expansion of the schematic shows the series of information transformations involved in producing VR; c) A transformational view of cartography (after Tobler, 1979; Robinson *et al.*, 1996).

geographical environment (Figure 2.1a). This simple model can be expanded to include two other critical factors common to each of our applications. These are the geographical information derived from the environment and the users themselves (Figure 2.1b). To readers who have a background in cartography, it will be apparent that if we replace the terms 'representation' and 'user' with 'map' and 'map image' respectively, our framework for VR corresponds to the traditional cartographic process viewed as a series of transformations (Robinson *et al.*, 1996, after Tobler, 1979; Figure 2.1c).

So far our attempt to create a framework for our VR applications has got us little further than mere semantics. Are we to define VR simply as a subset of cartography? Is there anything that serves to distinguish approaches such as VR from traditional and more established means of representing the world?

We argue that VR is distinct from traditional cartographic transformations. What makes it different is the nature of the relationship between the representation (map) and user (map image). In such a formulation the emphasis of VR, as a form of HCI, is on the process linking the representation and its user. This transformation involves high levels of interaction between user and representation. In our schematic this feature of VR is represented by a bi-directional arrow flowing between the map and the user rather than the single, unidirectional arrow of cartography. This relationship is also stressed by our alternative terminology in the constructs used in the transformational view of cartography. In our VR applications the emphasis is on a 'representation' defined more broadly than the traditional 'map', and the physical user, rather than their 'map image'. This is because the real world affordances that we provide in VR to facilitate the transformation between representation and map image (involving the processes of reading, analysis and interpretation) form the crux of our applications. What is more, unlike any other mode of cartographic representation, in VR the level of engagement between map and user can be varied. In our schematic this is indicated graphically by the length of the arrow relating the two. If a single feature can be said to characterise VR, it is the ability to embed the user fully within the representation, permitting the kind of real world representation desired by, but unavailable to, Tobler when he noted that

> Any given set of data can be converted to many possible pictures. Each such transformation may be said to represent some facet of the data, which one really wants to examine as if it were a geological specimen, turning it over in the hand, looking from many points of view, touching and scratching.
>
> (Tobler, 1979: 105)

Ways in which VR can achieve this interactive, real-world interface between recognised geographical information, representation and user are shown schematically in Figure 2.2.

The degree to which the user and representation are collapsed is dictated by the precise use to which a given application of VR is oriented. As this volume shows, the breadth of applications of VR even within a single discipline such as geography are enormous. They cover the whole spectrum of approaches and users from the initial exploration of a complex data set by an individual expert in an attempt to find patterns, through to the final graphical presentation of results to a wider audience lacking in the same level of expertise. For example, if a VR model is constructed to test the effects of alcohol intake on drivers or to train surgeons in delicate techniques, the level of user immersion in the virtual representation (i.e. the degree of collapse) must be high (Brodlie and El-Khalili, Chapter 4). In contrast, for the purposes of creating a gallery space in which to deploy

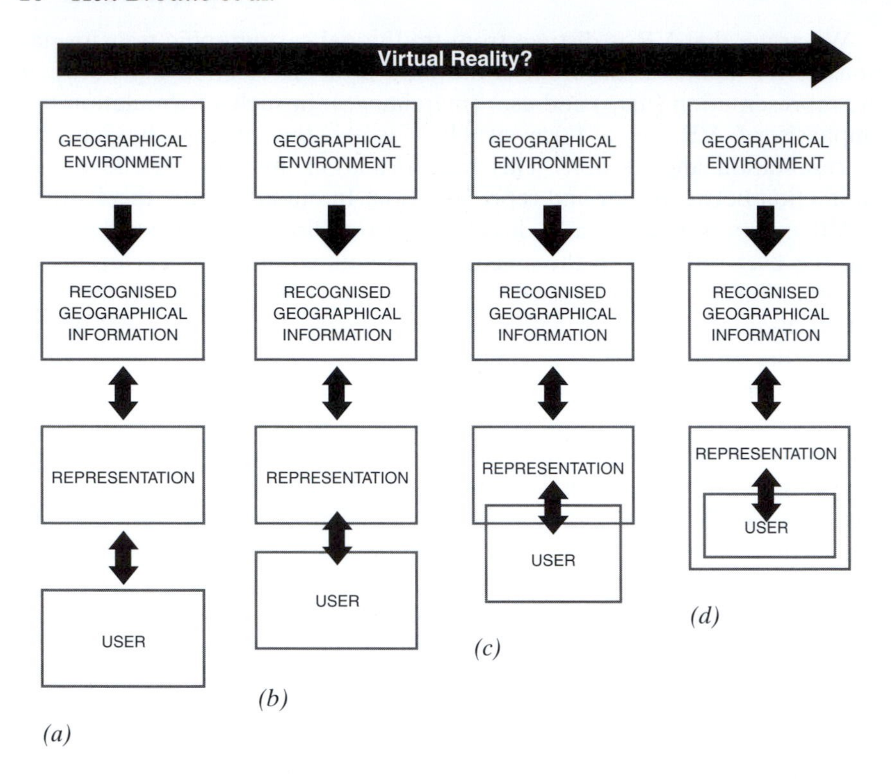

Figure 2.2 VR can be regarded as a continuum based upon levels of interaction and the real-world affordances used to support and facilitate the third transformation.
a) VR takes advantage of interactive visualization techniques where the user can interact with the data to vary transform 2; b) The interface uses some real-world spatial affordances – e.g. Chapter 7; c) The interface relies upon real-world spatial affordances and takes advantage of a strong sense of immersion – e.g. Chapter 12; d) The user is fully and physically immersed in the model and responds as if operating in the real world – e.g. Chapter 4.

virtual agents, the level of collapse can be negligible (Batty and Smith, Chapter 19).

The nature of the geographical information used to generate a given virtual construct can also vary enormously. In many instances this will reflect some aspect of the physical world, as shown by our schema that use 'tangible' data obtained by survey. A good example is the work of Lovett *et al.* (Chapter 9) who model sustainable landscapes. Other applications may rely on survey for their data, but record and thus visualize less tangible phenomena such as annual precipitation levels, social conditions or urban land use (Moore, Chapter 18). Equally, a given construct can be based on 'non-tangible' information that does not derive from the

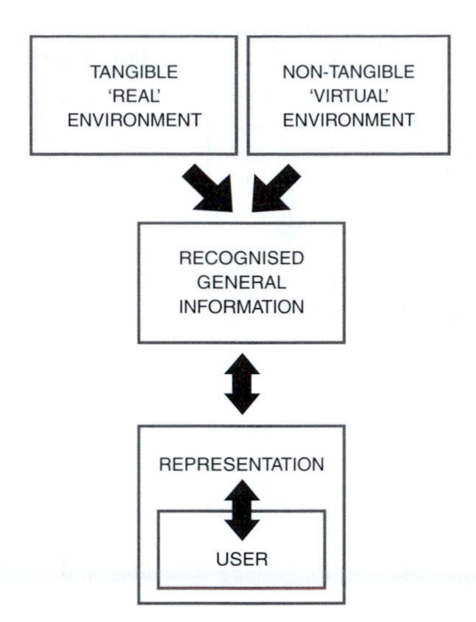

Figure 2.3 Applications can take advantage of a VR interface to present information that does not describe the physical environment of a real location.

physical world. In much the same way as a cartographer may seek to map out the fictional world of a novel, there are many instances where the virtual construct derives from the imagination, speculation or the realisation of abstract data spaces (e.g. Harvey, Chapter 22) as shown in Figure 2.3.

Taken together these factors go a long way towards explaining the diversity of definitions of VR the reader will encounter in this book. The framework outlined here allows a given VR representation to be defined and assessed not by any external criteria or pragmatic guidelines, but by its fitness for purpose. In this sense consider the following simple examples: the placement of a solid block in a simplified urban landscape; a series of geo-referenced images and linked data sets; and a simple series of coloured spheres that show rock types at different depths and locations. Each is as much an example of VR as a sophisticated reconstruction of a structure on a virtual brick-by-brick basis if it utilises spatial, real-world metaphors to enable the user to interpret the information by effectively engaging them with the representation. These examples might well enable residents visually to assess the impact of a proposed structure, students to learn about the geography of a region, and experts to identify geological structures from borehole samples.

As a result, rather than offering a single definition of VR, instead we propose a loose framework within which a series of task-specific

Gillings – Chapter 3

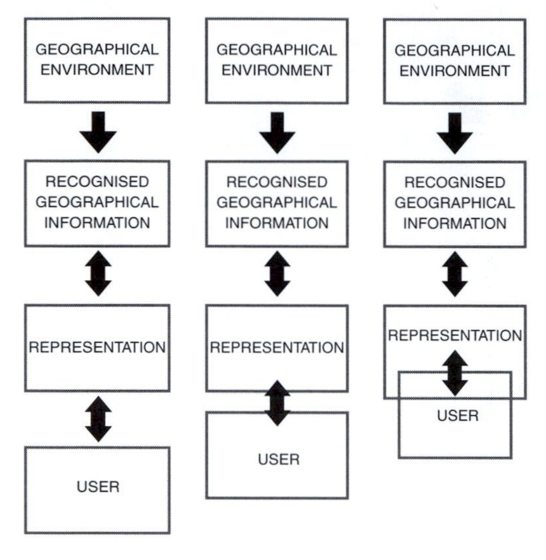

Transform 1

- Measurement and recording fragmentary material remains of the past using spades, trowels, total stations and cameras

Transform 2

- Empiricist measurements are blended with dominant disciplinary views on topic enriched through informed speculation and opened to peer discussion

Transform 3

- Degree to which user collapsed with representation dictated by guiding purpose of simulation

Brodlie – Chapter 4

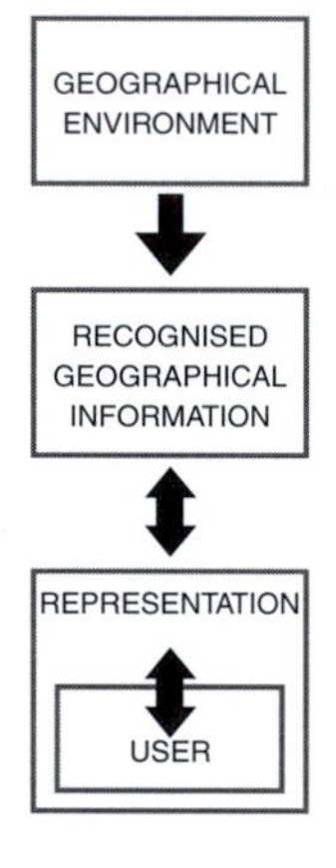

Transform 1

- CT Scan

Transform 2

- Lagrangian Dynamics

Transform 3

- High and realistic levels of VR interaction with haptic feedback

Kraak – Chapter 6

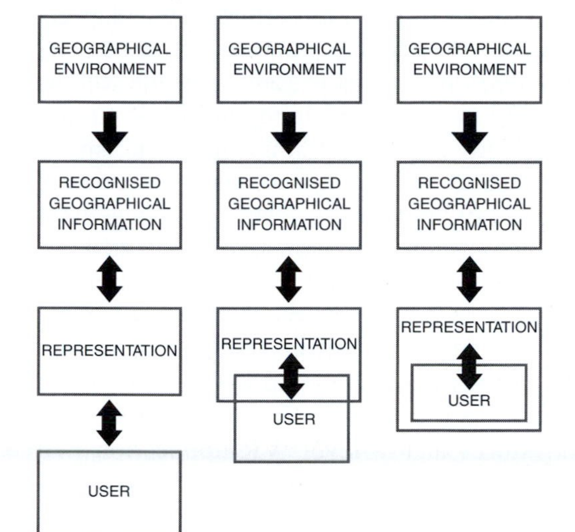

Transforms 2 & 3

- Development and use of tools for visual exploration to further understanding
- Schemas shown here represent plan, model and world views of data set
- Use of World Wide Web can increase accessibility and facilitate collaboration

Dykes – Chapter 7

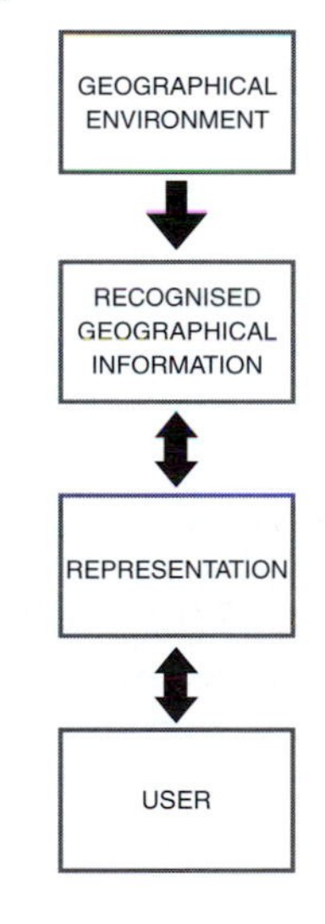

Transform 1

- Survey
- Secondary data
- Camera
- Image stitching

Transform 2

- Photorealism
- Spatial affordances

Transform 3

- Low-level 'VR'

Figure 2.4 Task-specific trajectories through our framework for selected contributions.

trajectories can be identified. Although the basic framework remains the same in each case, the processes involved in negotiating the various stages, and the degree to which user and representation are effectively collapsed, vary. To illustrate this we have sketched out a number of trajectories linked closely to the case studies and discussions presented in the following section. In Chapter 3, Gillings assesses the development of VR models to help the interpretation of archaeological information from the perspective of process. In Chapter 4, Brodlie and El-Khalili introduce a series of scenarios requiring highly-realistic rendering, high levels of engagement and sophisticated forms of interaction between representation and user. Chapter 5 presents the results of a survey by Haklay of applications in geography that further emphasises the diversity of work in the field. Kraak's consideration of the use of VR for the visual exploration of geographic data in Chapter 6 identifies the plan, model and world views which relate to distinct trajectories within the framework. Finally in this section, in Chapter 7, Jason Dykes introduces a 'low-level' VR application that combines photographic imagery and spatial information to produce a photorealistic virtual environment with spatial affordances for use in supporting student fieldwork.

These are summarised in Figure 2.4, where in each case the processes involved in traversing the stages and the desired/required degree of collapse between user and representation are clearly indicated.

Issues in VR

It should be appreciated that the creation and use of virtual constructs raises a number of important issues. Some are negotiable, some intractable, but all should be considered explicitly when creating and using virtual constructs. The first is that of accessibility. Are VR simulations to be created on egalitarian principles and made as accessible as possible, using perhaps an open standard, such as VRML or Java, rather than limited through the use of proprietary software? Linked to this are a number of commercial issues and technological considerations ranging from the restraints inherent in current hardware and delivery strategies through to the thorny issue of technological determinism which levels the accusation that we generate VR models largely because we can, only then deciding what we can actually do with them. This is an issue that is tackled head-on by Gillings (Chapter 3) and is noted by Haklay (Chapter 5).

Looking at the VR models and environments themselves, we have a number of issues relating to the precise relationship that is claimed between a given simulation and the source reality it purports to represent. In effect this poses the question: what is the virtual component of any given virtual reality and what is the real? Despite considerable rhetoric to the contrary, the answer to this basic question is often far from self-evident. A number of navigational issues persist in multidimensional, complex spaces. Which users and which applications might benefit from limiting users in a virtual landscape to real-world navigation and movement? For example, this might be appropriate in the mountain navigation exercises described by Purves *et al.* (Chapter 13), whereas the ability to teleport and float around the virtual world might be appropriate in other circumstances (e.g. Dodge, Chapter 21). What is clear is that the increased complexity of our virtual simulations requires ever more elegant solutions to the problems of interactivity, sometimes incorporating projected presence. The final issue that needs to be highlighted is the lack of people in our often highly-sophisticated virtual simulations. Navigating a virtual construct can often be a ghostly and unsettling experience, yet the introduction of avatars and virtual inhabitants poses considerable technical and representational challenges.

Geography in VR

Up to this point we have been discussing VR as a tool, or approach, that can be used by geographers to undertake investigations or present information in new and often challenging ways. The discussion has been centred upon the explicit role of VR in geography. However, such an emphasis neglects a whole field of study opened up by the combined collapse of user and representation, and the gentle blend of tangible and non-tangible data sources. We can undertake geography in VR, with virtual constructs becoming the objects of study rather than mere heuristics. Such a re-orientation can already be seen in the work that has generated satellite images and demographic trend maps for virtual worlds such as AlphaWorld (Chapter 21). Sociologies, histories and archaeologies will no doubt follow!

Influencing reality?

The relationship between reality and a given construct and the resultant 'reality' or 'authenticity' of any virtual simulation is a topic that has generated much heated discussion. 'Virtual reality is as real as a picture of a toothache' (Penny, 1993: 19). It is discussed at length in a number of the chapters in this book. What has prompted less discussion is the impact VR constructs can have upon the reality from which they are derived. From discussions of the hyper-real, whereby the only reality at stake is that generated by the construct (Gillings, Chapter 3), through to the augmented reality applications (Cheesman and Perkins, Chapter 24; Jacobson *et al.*, Chapter 25), there are a growing number of applications where the virtual construct serves to enhance, if not define, the world traditionally thought of as 'real'. The developing fields of robotic and telepresence will undoubtedly serve to further develop this theme.

Conclusion

In seeking to introduce and associate the contributions to this section we have highlighted diversity as one of the defining features of the technology. Instead of a prescriptive straitjacket we outline a flexible framework within which applications of VR can be developed and identified in their contexts. What VR is has as much to do with what we do with it as it has with sterile computational or theoretical definitions. In saying this, a number of critical issues and limitations exist which researchers must acknowledge and address if VR is to have the impact on the discipline it deserves. The chapters in this opening section seek to highlight and explore precisely these issues.

References

Boyd-Davis, B.S., Lansdown, J. and Huxor, A. 1996. *The Design of Virtual Environments.* SIMA Report No. 27.

Penny, S. 1993. Virtual bodybuilding. *Media Information Australia*, 69, 17–22.

Robinson, A.H., Morrison, J.L., Muehrcke, P.C., Kimmerling, A.J. and Guptill, S.C. 1996. *Elements of Cartography: Sixth Edition*. New York: Wiley, 674 pp.

Tobler, W.R. 1979. A transformational view of cartography. *The American Cartographer*, 62, 101–6.

3 Virtual archaeologies and the hyper-real

Or, what does it mean to describe something as *virtually*-real?

Mark Gillings

Introduction

> At the time of writing, virtual reality in the civilian domain is a rudimentary technology, as anyone who has worn a pair of eyephones will attest. That the technology is advancing rapidly is perhaps less interesting than the fact that nearly all commentators discuss it as if it was a fully realised technology. There is a desire for virtual reality in our culture that one can quite fairly characterise as a yearning.
>
> (Penny, 1993: 18)

Over the last five years, the term 'virtual reality' (VR) has become ubiquitous within all aspects of contemporary western society, synonymous with a developing generation of photo-realistic and fully interactive computer-generated environments. VR models are being used in a bewildering variety of contexts: from analysing the effects of alcohol intake on the drivers of automobiles to training surgeons and astronauts; from calming nervous dental patients to creating utopian worlds within which individuals can socialise and gather (Brodlie and El-Khalili, Chapter 4; Kitchin and Dodge, Chapter 23). In this chapter I argue that despite this widespread and growing interest in VR, researchers and commentators have not yet begun to grapple adequately with the question: What does it actually mean to describe something as *virtually* real? It is my contention that until they do the unique potential VR has to change the way we approach, study and think about the physical world will not be fully exploited.

Academic disciplines have been quick to register an interest in these innovative new approaches. For example, my own discipline, archaeology, has been actively exploring the potential of VR since the late 1980s through a number of expensive and highly-sophisticated collaborative projects. These include IBM's work on Winchester Cathedral and Roman Bath and more recently English Heritage's much-publicised Virtual-Stonehenge (Burton *et al.*, in press). As early as 1991, a clear blueprint was laid down for the future realisation of what was grandly termed a 'virtual archaeology' (Reilly, 1991). In 1996 the first popular text on virtual archaeology was published (Forte, 1996) and at the 1998 *International Computer Applications in Archaeology* conference, six full sessions were devoted to the topic of archaeology, heritage and VR. One clear and consistent trend can be identified in this explosion of interest in all things virtual. VR appears to have achieved widespread acceptance without having any history of its own production, or any clear definition as to what precisely it is. The adoption of VR techniques and applications within archaeology, as within many other disciplines, has been largely uncritical. VR is seen as self-evident and researchers rarely stop to consider what it actually means to define a given model or simulation as 'virtually real'. Related to this, and implicit in it, is a sense of ambiguity as to what we are supposed to do with VR models once we have created them. As Penny noted, there is a very strong tendency for commentators to talk about VR not in terms of what it is currently capable of doing, but instead in terms of what they expect, or desire, it to be able to do (Penny, 1993: 18).

Equating VR with fully-immersive, wholly-convincing (i.e. realistic) virtual worlds leads inexorably to the assumption that VR models should act as surrogates, digital replacements for original artefacts, structures and landscapes. The faithfulness with which these surrogates approximate their original referents is regarded as a direct correlate of the quality and volume of data that has gone into their creation. This has certainly been the case within archaeology, where the developmental blueprint laid down in 1991 defined VR models in precisely this way – representations that should strive to imitate a given source reality. The more painstaking the attention paid to the detail incorporated within a given model, the better the process of imitation, and thus the overall model, would be.

This uncritical orthodoxy has resulted in a developmental trajectory characterised by a continual drive towards methodological refinement. Ever more powerful computing resources are used as part of a relentless quest for the elusive grails of photo-realism and precise visual concordance, or verisimilitude. This has always been driven and justified by the belief that this is, and has always been, what VR is all about. As an aside, it is interesting to note that this spiralling reliance upon ever more powerful computer resources has tended to come at the direct expense of any attempt to develop a suite of reproducible, accessible, practical methodologies that can be routinely used throughout the discipline (Gillings, 2000; Pollard and Gillings, 1998).

As a result of these factors, archaeological use of VR is at present all about the creation of pictures. Ingenious pictures undoubtedly, but pictures nevertheless. Sophisticated VR models are created largely because we can, and are always generated as finished and freestanding, in effect, end products. Only *after* they have been generated does attention turn to the uses to which such models can be put. We are all too often encouraged to sit back and admire the resultant models than to do anything useful with them. Needless to say, when they are utilised for something other than purely decorative purposes, the levels of ingenuity and sophistication employed fall far short of the levels expended in their creation. A good example of the latter can be seen in the increasingly sophisticated approaches being advocated in archaeology for the solid modelling of complex architectural structures. Rather than creating VR models using simple geometric shapes and surfaces, for example a single, flat plane to represent the wall of a building, such approaches encourage instead the construction of models on a virtual stone-by-stone basis. In effect this is a virtual re-building. The levels of technical detail and sophistication that go into the creation of such a virtual reconstruction are breathtaking but the benefits of such a painstaking and technologically-intensive approach are minimal. Questions relating to the volume, mass and centre of gravity of the structure can be answered, and factors relating to the economics of construction, such as quantities of material and labour cost per cubic metre, can be explored (for an enthusiastic discussion of such approaches see Daniel, 1997). The object of analysis becomes reduced to a sterile architectural shell. Such questions, whilst no doubt interesting, seem scant reward for the enormous efforts expended in the creation of the model. In addition, it can also be argued strongly that people do not need to know or understand the mass, or centre of gravity, of a building to be able to dwell within it and imbue it with meaning.

This chapter takes a step back and challenges this orthodoxy by making a concerted attempt to answer the question posed in the opening paragraph: what does it actually mean to describe something as *virtually* real? This will entail a detailed and explicit discussion of a wide range of issues and concepts, from taken-for-granted notions such as realism and authenticity, through to an investigation of the relationship between archaeological applications of VR and the hyper-reality of Jean Baudrillard. If this does not result in a clear redefinition of VR, for reasons that will become clear in due course it will hopefully serve explicitly to identify the key issues and building blocks upon which a critical re-fashioning of VR can proceed.

To summarise, in adopting and embracing VR, like so many other disciplines, archaeology has failed adequately to foreground and explore a number of fundamental epistemological issues: what is VR, what is its potential, and how can this potential be realised? At present VR represents a rather taken-for-granted technology, shaped by desires and expectations rather than any detailed understanding and careful evaluation. As long as this remains unchallenged, VR runs the risk of marginalisation, being perceived not only as specialist and resource intensive, but ultimately of use only in the creation of what amount to little more than esoteric galleries. An expensive form of 'eye-candy', there to be visually devoured and little more. As long as individuals continue to characterise VR in terms of the expected, or *yearned for*, we run the very real risk of failing to realise the enormous potential it has in creating for the very first time a set of heuristic devices for the examination and exploration of the sensuous and physical three-dimensional world in a flexible and fundamentally three-dimensional fashion.

A question of definition: so what is VR?

An extremely useful framework for thinking about this question has been offered by the historian of science Noel Gray who identified two basic strategies which he elegantly summarised in his mutually-exclusive definitions of 'deficiency' and 'intensity' (Gray, 1995). The first of his categories defines VR models as 'manufactured deficiencies' where the model is seen to stand in an inferior position to an original referent. The degree of closeness, or similarity, between a model and the original is related directly to the quantity and quality of information put into the former. This parallels closely the dominant archaeological view of VR models, where the model is portrayed as a *döppelganger*, or surrogate, that is as faithful a replica as possible of some objective external reality, but always somewhere, somehow, lacking.

In contrast, Gray's second category positions VR models as 'manufactured intensities'. Here a virtual reality is a reality that is *more* intense and concentrated than so-called everyday nature, in effect more real than real. This idea of a reality more real than real can be difficult to visualize. A useful concept in helping to shed light upon this notion of manufactured intensity can be found in the term 'hyper-reality', introduced by the critical theorist Jean Baudrillard (1983). Baudrillard's hyper-reality relates directly to the loss of the real in contemporary society, where the barrage of images presented by media such as advertising and television have led to a disappearance of the distinction between factors such as real and imagined, reality and illusion, surface and depth (Barry, 1995: 87). The important theme in this is the recurring characterisation of the hyper-real in terms of simulations that are generated of the real, without origin or reality – the only reality at stake being that generated by the simulations themselves (Baudrillard, 1983: 146; Horrocks and Jevtic, 1996: 109). Like many of Baudrillard's more useful concepts, the term hyper-reality works tirelessly to elude concise definition. Fortunately, a number of excellent examples of hyper-reality exist and serve as useful exemplars for a careful exploration of the concept. One such example is Walt Disney's Animal Kingdom located in Florida. Animal Kingdom opened at 6.00 a.m. on the 22nd of April 1998. According to contemporary news reports, it had admitted its maximum of 28,000 paid visitors by 6.45 a.m. One important component of this complex theme park is the 'Kilimanjaro Safari'. In Disney's Africa, visitors are first encouraged to wander around the entry village of Harambe, which painstakingly re-creates the Swahili architecture of a well-established east African port city, complete with centuries-old peeling paint and worn benches. Having savoured the sights and smells of the village they are then taken off to experience the excitement of safari, where the precise flora and fauna of the African savannah have also been painstakingly and carefully re-created in this corner of Florida. It is only when we begin to look in detail that the essential hyper-reality of Animal Kingdom is laid bare. Despite the microscopic attention to detail, this is an Africa free of danger, disease, threat and disorientation. Visitors need no visas, inoculations or travel insurance. None of the animals come from the wild, most are from American zoos, and the village of Harambe serves cappuccino, ice cream and cocktails. Despite the painstaking detail and accuracy that have gone into the construction of 'Africa' the only Africa experienced is that generated by the simulation itself. The hyper-reality of Animal Kingdom is perhaps best summed up by a quote from the website advertising which stated that 'Asia is still under construction and won't be open until 1999' (see http://www.disney.com, accessed July 1998).

A second, and perhaps more intriguing example of the hyper-real has been presented by Staniszewski in her discussions regarding the creation of the culture of art. Here she describes the unique hyper-reality of the 1988 Winter Olympics in Calgary, Canada, where the 'snow' that featured in the opening ceremony was, in fact, sand. The principal reason for this substitution was that when shown on television sand looks more like snow than real snow. In addition it was uniform, could be smoothed and raked between broadcasts, and would not melt, get dirty, or reflect as much light into the camera. The sand looked like snow but didn't have any of the problems of snow. In a sense, the only 'real' Olympics were those viewed and accessed via the medium of the television screen (Staniszewski, 1995: 74).

Authors disagree as to whether hyper-reality is best treated as a concept or a tool, or whether it represents a tangible thing that resides in places, or a process that exists between subjects and objects (for contrasting views see Eco, 1986; Rodaway, 1994, 1995). It is perhaps most profitably thought of as representing aspects of both tool and concept, being embodied in particular locations and situations that are themselves part of a broader, on-going process of experiencing the world.

To summarise: in attempting to define VR, two contrasting, and mutually exclusive, definitions have been identified. These differ in the precise relationship that is claimed to exist between a given VR model and reality. On the one hand we have the notion of a surrogate, straining to replicate as authentically as possible some tangible, objective, external reality. On the other we have a simulation where the only reality at stake is that internal to the model itself.

Archaeology and VR

At this point, it is useful to compare the accepted archaeological definition of VR with Gray's categories. The first point to note is the remarkable similarity between the notion of manufactured deficiency and the current orthodoxy and developmental trajectory in archaeological VR. In such a portrayal, authenticity, expressed as visual approximation of form, is everything, and it is at precisely this point that we encounter a major stumbling block inherent within our accepted classification.

As has been discussed, underlying this notion of deficiency is the idea of an attainable, tangible reality to which the virtual surrogate aspires, and therefore against which the VR representation can in some way be compared and tested. In the case of a complete artefact, extant structure, or unaltered landscape, you could argue that a tangible original exists against which a model can be evaluated. However, with the majority of archaeological representations, whether of artefacts, built structures, or the very landscape itself, there are few, if any, wholly intact and unaltered referents. As archaeologists we base our models on primary data, however fragmentary, and are situating them within a well-developed body of academic knowledge. Any given model is based on a combination of rigorous empirical observation and informed archaeological inference. This latter is speculation shaped by the current, dominant disciplinary views on the topic that are always open to change, revision and re-negotiation. By uncritically accepting the path of manufactured deficiency we find ourselves continually confronted by the question: in what sense are our archaeological representations and reconstructions lacking and how are we to evaluate them? As the phenomenologist Kimberly Dovey has argued, the tendency to characterise VR representations as imperfect surrogates tends to engender an investigative operation on the part of the observer. This comprises a search for clues indicating authenticity, which itself leads to a sense of empirical testing rather than exploration or sense of experience of the represented place. Viewers are encouraged to *evaluate* rather than *negotiate* the model. This *attitude of mistrust* in turn leads to a sense that virtual representations have to become increasingly sophisticated in order 'to thwart investigations and capture real meanings' (Dovey, 1985: 38–9). Rather than treating VR representations as flexible interpretive devices, the tendency is to employ ever more powerful computers to create ever more sophisticated models to fool ever more suspicious viewers. This is precisely the pathway I would argue that archaeological-VR is currently following.

In the case of treating VR models as manufactured intensities, taking our discussion of hyper-reality to its logical conclusion, we could follow Baudrillard in arguing that when dealing with the hyper-real such questions have no meaning. 'Illusion is no longer possible because the real is no longer possible' (quoted in Rodaway, 1995: 246). Now to deny wholly the existence of the real may be provocative and is undoubtedly intellectually stimulating, but in following such a pathway there is the very real risk that we will achieve little more than the concealment of issues such as faithfulness and authenticity beneath a thick veneer of postmodern rhetoric. This is not to say that the points Baudrillard raises do not have a profound resonance within contemporary media culture. It is simply to acknowledge that in archaeology, as in so many disciplines, we do have an empirical commitment to the material remains we painstakingly recover, curate and analyse.

Approaching the issue from such a perspective it would appear that the requirements of a truly useful VR hover somewhere between Gray's mutually exclusive categories. In creating a given model there is both a commitment to the material remains of the past, and the ever-present realisation that the final model will always incorporate a degree of creativity and speculation. Archaeology is not unique in this ambiguity. We can identify applications of VR that have a clear commitment to some form of tangible, external reality, in a wide variety of contexts. Examples are the production of virtual simulations of existing structures for the purposes of heritage and tourism (Ogleby, Chapter 17), or for the complex analysis of patterns of internal visibility and visitor flow (Batty and Smith, Chapter 19). A perhaps more familiar example is in the simulation of artefacts themselves, paralleling the use of holograms and photographs to give a sense of experience of an otherwise inaccessible object. The creation of medical or driving simulators also demands a clear commitment to an external reality, whether the behaviour of a car on an icy road or the human body to a surgical intervention (Brodlie and El-Khalili, Chapter 4), as do augmented-reality applications for the visually impaired (Cheesman and Perkins, Chapter 24).

Equally, we can highlight applications where there is a much more tenuous relationship to a tangible external reality, if any relationship at all. An example is the creation of museums of the imagination. These can be idealised structures, such as the Virtual Museum of Art of Uruguay. This was designed by a consortium of architects to fulfil the pressing need to provide gallery space for Uruguay's artists, whilst avoiding the economically-crippling costs of creating a new physical structure (see http://www.diarioelpais.com/muva/). They can also be high-concept structures that were never realised. One example of the latter is the unique circular museum proposed in the nineteenth century by the pioneering archaeologist Pitt-Rivers to encapsulate and embody the process of evolution. A more abstract example is in the use of VR tools for the navigation of complex data spaces, in effect making intangible issues, relating to factors such as data quality, tangible (Harvey, Chapter 22). As with the vast majority of archaeological models, the precise relationship claimed between a VR model and 'reality' differs widely. We have applications that have a strong commitment to verisimilitude, and others which acknowledge no tangible original referent.

A question of definition again: defining VR?

This diversity and ambiguity begs the question of whether or not the difficulty we find in defining VR lies not with any inherent slipperiness on the part of the term itself, but instead with our desire to assign a stable and inflexible label. In his discussions regarding the impact of mechanical reproduction upon works of art, the philosopher Walter Benjamin asserted that terms such as 'art', 'perception' and 'reality' were inherently universal in application and thoroughly indeterminate in meaning. For Benjamin, a term such as 'reality', along with the standards we use to judge representations against it, gains meaning only in specific historical circumstances and in the context of specific tasks and uses (Snyder, 1989). Might we suggest that the same is true of VR? Any adherence to a single, orthodox classification, or search for any stable and emphatic definition is not only inherently futile, but also profoundly restrictive. What a given VR model *is*, is dictated more by the context and purpose to which it is put. There are instances where precise visual concordance is paramount. In an archaeological context this might comprise an attempt to explore the disorienting effects of flickering light and rhythmic chanting when peering through smoke and gloom at a cave painting. It could equally be concerned with the behaviour of a blood vessel wall to the impact of a surgical instrument (Brodlie and El-Khalili, Chapter 4). There are others when visual concordance is not an issue. For example, a series of models have been created of a suspected Roman lookout station on the second century AD frontier of Hadrian's Wall. These models were created solely to answer the question that if a rectangular tower, connected to a continuous stretch of walling was located here, what could be seen from it? They were not driven, or structured, around any desire to re-create a stretch of Hadrian's Wall (Gillings and Goodrick, 1996). Following this argument to its logical conclusion, rather than a firm definition, what is needed is a broad conceptual framework within which the full range of meanings encompassed by VR can be accommodated, including both the traditional sense of manufactured deficiency (visual approximation) and this new and potentially liberating notion of manufactured intensity (creative simulation). Before we go on to consider one such framework, I would like to look in more detail at the issues of reality and authenticity and their complex relationship with all things virtual.

Authenticity and realism

> A camera is a machine constructed to produce an image based upon artificial perspective. Only if one accepts the claims of the naturalness of renaissance artificial perspective can we accept photography as a mimetic representation of the world
>
> (Duncan, 1993: 43)

It is fascinating to note that when the *realism* of a VR model is judged, it is rarely against the physical world of things and processes, but instead against photographs. In the case of VR models, for *realism* one is actively encouraged to read *photo-realism*. As a number of authors have pointed out, photographs are by no means an unproblematic and objective representation of the world. Far from simply representing, photographs teach a particular way of seeing and as a result have to be carefully read (Ihde, 1993: 93). The anthropologist Nigel Barley gives a vivid illustration of this. Desperate to learn the correct names for lions and leopards in the language of the Dowayo (a group living in the north of Cameroon), he hit on the idea of passing postcard images of the animals around the village elders. He watched patiently as the elders stared at the photographs, pondered them, and turned them through all angles before handing them back with the phrase, 'I do not know this man' (Barley, 1983: 96–7).

This tendency to judge the *reality* of VR models on the basis of photographs is ironic given the debates that raged in the early decades of the twentieth century about the relationship between photographs themselves and the assumed reality of the external world (for a comprehensive discussion see McQuire, 1998). Walter Benjamin was convinced that the advent of photography and the widespread dissemination of photographic images demanded a new set of standards against which the reality of a given reproduction could be assessed. Photography was seen to represent a new vision of reality that pre-existing notions as to what constituted the real had no capability to test. Benjamin went on to argue that there existed no a priori natural standard against which *all* representations could be assessed. Instead the standards we use to judge the veracity of any claim to realism are fluid and inherently historical. The advent of photography demanded one such point of change in the historical transformation of these standards (Snyder, 1989). Following Benjamin, might we suggest that the introduction and widespread exploitation of VR represents another such point. Like photographs, VR models demand new ways of seeing and, also like photographs, they have to be carefully read.

Looking now to the issue of authenticity, and in acknowledging that it is central to any discussion of VR, it may be that many of the conceptual problems we associate with the issue are as much to do with our understanding of the concept of authenticity itself, as with its relationship to virtual simulations. Rather than stressing the issue of visual approximation, intimately bound-up in ideas of manufactured deficiency, we might instead follow Dovey in asserting that authenticity is not solely a property of form, but is instead a property of process and a relationship, or connectedness, between people and their world. Central to this argument is the assertion that authentic meaning cannot be created through any manipulation of mere form, as authenticity is the very source from which form gains meaning (Dovey, 1985: 33). The critical difference between an original and its representation is not in any detail of form but in the richness of environmental, or experiential, depth that attend the object or landscape. Dovey illustrates this idea of environmental depth through the discussion of a 'fake' beach constructed in the Arizona desert. The basic form of the beach is reproduced in so far as water impacts rhythmically upon a strip of sand. However, despite the fact that the beach has waves and sand, it has no crabs, sharks, undertow, driftwood, shells to be found, rock pools to explore, sea breezes or salt air (Dovey, 1985: 39–40). Whereas the original is a learning environment that fully embodies a sense of encounter, experience and process, the representation lacks spatial and historical depth, diversity and variation. Looking to VR, if we take the sophisticated VR model of a building, for example the Byzantine basilica church of the Alicami in southern Turkey (Bayliss, 1997), however architecturally faithful the representation, the virtual-stones do not carry mason's marks, the virtual-floor the shine of a thousand footfalls, nor is the virtual-interior cluttered with the accumulated bricolage of everyday social practice. And the critical point is that upping the processing power and adding them to the model will make no difference to this issue of authenticity. It is perhaps no surprise that, during a public demonstration of English Heritage's Virtual-Stonehenge, the audience became aware of the painstaking attention to detail and accuracy that had gone into its production, not when moving amongst the meticulously measured stones, but instead when they encountered the everyday familiarity of the underpass, information boards, turnstiles and post-box of the visitors' centre. These are the features they had already repeatedly encountered at Stonehenge and in a host of other routine daily contexts. (The demonstration was held at the 1997 Computer Applications and Archaeology (CAA) conference held at Birmingham, United Kingdom. A simplified version of the model can be accessed at http://www.intel.com/cpc/explore/stonehenge/index.htm.)

This notion of experiential depth has a parallel in Benjamin's ideas regarding what he termed the 'aura' of a work of art (Caygill *et al.*, 1998: 134–7; Snyder, 1989: 162–3). The aura is the facet which is lost when the work of art is reproduced. It relates to factors such as its location and history and the way in which it has gradually become embedded in social practice. A reading of Heidegger's notion of 'gathering' is also pertinent, where it is claimed that things and locales gain meaning through the unique way each gathers the world around itself (Thomas, 1996: 49). A good example of this idea of aura can be seen in the programme of VR modelling being undertaken at the site of Avebury in the United Kingdom. The monument at Avebury is an immense late Neolithic bank and ditch containing a circle of large standing stones, the largest stone circle in Europe. In a detailed discussion of one of the standing stones making up the monument, researchers have contrasted the long and detailed biography, or 'life', of the stone itself with the release into cyberspace of its virtual simulacrum. Whilst the latter is a carefully measured and highly accurate simulation, it is ultimately a hollow shell of pure form, stripped of its aura and the other stones, earthworks and landscape features it gathers to itself (Gillings and Pollard, 1999). The critical point to emphasise is that, however stunning and detailed their appearance, VR representations can *never* be wholly authentic. Stating that a given re-creation is inauthentic is not to say that it is either actively deceiving us or that it is not useful. A beach located in the middle of a desert is not in a position to fool anybody but it does not stop it being a nice place to spend an afternoon. Nor is it to say that VR models cannot in time generate new and wholly different auras as they are experienced, altered and incorporated into a host of virtual settings and social contexts, gathering new things and worlds around them. This is certainly the hope for the Avebury stone simulation (Gillings and Pollard, 1999). Virtual representations can serve to facilitate new modes of engagement and interpretation through emphasising process rather than the critical appraisal of mere form. In this sense the frustration encountered in a model at not being able to see over a section of wall to the other side becomes as important as the attention that has gone into realistically weathering the texture of the stones blocking the view.

A flexible framework for the development of VR: mimesis

> The Greeks did not suppose that a poet could create something out of nothing by words, which are only symbols of reality. They considered the poet created an artificial imitation of reality, a mimesis. For Plato the poet is essentially a man who mimics the creations in life in order to deceive his hearers with a shadow-world. In this the poet is like the Demiurge, who mocks human dwellers in the cavern of life with shadows of reality.
>
> (Cauldwell, 1946: 48)

As was implied earlier, rather than any set of restrictive classifications, a broader conceptual framework is needed to cope with the inherent fluidity, contextuality and historical contingency of any given definition of VR. What is needed is a framework within which the full continuum of meanings, stretched out between the deficiency of the surrogate and intensity of the simulation, can be accommodated and nurtured. Fortunately such a framework already exists that has had a long, if chequered, history within academic thought. This is encompassed by the term 'mimesis'. Mimesis has its origins in the Greek world, achieving prominence in the fifth and fourth centuries BC. In its original formulation the term was awarded a very general set of definitions, encompassing such characteristics as (Gebauer and Wulf, 1995: 28):

- miming – the direct representation of the looks, actions, and/or utterances of animals or men through speech, song, and/or dancing.

- imitation – the imitation of the actions of one person by another.
- replication – the production of an image or effigy of a person or thing in material form.

With Plato, and more markedly Aristotle, these broad definitions were replaced by a much more restricted understanding of the term. This laid special emphasis upon the notion of imitation in the aesthetic sphere and it is this definition that has persisted most forcibly into the present. In recent years the concept of mimesis has been foregrounded within twentieth-century humanistic geography. In discussions regarding the nature and history of the representation of geographical reality, Duncan and Ley have suggested that the question as to *how* the world is to be represented has been largely taken as given. This process of representation has been based upon what the authors identify as a fundamentally mimetic notion: the belief that individuals, in this case geographers and social scientists, should aim to produce as accurate a reflection of the world as is possible (Duncan and Ley, 1993: 2). In arguing this point, four dominant, and inherently chronological, modes of representation were identified which were termed 'descriptive fieldwork', 'positivist science', 'postmodernism' and 'hermeneutics'. These were then grouped on the basis of whether they adhered to, and hence reinforced, or directly challenged, and thus undermined, the notion of mimetic representation. These modes are summarised in Table 3.1.

We can easily position archaeological, and broader disciplinary, applications of VR in this classificatory schema. The current orthodoxy, characterised as it is by spiralling technological sophistication and the quest for photo-realistic approximation, straddles the first of Duncan and Ley's categories. In constructing models, a fastidious empiricism is allied to the abstraction, rigour and testability of positivist science. In each case the mimetic mode of representation is adhered to rather than challenged or questioned. Through painstaking attention to detail the aim has been to generate a series of representations which repeat one of what Duncan has called the central tropes of mimesis, the 'persuasive claim to represent the nature of a place as accurately and objectively as is possible' (Duncan, 1993: 40). What is interesting to note is that in these discussions mimesis is portrayed as an inherently flawed and restrictive concept, shackled to the notion of a single, neutral world-out-there, an objective reality against which all representations can be unequivocally evaluated, Benjamin's a priori universal standard. Taking such debates at face value, the implication for VR is that, by adhering to such a representational mode, current trends have side-stepped, or wilfully ignored, a considerable body of productive theoretical debate that since the early 1980s has raged in the wider social sciences and humanities.

Table 3.1 (adapted from Duncan and Ley, 1993: 2–3)

Mode	*Correlate in twentieth-century archaeological thought*	*Characteristics*	*Relationship to mimesis*
Descriptive	Culture–historical	Empiricist, observational	Profoundly mimetic.
Positivist science	New archaeology	Abstract, reductionist	Profoundly mimetic.
Postmodernity	Post-processual archaeology	Anti-foundational, relativist	Radical attack on mimesis.
Hermeneutics	Post-structuralist archaeology	Interpretation as dialogue between data and researcher	In acknowledging the role of the interpreter it rules out mimesis in the strict sense of the term.

An enriched notion of mimesis

Following this argument to its logical conclusion, it would seem that as a working concept, mimesis has little to offer, bound up as it is with the idea of manufactured deficiency that I have argued is ultimately counter-productive. Rather than a flexible framework, mimesis offers little more than another classificatory straitjacket. In seeking to retain mimesis, it has to be realised that the definition of the concept expressed by Duncan and Ley represents only a partial, reductionist and highly-specific reading of the term. This is a reading that, following Aristotle, emphasises solely those aspects which concern fastidious copying and mimicry. As a number of researchers have noted, such understandings neglect to acknowledge the complex nature of mimesis and the wide range of meanings that have been afforded to it during the course of its historical development (Auerbach, 1953; Gebauer and Wulf, 1995: 1; Taussig, 1993: 70). Even Aristotle, whose characterisation has proven so lasting and influential, appears to have assigned a dual meaning to the term. The first emphasised the creation of images whereas, more interestingly for us, the second concerned the creation of a plot or fable. Any emphasis on the first meaning alone, as is the case with the definition offered by researchers such as Duncan and Ley, neglects to acknowledge that '(f)or Aristotle, the critical point is that mimesis produces *fiction;* whatever reference to reality remains is shed entirely of immediacy' (Gebauer and Wulf, 1995: 55, italics in original).

If we accept that as a concept mimesis is far richer than recent debates suggest, the 'myth of mimesis' highlighted and critiqued by Duncan and Ley begins to take on more than a whiff of straw. Whilst accepting the force of the arguments concerning the nature of representation, I argue that to reject mimesis out of hand is to throw out not only the baby with the metaphorical bath water, but also the bath. A close reading of the history of use of the concept reveals three key factors that serve to take us far beyond any convenient correlation with simplistic processes of objective imitation:

1 The first is that, as a notion, mimesis is best thought of as two-layered. In the first instance you have imitation, in the second you have what has been referred to as a palpable and sensuous connection between the perceiver and that being mimetically reproduced. This is, to quote Taussig: 'the two-layered character of mimesis: copying, and the visceral quality of the percept uniting viewer with the viewed' (Taussig, 1993: 24). This two-layered definition is also evident in the traditional mimetic theory of art, which asserts that not only is a work of art something that copies or imitates an object, but it is also something that acts as a source of knowledge of the object copied in order 'to provoke in viewers the feelings provoked in them by the thing copied' (Bailey, 1998).

2 The second factor that emerges is that mimesis is a highly-fluid term. It is firmly embedded in practice and gains its immediate meaning largely through specific historical context, rather than any unambiguous and normative definition (Benjamin, 1979: 160–3). This is most clearly expressed by Gebauer and Wulf when they claim that:

> Mimesis is deeply entangled in society. Its respective historical positions are defined by authors, painters, musicians, architects, historians and philosophers; they offer designs of how it might be possible, under the conditions of their time, to make artistic and other worlds.
>
> (Gebauer and Wulf, 1995: 6)

3 The third is that mimesis is inherently creative, as hinted by Aristotle's second meaning sketched out above (Gebauer and Wulf, 1995: 17). Modern characterisations, such as that adopted by Duncan and Ley, have sought to deny or deprive it of this creative element.

It is in this much richer and less restrictive set of definitions that I believe the term has a constructive role to play in helping us to reformulate our understanding of the role of VR within archaeology and the broader social sciences. If a criticism can be levelled at VR applications to date, it is not that they have been mimetic per se, but that their mimetic faculty has been focused exclusively on issues of mimicry and imitation.

Consolidating earlier points and debates regarding hyper-reality and the nature of authenticity in representation, the adoption of mimesis as a framing concept provides us with an opportunity to concentrate not only on verisimilitude, but also upon aspects concerned with emotional engagement and creativity. This entails not only a re-appraisal and broadening of our existing theoretical foundations, but also a re-expression of the term mimesis itself. We must question the use of any ossified definitions that force the change in media heralded by VR to conform to a pre-existing and tightly-defined set of normative rules to be adhered to or confronted. Instead, the challenge must be to realise that the change of media initiated by VR, and developments such as the Internet, herald a new mimetic moment, uniting the threads of hyper-reality, deficiency, intensity, realism and authenticity I have discussed above. Our challenge, ultimately, will be to add archaeologists and geographers to Gebauer and Wulf's list quoted above.

Conclusion

If VR is to achieve and perhaps even exceed its undoubted potential, it must be realised that the generation of ever more photo-realistic models is but one application amongst many. Although the dominant approach taken at present, in the final analysis it may prove to be the least useful to disciplines such as archaeology and geography. Before the potential of VR can be realised, researchers must confront a number of fundamental questions regarding the nature of VR and its relationship to reality. They must also be prepared to highlight and challenge existing orthodox views. Through a consideration of the nature of representation and a detailed discussion of concepts such as deficiency, intensity, hyper-reality, authenticity and realism, the aim of the present discussion has been to provide a firm foundation upon which such a re-negotiation of VR can proceed.

Through the introduction of the richly-nuanced concept of mimesis, a constructive, yet flexible, framing device has been offered for future developments and discussion. The conceptual framework it offers emphasises not only imitation, but also sensual involvement, creativity and an emphasis upon the critical relationship between representational schema and social practice.

What is clear is that many challenges remain. VR models tend to be ghostly and unsettling places, notable not so much for their sense of embodied perspective and 'being-there', but instead for the lack of people (humanity) in them. At present, the 'body' of the embodied participant/observer in a given VR situation is also seen as unproblematic. It can be argued that, whilst giving the appearance of direct participation, the medium for engagement with a given simulation is more subtly that of a neutral and idealised observer. Following the work of, amongst others, Haraway (1991) and Penny (1993) an important challenge will be to explore more fully the idea of the situated and gendered social body.

References

Auerbach, E. 1953. *Mimesis: The Representation of Reality in Western Literature*. Princeton: Princeton University Press.

Bailey, G.W.S. 1998. *Art as Mimesis*. (http://ecuvax.cis.ecu.edu/~pybailey/mimesis.html: browsed 29/4/98).

Barley, N. 1983. *The Innocent Anthropologist: Notes from a Mud Hut*. London: Penguin.

Barry, P. 1995. *Beginning Theory: An Introduction to Literary and Cultural Theory*. Manchester: Manchester University Press.

Baudrillard, J. 1983. *Simulations*. New York: Semiotext(e).

Bayliss, R. (1997). The alacami in Kadirli: transformations of a sacred monument. *Anatolian Studies*, XLVII, 57–87.

Benjamin, W. 1979. *One-way Street and Other Writings*. Frankfurt: Suhrkamp Verlag.

Burton, N.R., Hitchen, M.E. and Bryan, P.G. (in press). Virtual Stonehenge: a fall from disgrace? In Gaffney, V. *et al.* (eds) *Computer Applications and Quantitative Methods in Archaeology 1997. Proceedings of the 25th Anniversary Conference, Birmingham*. British Archaeological Reports International Series.

Cauldwell, C. 1946. *Illusion and Reality: a Study of the Sources of Poetry*. London: Lawrence and Wishart.

Caygill, H., Coles, A. and Klimowski, A. 1998. *Walter Benjamin for Beginners*. Cambridge: Icon Books.

Daniel, R. 1997. The need for the solid modelling of structure in the archaeology of buildings. *Internet Archaeology* 2, 2.3 (http://intarch.ac.uk/journal/issue2/daniels_index.html).

Dovey, K. 1985. The quest for authenticity and the replication of environmental meaning. In Seamon, D. and Mugerauer, R. (eds) *Dwelling, Place and Environment: Towards a Phenomenology of Person and World*. Dordrecht: Martinus Nijhoff Publishers, pp. 33–49.

Duncan, J. 1993. Sites of representation: place, time and the discourse of the other. In Duncan, J. and Ley, D. (eds) *Place/Culture/Representation*. London: Routledge, pp. 39–56.

Duncan, J. and Ley, D. 1993. Introduction: representing the place of culture. In Duncan, J. and Ley, D. (eds) *Place/Culture/Representation*. London: Routledge, pp. 1–24.

Eco, U. 1986. *Faith in Fakes: Travels in Hyperreality*. London: Minerva.

Forte, M. (ed.). 1996. *Virtual Archaeology: Great Discoveries Brought to Life Through Virtual Reality*. London: Thames and Hudson.

Gebauer, G. and Wulf, C. 1995. *Mimesis: Culture – Art – Society*. Berkeley: University of California Press.

Gillings, M. (2000). Plans, elevations and virtual worlds: the development of techniques for the routine construction of hyperreal simulations. In Barcelo, J.A., Fork, M. and Sanders, D.H. (eds) *Virtual Reality in Archaeology* (Oxford: Archaeopress), pp. 59–70.

Gillings, M. and Goodrick, G.T. 1996. Sensuous and Reflexive GIS: exploring visualization and VRML. *Internet Archaeology*, 1 (http://intarch.ac.uk/journal/issue1/gillings_index.html).

Gillings, M. and Pollard, J. (1999). Non-portable stone artefacts and contexts of meaning: the tale of grey weather (www.museums.ncl.ac.uk/Avebury/stone4.htm). *World Archaeology*, 31, 2.

Gray, N. 1995. Seeing nature: the mathematisation of experience in virtual realities. *History of European Ideas*, 20, 1–3, 341–8.

Haraway, D.J. 1991. *Simians, Cyborgs and Women: the Reinvention of Nature*. London: Free Association Books.

Horrocks, C. and Jevtic, Z. 1996. *Baudrillard for Beginners*. Cambridge: Icon Books.

Ihde, D. 1993. *Postphenomenology: Essays in the Postmodern Context*. Evanston, Illinois: Northwestern University Press.

McQuire, S. 1998. *Visions of Modernity: Representation, Memory, Time and Space in the Age of the Camera*. London: Sage.

Penny, S. 1993. Virtual bodybuilding. *Media Information Australia*, 69, 17–22.

Pollard, J. and Gillings, M. 1998. Romancing the Stones: towards an elemental and virtual Avebury. *Archaeological Dialogues*, 5.2, 140–64.

Reilly, P. 1991. 'Towards a Virtual Archaeology'. In Lockyear, K. and Rahtz, S. (eds) *Computer Applications and Quantitative Methods in Archaeology 1990*. Oxford: British Archaeological Reports (Supp. Series 565), pp. 133–40.

Rodaway, P. 1994. *Sensuous Geographies: Body, Sense and Place*. London: Routledge.

Rodaway, P. 1995. Exploring the subject in hyper-reality. In Pile, S. and Thrift, N. (eds) *Mapping the Subject: Geographies of Cultural Transformation*. London: Routledge, pp. 241–66.

Snyder, J. 1989. Benjamin on reproducibility and aura: a reading of 'The Work of Art in the Age of its Technical Reproducibility'. In Smith, G. (ed.) *Benjamin: Philosophy, Aesthetics, History*. Chicago: University of Chicago Press, pp. 158–74.

Staniszewski, M.A. 1995. *Believing is Seeing: Creating the Culture of Art*. New York: Penguin Books.

Taussig, M. 1993. *Mimesis and Alterity: a Particular History of the Senses*. London: Routledge.

Thomas, J. 1996. *Time, Culture and Identity*. London: Routledge.

4 Web-based virtual environments

Ken Brodlie and Nuha El-Khalili

Introduction

The interaction between human and computer has always been important. In the early days of computing, Ivan Sutherland envisioned a world in which the user could be immersed in the computer output (see Kalawsky, 1993). As computer technology has developed, so Sutherland's dream has become a reality: virtual reality (VR).

VR is now a major branch of computing, covering a broad spectrum of human interaction with computers. It ranges from interaction with real environments, to interaction with synthetically-generated worlds. Figure 4.1 shows this spectrum. Teleoperation at the left end involves a 'machine that operates on an environment and is controlled by an operator at a distance' (Barfield and Furness, 1995). Augmented reality involves a representation which mixes part of the real world – either captured by video or directly perceived – with synthetic images. A virtual environment is a wholly-synthetic world in which the user interacts.

In this chapter we shall be primarily concerned with virtual environments. As previous chapters have demonstrated, attempting a definition is difficult, but from our perspective we can identify two essential components: simulation and presentation. Simulation is the process whereby the behaviour of people and objects within the world is modelled – so that objects fall to the ground under gravity when we drop them. Presentation is the process of displaying the synthetic environment to the user – either with immersive technology (such as head-mounted display), or panoramic

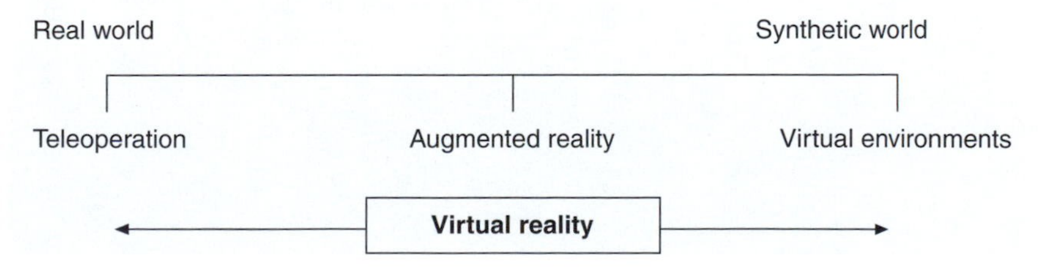

Figure 4.1 The spectrum of VR.

technology (such as stereo projection), or simply on the desktop. Taken together, the simulation and presentation must be realistic enough to engage the user in the environment.

A nice example of a virtual environment is the Leeds Advanced Driving Simulator (Driving Simulator, 2000) shown in Figure 4.2. Here the user sits in a real car, with the steering wheel, accelerator, braking and gears operated by the driver exactly as normal. However, the car itself sits in a virtual environment, projected on a screen in front of the driver. As the driver turns the steering wheel, for example, the torque applied is passed to a simulation model running on a powerful Silicon Graphics computer, which calculates the new heading direction of the car. This, in turn, is passed to a graphics process that calculates the new scene to be projected in front of the driver. The overall experience is highly convincing; the simulation and presentation work in tandem to engage the user in the virtual environment.

Figure 4.2 Leeds Advanced Driving Simulator.

This is an example of a highly-sophisticated virtual environment. A powerful, dedicated computer drives the simulation and is necessary to achieve real-time performance of thirty or more frames per second, responding without lag to user input. Interaction is provided in the most realistic way possible, by using a real car, not an arcade mock-up. This is characteristic of many traditional virtual environment applications making use of dedicated equipment at a fixed location with powerful and expensive computing equipment.

The World Wide Web offers a quite different approach to virtual environments. This began as a distributed information repository, but it has evolved into a powerful distributed computing environment. This environment hosts a range of applications across a range of disciplines, from the Virtual Field Course (VFC, 2000) in geography, through air quality visualization (Wood *et al.*, 1996) in environmental science, to physical modelling and simulation (Mann and Waldmann, 1998) in engineering. Thus there has emerged a Web society that supports our everyday lives.

A key aspect of the Web is its accessibility. With a PC and a connection to an Internet Service Provider (possibly by way of a wireless connection), a user can access the Web and enter the Web society. Technology exists in order to create simple virtual environments. Java, a programming language for Web applications, can be used to create simulations and the Virtual Reality Modelling Language (VRML), a standard for defining 3D 'worlds' on the Internet, can be used to create the presentation. This is a very different VR from the traditional applications typified by the driving simulator. It is ubiquitous rather than dedicated and cheap rather than expensive. The cost inevitably is the level of engagement that can be offered to the user.

In this chapter we study Web-based virtual environments. There are a number of different approaches that can be adopted, exploiting the client–server nature of the Web to distribute the computing in different ways. To improve our understanding, we develop an architectural model (pp. 38–39) that identifies the components of Web-based virtual environments, and can be used to compare and contrast different approaches. Our focus will be much broader than geography, because the issues we shall study are generic and can be applied to any discipline.

On pages 39–45 we look at a number of successful examples of Web-based virtual environments. These serve to illustrate the model and how it can distinguish different approaches. The examples are deliberately taken from an area other than geography – in fact, medicine – in order to demonstrate the generality of the approach. The geography reader is invited to express examples of Web-based virtual environments with which they are familiar in this same framework. A final example, a bicycle simulator, highlights what is now achievable.

This chapter concludes with a summary of what is possible, and what the limitations are, with pointers to how these might be alleviated in the future.

A model for Web-based virtual environments

Over the next few pages we construct an architectural model for Web-based virtual environments. At the heart of the model are two components, representing the two aspects we identified in the Introduction: simulation and presentation. As Figure 4.3 shows, there is communication between these components. Simulation results concerning the state of the environment are fed to the presentation component, and interaction from the user is returned from the presentation component to the simulation component. This is a continuous process.

The simulation component is constructed from two models. The geometry model describes the environment as a set of objects. The behaviour model describes the way in which objects in the environment respond to interaction with the user, to interaction with each other (for example, in collision), and to interaction with the environment (for example, under gravitational force).

The presentation component interfaces to the human. The output from the component is a sensory experience, reaching out to our senses of vision, hearing and touch. The input to the component is usually kinesthetic – for example, our hand moving a mouse – but might be oral in a speech-driven system.

We can use this model to characterise a Web-based virtual environment. Almost any simulation component will include a geometry model, but this can vary in sophistication. The more complex it is, the harder it will be to render in real-time. As we shall see, the behaviour component may or may not exist, and if it does, may vary in sophistication. The presentation component may appeal to only certain of our senses. The examples in the following pages will illustrate these differences.

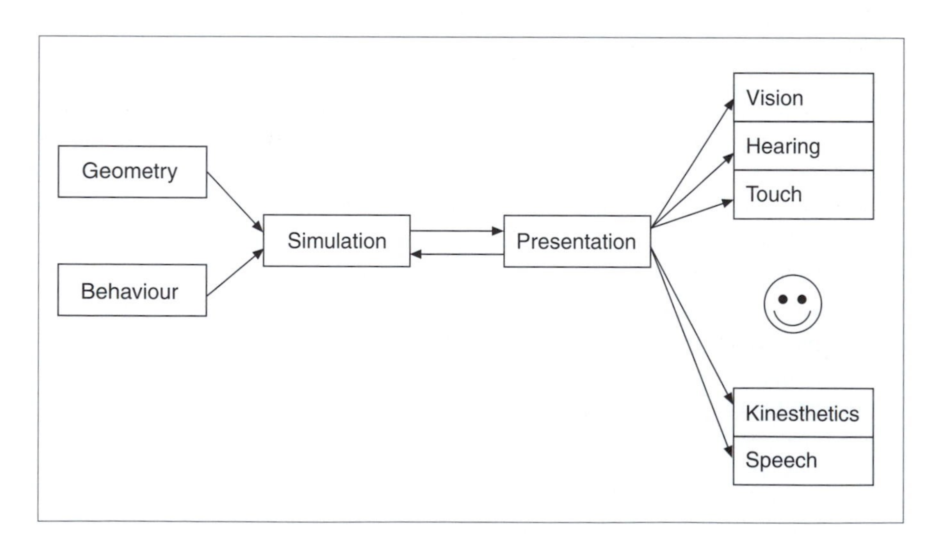

Figure 4.3 A model for Web-based virtual environments.

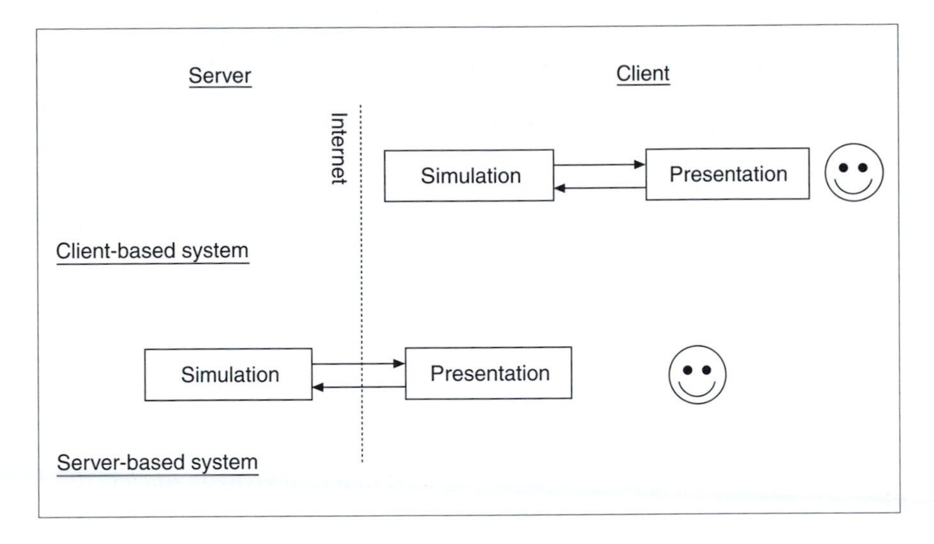

Figure 4.4 Client–server architectures.

The model can also be used to distinguish different system architectures of a Web-based application, as shown in Figure 4.4. The presentation component, facing the user, will always run on the client system. In terms of technology, this will be in the form of a VRML browser, such as CosmoPlayer™, or a Java applet. However, the placing of the simulation component divides applications into two classes. In a client-based system, the simulation component runs on the client. The technology to enable this is Java, with its ability to provide portability. Java code can be incorporated as an applet within a Web page, and can communicate with a VRML presentation component through the External Authoring Interface (EAI). In a server-based system, the simulation component runs on a server process and communicates with the client through a socket connection, or some other means. There are strengths and weaknesses of each approach. In a client-based system, there is close coupling of simulation and presentation that fosters good interaction. However, if the simulation is compute-intensive, there is attraction in using a server-based system where a powerful server may be shared between several users.

Examples of Web-based virtual environments

A patient monitoring system

This is a very simple, but instructive, example. At the Children's Hospital in Boston, a project has looked at the monitoring of patient data over the Web (Wang *et al.*, 1996). Data from bedside monitors is fed to a central Oracle database. A Web page is created for each patient, and the bedside

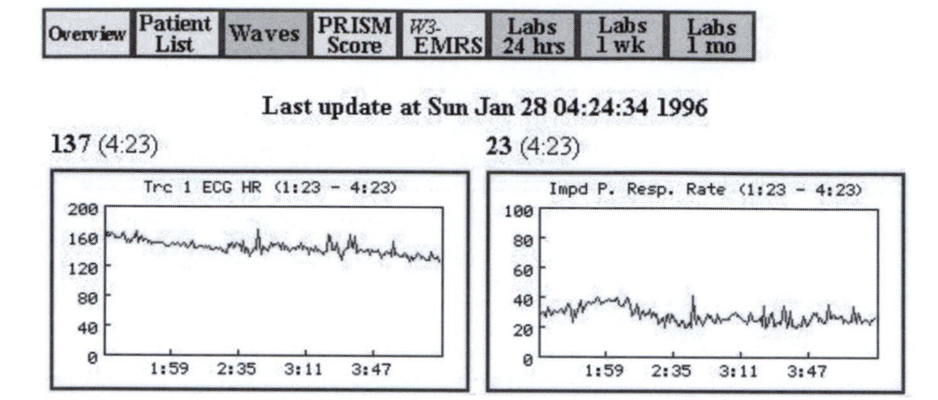

Figure 4.5 Patient monitoring system.

data are fed continuously from the database to the Web page, using server-push technology. The effect is that the bedside monitors are available for access by clinicians over the Web as shown in Figure 4.5.

Is this a Web-based virtual environment? Most certainly. The simulation process is mimicking the bedside monitors. The geometry may only be two-dimensional (2D), but a monitor screen is 2D so there is absolutely no need for three dimensions. The behaviour is the graph of the quantity being monitored, updated continuously over time. The presentation component receives the updated imagery from the server and displays on the Web page.

In terms of our architectural model, we see both geometry and behaviour models in the simulation component; one-way communication between simulation and presentation; and presentation appealing only to the visual sense. It is a server-based system, since the simulation is created on the server. This is highlighted in Figure 4.6.

Neurosurgery training

This is an example of a simple Web-based environment for surgical training, developed by John *et al.* (1999). A 3D anatomical model of a human head, and a model of a surgical instrument called a cannula, are downloaded from server to client as a VRML model, and viewed within a browser such as CosmoPlayer™. The trainee neurosurgeon manipulates the cannula into the skull using directions given on the Web page. There is no behaviour involved: there is no testing of collision between cannula and skull, for example. This is shown in Figure 4.7.

In terms of the architectural model, this is an example of a client-based system. The simulation only has a geometry component, while the presentation interfaces to our visual senses, and receives kinesthetic input as we use the mouse to navigate the cannula (see Figure 4.8). Note that the

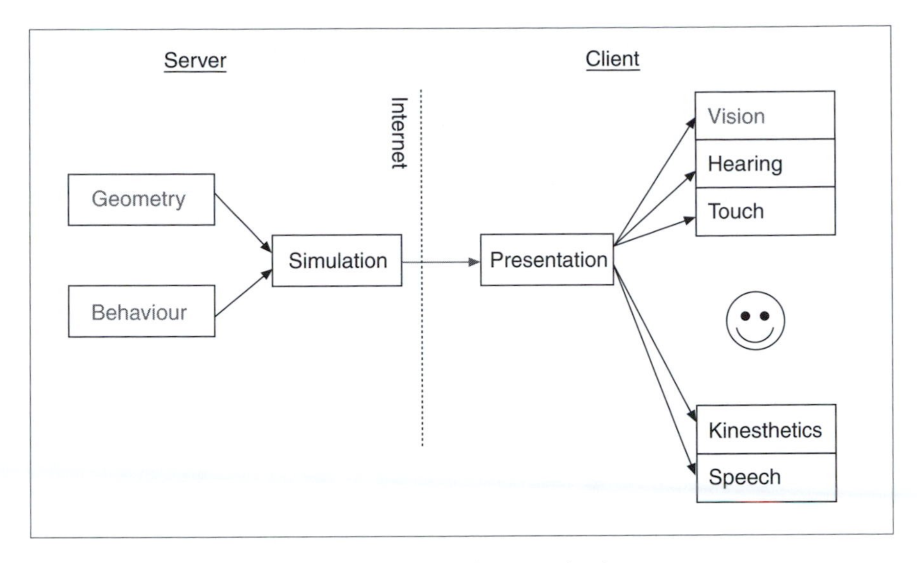

Figure 4.6 Architectural model for the patient monitoring system.

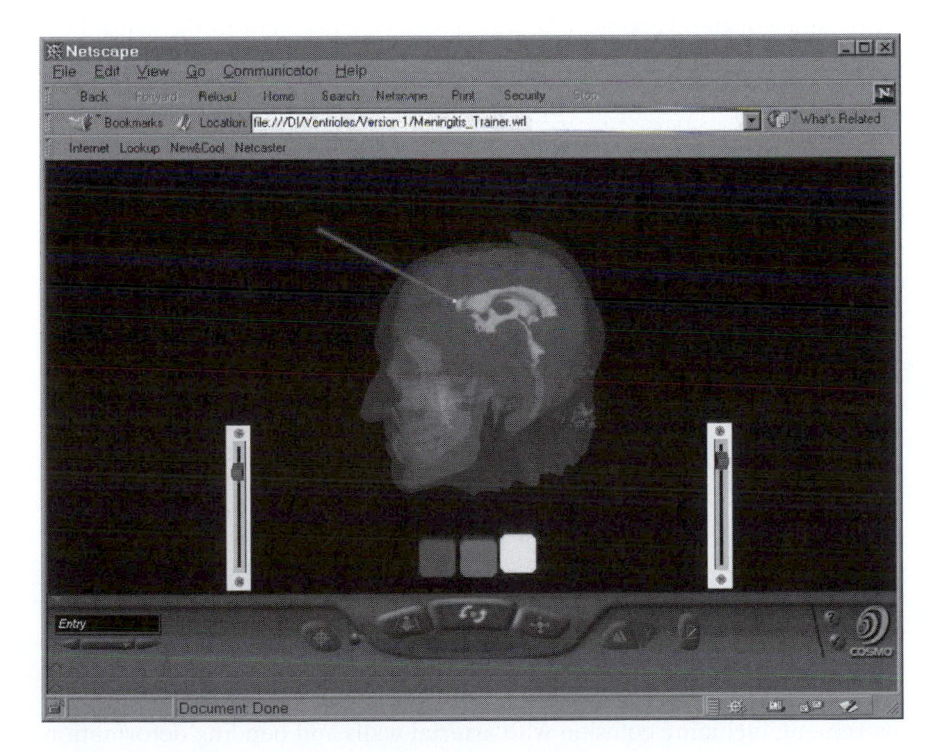

Figure 4.7 Neurosurgical training system.

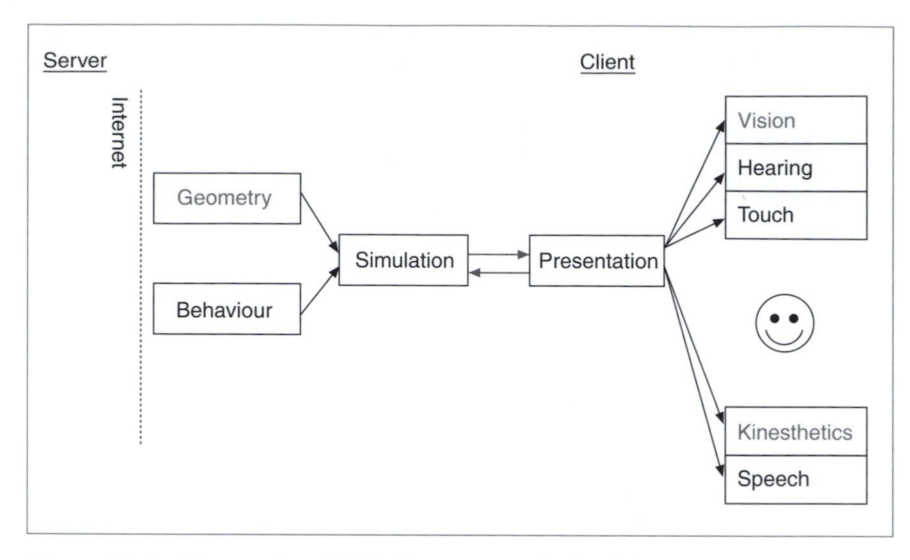

Figure 4.8 Architectural model for the neurosurgical training system.

Virtual Surgery website (Virtual Surgery, 2000) now contains in addition a more sophisticated version, with a behaviour component in the sense that simple collision detection is supported.

Interventional radiology training

This is a more complex example of surgical training. It was developed to help train radiologists in manipulating catheters and guidewires through complex arterial structures in the human vascular system. This is needed, for example, in the treatment of abdominal aortic aneurysm. This is a swelling or dilation of the aorta; if it bursts, the consequences are life-threatening. Thus a stent, or collar, is inserted in the aorta to protect the flow of blood. This is achieved by manipulating a catheter from an incision in the thigh, through the tortuous femoral artery and into the aorta. The catheter bends under collision with the arterial wall – indeed this bending is necessary in order to navigate it to the required position. This is shown in Figure 4.9.

The simulator allows the trainee to select a particular catheter or guide wire, insert it into a 3D model of the aorta and manipulate it using push, pull and twist forces. The aorta model, and the surgical tools, are downloaded as VRML geometry. The operation of the tools is simulated using a virtual representation of the real-life control handle. The forces applied to this virtual tool are communicated to a server process, which (like the driving simulator) calculates the effect of these forces on the surgical instrument, including collision with arterial walls and bending deformation which results. The simulator is shown in Figure 4.10. The screen is divided

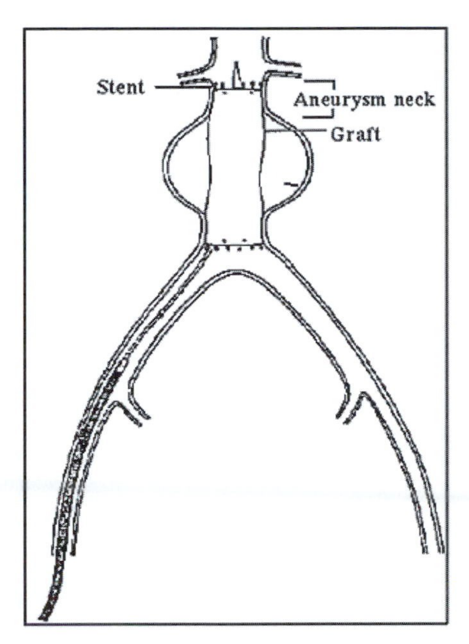

Figure 4.9 Stent deployment.

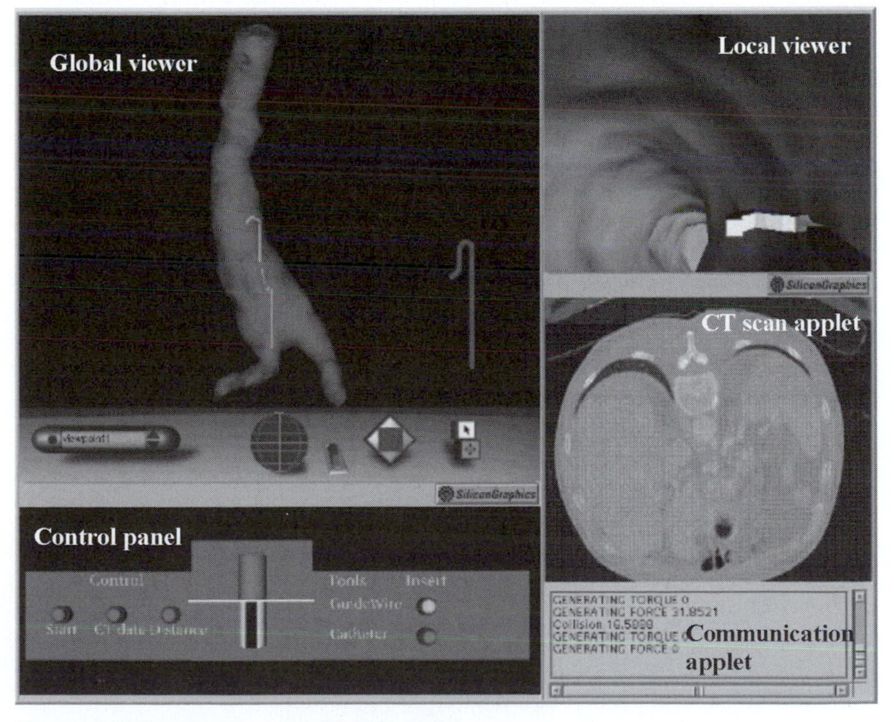

Figure 4.10 Interventional radiology training.

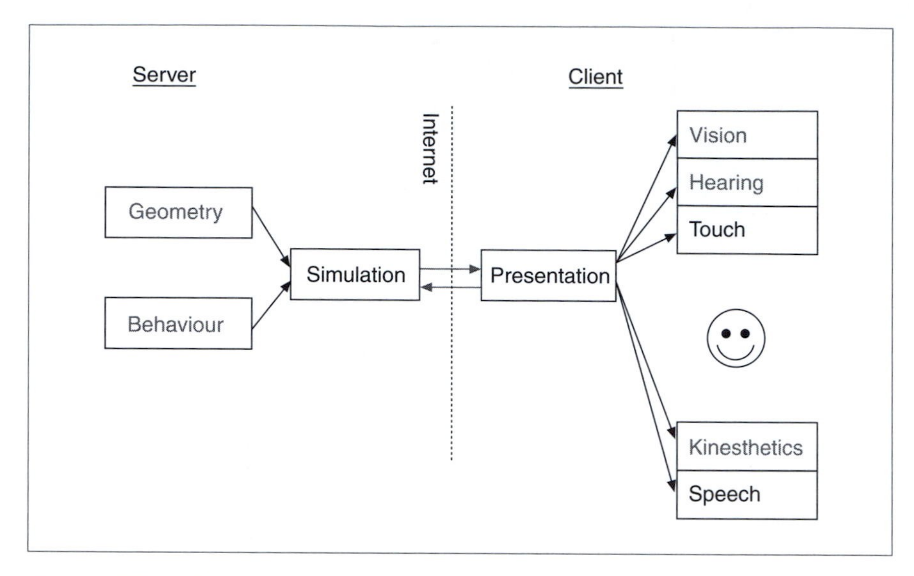

Figure 4.11 Architectural model for the interventional radiology training system.

into different parts: the main view (top left) shows the catheter within the aorta; top right shows a view inside the aorta from the catheter tip; lower left is the control panel with virtual tools; middle right are CT slices through the aorta to help orient the radiologist; and finally, lower right is a Java applet window with control messages.

In terms of the architectural model, this is a server-based system. The simulation has both a geometry and behaviour component. The presentation appeals to the visual sense, and reacts to kinesthetic input. A major distinction from the patient monitoring on pages 39–40 is the two-way communication between simulation and presentation, requiring a sophisticated behaviour model (see Figure 4.11).

Bicycle simulator

The Peloton bicycle simulator currently represents the ultimate in Web-based virtual environments. Indeed it suggests what is likely to be an eventual convergence between traditional virtual environments, and the newer Web-based approach. It was developed at Bell Laboratories by Carraro *et al.* (1998). The user sits on a real bicycle (a RacerMate Computrainer™ to be precise), operating real pedals (similar to the driving simulator). A fan simulates the effect of the breeze! The scene in front of the cyclist is calculated in response to the pedal action and displayed on a screen. It is multi-user, so that several cyclists can participate in a race.

In terms of the architectural model, this is essentially a client-based

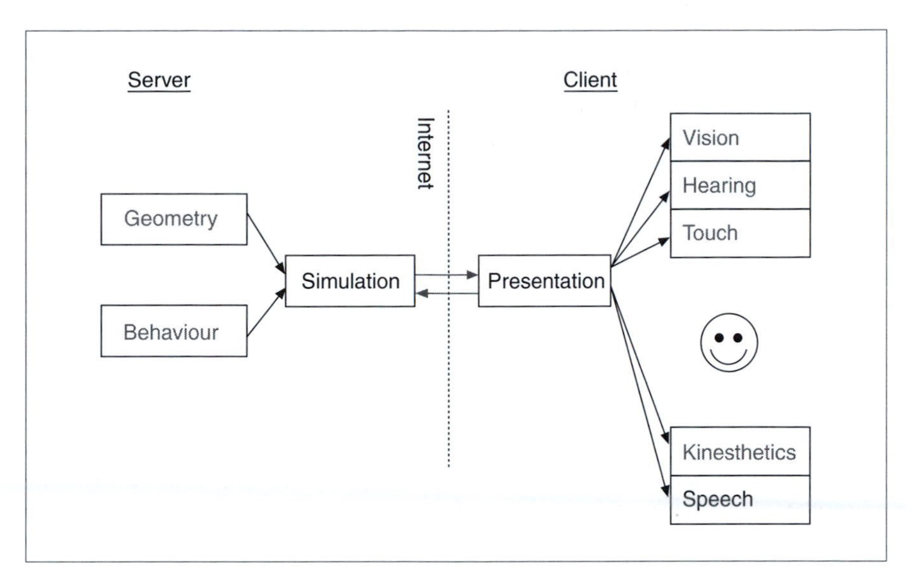

Figure 4.12 Architectural model for Peloton bicycle simulator.

system in that the simulation component runs locally. However, a server is needed to manage the multi-user involvement, which will always be the case in collaborative environments, and so it might be described as a mixed client-based/server-based system. The simulation component has both geometry and behaviour; the presentation component appeals to all senses, and reacts to kinesthetic input from the cyclist (see Figure 4.12).

Conclusions, limitations and the future

Web-based virtual environments offer a simple, accessible, low-cost alternative to traditional approaches to virtual environments. A number of these have already emerged on the Web and make a very useful contribution to the Web society. This chapter has developed an architectural model for Web-based virtual environments, and applied it to a number of examples largely from the medical field. The differences in the examples are highlighted with reference to the model.

Of course there are limitations. Despite its name, VRML was not really designed for VR. For example, it supports collision detection only between viewer and object, not inter-object collision. This was one reason the interventional radiology training system was forced to be server-based. It only considers the geometry of objects, not their physics. Thus we know nothing about mass or elasticity, information we need to do accurate physical modelling. There is tension too between the ubiquity of the Web which forces common denominator solutions, such as using the 2D mouse for all interaction.

However, we can look forward to progress in the future. Java3D, the additional set of classes for 3D graphics programming in Java, will support inter-object collision, so client-side simulation will become more feasible. The Peloton project has already shown how, albeit with some difficulty, a 2D mouse can be replaced by a bicycle for input.

We can look forward, therefore, to a new era in computing where the combination of virtual environments and the Web can deliver a new reality to a mass market: Web-based VR.

References

Barfield, W. and Furness, T.A. 1995. *Virtual Environment and Advanced Interface Design*. New York: Oxford University Press.

Carraro, G., Cortes, M., Edmark, J. and Ensor, J. 1998. The Peloton bicycling simulator. In Spencer, S. (ed.) *Proceedings of VRML '98*. Monterey, California: ACM, Inc, 63–70.

Driving Simulator. 2000. http://mistral.leeds.ac.uk/, visited August 2000.

John, N., Phillips, N., Vawda, R. and Perrin, J. (1999). A VRML simulator for ventricular catheterisation. In *Eurographics UK '99*. Cambridge, UK, April 1999.

Kalawsky, R. 1993. *The Science of VR and Virtual Environments: A Technical, Scientific and Engineering Reference on Visual Environments*. Wokingham: Addison-Wesley.

Mann, H. and Waldmann, L. 1998. Web-based modelling and simulation of multidisciplinary engineering systems. In Fishwick, P., Hill, D. and Smith, R. (eds) *The International Conference on Web-Based Modelling and Simulation*. San Diego, California: The Society for Computer Simulation International, 183–8.

VFC. 2000. Virtual Field Course. Department of Geography, University of Leicester, http://www.geog.le.ac.uk/vfc/, visited August 2000.

Virtual Surgery. 2000. http://try.at/virtual.surgery, visited August 2000.

Wang, K. *et al.* 1996. *A Real Time Patient Monitoring System on the World Wide Web*, http://www.emrs.org/publications/amia_icu.html, visited August 2000.

Wood, J., Brodlie, K. and Wright, H. 1996. Visualization over the World Wide Web and its application to environmental data. In Yagel, R. and Nielson, G.M. (eds) *Proceedings of IEEE Visualization '96,* San Francisco: ACM Press, 81–6 (see http://cself27.leeds.ac.uk:8010/jason/html/pollution.html, visited August 2000).

5 Virtual reality and GIS

Applications, trends and directions

Mordechay E. Haklay

Introduction

During the last decade, research projects that merge Geographical Information Systems (GIS) with virtual reality (VR) systems have become popular. Several developments in computing can be seen as the basis for this trend. These include the widespread use of powerful desktop computers (Hughes, 1996), the re-evaluation of VR as a simulation of reality rather than imitation of it (Gillings and Goodrick, 1996), and the rise of the Virtual Reality Modelling Language (VRML) standard coupled with a range of software tools that accompany it. It is now possible to locate and classify the research areas that focus on applications of Virtual Reality GIS (VRGIS) and to envisage the future directions in these fields.

To anchor this chapter in the context of this volume as a whole, there is a need to define the sort of VR that it discusses. In this chapter the term is used to describe computerised systems that provide the end-user with interactive, three-dimensional Computer-Generated Imagery (CGI). This definition contextualises VR as a Human–Computer Interaction (HCI) technology. Using this definition, computer-generated animated movies where the user cannot control the sequence of images will not be included, but if the user can manipulate the imagery and set the vantage point, then it is VR.

It follows that we can position virtual reality/geographical information systems in the context of geographic visualization. VR can be seen as the uppermost level in a hierarchy that starts with the traditional two-dimensional map. Most GIS applications and software still rely on this map representation as the main access method to present and manipulate spatial data, and there is evidence that it is also useful in a VR environment (Navas *et al.*, 1997; Verbree *et al.*, 1998). The next level is the perspective view, also known as '2.5D' representation. This is a static representation that uses the laws of perspective to give the viewer the feeling of distance and depth. By packaging a sequence of 2.5D representations, it is possible to create an animated movie. The movie gives a more realistic notion of depth, but the user cannot interact with it.

In a 2.5D/3D interactive environment, the user can control the way that the representation is seen, and have control over the viewpoint. The interface to this environment is still a flat screen with a limited field of view and the user does not need any special-purpose devices to see the images. The final level is when the user is immersed into the GIS environment through the use of various enabling devices, from Head Mounted Displays (HMD) to systems like the CAVE (for a discussion on the CAVE and other display mechanisms, see Verbree *et al.*, 1998).

In this chapter, the term VRGIS will be used to refer to systems that exhibit elements of the last two levels, as it is possible to find common properties and attributes to all those systems. This chapter can be read as an attempt to locate and state them. It is based on literature and Internet surveys undertaken from November 1997 and November 1998, followed up by questionnaires that were sent at the beginning of January 1998 and December 1998 to all the centres with relevant projects in the field. The Internet survey was based on Web pages, reports in USENET groups, and printed literature on the progress in combining GIS and VR.

A short history of VRGIS

Although the roots of both GIS and VR can be easily traced back to the 1970s, the first documented successful fusion of GIS and VR was in the early 1990s with a system that depicted the Georgia Tech campus area (Faust, 1995). Since then, the number of application and research projects that involve VR and GIS have increased dramatically. Because of the geometrical computation in GIS and heavy rendering computation in VR, the first wave of VRGIS applications were based on high-end workstations, even supercomputers.

From the mid-1990s, a re-evaluation of the potential of VR has led to a major change in this research agenda. The result has been that VR is not seen as a method to *imitate* reality but rather to *simulate* aspects of it as a sensual form of communication (Gillings and Goodrick, 1996). This change paved the way for the next major development, that of the Virtual Reality Modelling Language. This emerged in 1995 onwards (Bell *et al.*, 1996) and opened up a new direction for the development of VRGIS. By using VRML, the economics of deploying a VRGIS dropped enormously. This development also coincided with the emergence of low-cost, yet powerful, desktop GIS packages that enabled workers to adopt a low-cost environment in which to experiment with VRGIS. According to the survey, over half of total applications in the field use VRML. It is noteworthy that the original ideas and concepts behind VRML were not toward a common format for VR applications but rather as an attempt to create a three-dimensional interface to abstract information on the World Wide Web (Pesce *et al.*, 1994). Furthermore, VRML was not designed to handle large-scale scenes or to use a geographic coordinate system and these

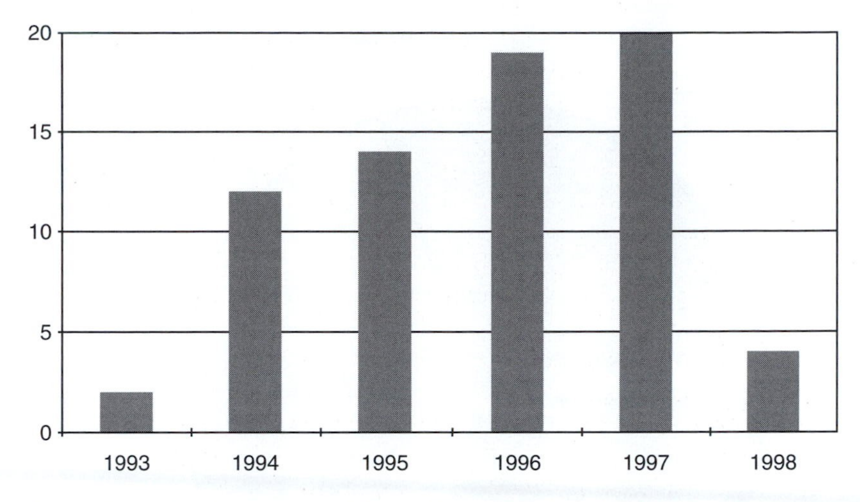

Figure 5.1 Virtual reality and GIS projects, initiations 1993–8.

deficiencies have fuelled efforts to steer its further development towards geographic applications (Rhyne and Fowler, 1998).

Figure 5.1 shows the number of project initiations between 1993 and 1998. There was a rapid increase in project numbers in 1994 and 1995 and a steady expansion since then. Since most of the reported projects are of more than a year's duration, this implies that the number of active projects in any year steadily increased through to 1998. A possible explanation for the sharp decrease in 1998 is the integration of VR since then into standard software packages which has reduced the justification for specialised research projects, as more and more users of these packages have used VR as just another visualization tool.

A taxonomy of VRGIS applications

Though it is possible to classify VRGIS applications and solutions according to technical aspects such as the level of immersion or computer platform used, in this chapter the focus is on the research areas that are currently active in VRGIS. The classification is used to identify the motives that have led to the current practice in this field. Figure 5.2 shows the division of the reviewed projects according to their main research area. Of course, not all the projects that have been reviewed belong to a single category. Considering the costs of a high-end VRGIS environment, it is not surprising to discover projects that have multiple agendas and, at least in one project, environmental evaluation was combined, for example, with military simulation.

According to the survey, the following research areas are experimenting the possibilities of VRGIS.

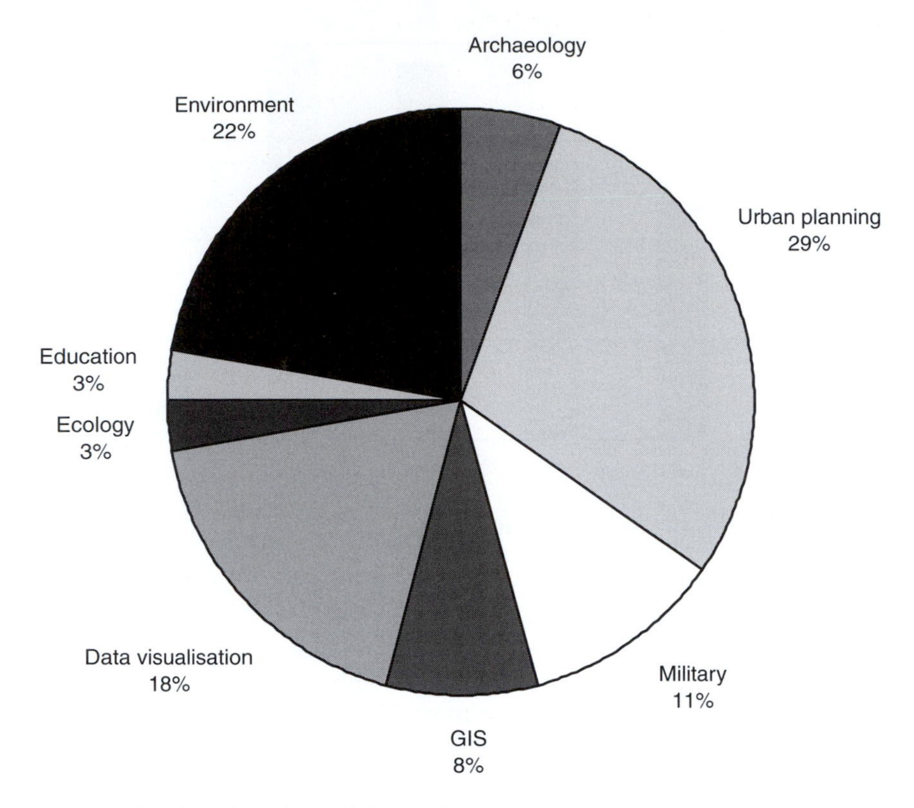

Figure 5.2 Virtual reality and GIS projects by research area.

Urban and regional planning

Applications in this field range from attempts to produce large-scale urban models (Liggetti *et al.*, 1995), to explorations of collaborative environments for design and planning (Dodge *et al.*, 1997). In these applications, the emphasis is on assessing the visual impact of different planning schemes and on enabling a group of planners to communicate through the planning process. These have the largest share (about a third) of the total number of applications. Taking into account the long tradition of using scale models in architectural design and the common use of highly-visual communication in these areas, it should not come as a surprise that implementation of VRGIS in this field was easy, at least at the conceptual level. Some anecdotal evidence suggests that between the two elements, VR is the more important with GIS playing a secondary role.

Environmental planning, modelling and impact assessment

Here, the visualization problems that lead to VRGIS range from assessing the visual impact of a planned forest (Buckley and Berry, 1997) to visualization of abstract phenomena like land contamination, or air pollution (Bishop and Karadaglis, 1996). Environmental modelling can produce voluminous, multi-variable data. The need to perform exploratory analysis is well known in this field and therefore results in the need for rich visualization environments. Unlike the previous group, here GIS is the dominant tool and it is more common to find questions about analysis and manipulation of the VR display in this field.

Scientific/geographic visualization

As MacEachren (1998) has noted, scientific visualization has emerged in the last decade as a response to the growing need for exploration and analysis of very large data sets. This is a common problem in many scientific fields, but the need to relate various data sets into a common geo-referenced framework raises specific issues that should be considered in this context. It is usually geographic visualization potentials that attract researchers in this. However, there are research projects that target specific questions about scientific and geographic visualization. The emphasis is on the visualization method as a goal in itself. In this context, the ICA commission on visualization (and the related Carto project of the ACM SIGGRAPH) is noteworthy.

Military simulation and intelligence applications

Military simulation, and more specifically flight simulation, has been a major driving force for many VR implementations. One of the goals in military VRGIS is to enable a 'virtual rehearsal' of future manoeuvres. Recently, simulation of this kind was used by the mission force in Bosnia (SGI, 1996) and it has been seen in TV broadcasts as part of the preparation for attacks on Kosovo and Iraq. This system (Powerscene™ from Cambridge Research Associates, MA, USA) provides a combination between physical features, such as satellite imagery draped on Digital Terrain Models (DTM), and tactical information such as the location and range of anti-aircraft weapons. A similar system is used by the British forces in Bosnia (Almond, 1997). Terrain modelling and visualization seem to have a major role in these applications.

Geographic information science

Applications in geographic information science stem from two main directions. The first is in attempts to advance the development of GIS through the use of the latest development in computing technology. Since

VR presents the latest development, to the extent that special purpose 3D functionality is now embedded into computer CPUs and graphic cards, researchers in this field are exploring the appropriate methods to incorporate VR capabilities into GIS. The second direction is the utilisation of VR in pursuit of classic problems in the field, such as the representation of data accuracy or the construction and representation of DTM.

Archaeological modelling

Most of the applications in this area deal with the reconstruction of landscape and structures. The applications that rely on GIS usually use their capabilities to handle a Digital Elevation Model (DEM), and then transfer the surface into a VRML model. The emphasis is on visibility analysis and hypothesis validation (Gillings and Goodrick, 1996).

Education

In education the idea of a virtual field course has become popular in a number of disciplines, at least in the research community. Examples include the Virtual Field Course initiative developed at Birkbeck College and Leicester University, and the Virtual Canyon developed by the Monterey Bay Aquarium Research Institute. Here the goal is to enrich the learning process through visualization of distance or 'hard to reach' places and it is the multimedia possibilities of VR that play an important role. The ability to add information to a VR scene is considered to be one of the important aspects of this environment.

Ecology

Applications in ecology are closely related to those in environmental modelling and exhibit similar features. GIS and remote sensing have been widely deployed in this field.

Motivations

By inspecting this range of applications and fields, it is possible to draw some observations on the incentives for implementing VRGIS. Some of the motives for developing VRGIS are explicit and the main one is to (re)construct landscape and urban settings that do not exist, either 'no more' or 'not yet'. The 'no more' case occurs when an archaeologist wants to explore an ancient landscape in 3D. In contrast, the 'not yet' case plays an important role in urban and environmental planning, especially for studies of the visual impact of planned buildings and other construction projects such as roads, dams and bridges. While many commercial GISs are able to conduct a visual analysis, the two-dimensional result is hard to grasp and perspective views, fly-through movies and animation are too rigid, enforcing the user to view from a specific point. In these applications, VRGIS adds an important freedom for the end-user.

In military applications, simulation is the driving force. Here, the actual mission is dangerous, and by familiarising the trooper/pilot with the mission area, it is possible to reduce uncertainty. Note that this is the only major research field that still strives to imitate reality as closely as possible. The third motive is the visualization of abstract variables and these types of application are gaining in popularity. However, in VRGIS these applications tend to create a hybrid representation of the 'virtual variable' on a background of 'real variable'. The abstract variable is usually an environmental variable (such as air pollution level) though there are sparse examples for socio-demographic or socio-economic variables.

A fourth motive is to improve communication of ideas and concepts in a collaborative process such as architectural planning. In this type of process, the VRGIS acts as a mediator and transmitter of ideas between the participants (Campbell and Davidson, 1996). In the GIS realm, the goal is to support users who are 'overwhelmingly map illiterate' (Jacobson, 1995). This in turn leads to the fifth motive, which is the search for an intuitive interface for spatial technology.

Finally, and more controversial, it is possible to identify an implicit motive in much work in VRGIS which is the 'Hammer looking for a nail'. GIS captures and stores spatial data, and, since many developers of VR applications are looking for applications that will be used as a case study, it is not surprising that GIS is used as a test case. Examples for this type of application appear in scientific visualization, or the CAVE GIS. Those projects tackle a real problem, but they serve primarily as examples of future implementations of the technologies rather than as contribution to solving the problem itself.

As an aside, Figure 5.3 charts the initiation year and the country of origin of the projects surveyed.

It can be seen that the overwhelming majority of VRGIS projects are taking place in highly-developed countries. This is perhaps clear evidence that VRGIS development is still costly, requiring resources that are beyond the reach of many other countries.

Properties of VRGIS

Faust (1995) describes the ideal VRGIS as having the following features:

- very realistic representation of real geographic areas. This can be based on remotely sensed imagery or through CGI. This, of course, requires fast rendering at different scales and resolutions;

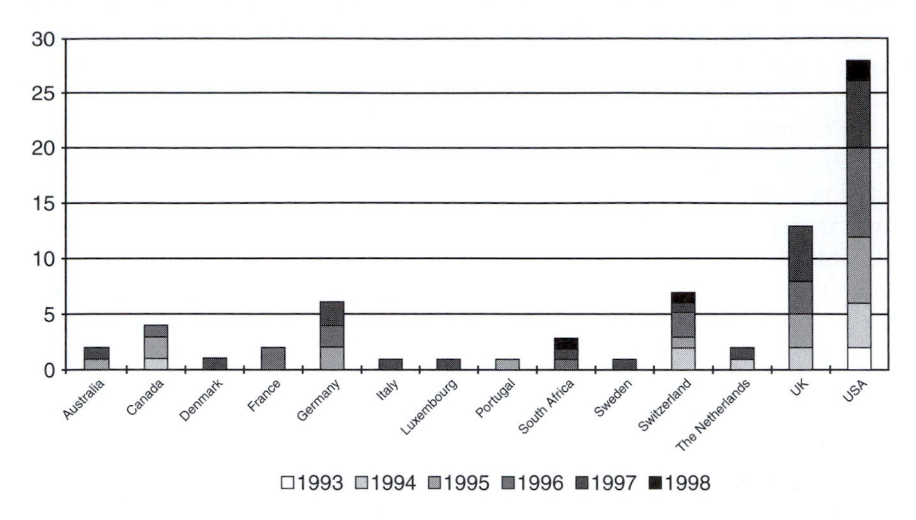

Figure 5.3 Country of origin and initiation year.

- free movement of the user within and outside the selected geographic terrain;
- standard GIS capabilities (query, select, spatial analysis, etc.) in a 3D database;
- the visibility function should be a natural and integral part of the user-interface;
- on the basis of more recent analysis (such as Gahegan, 1999; MacEachren, 1998; Verbee *et al.*, 1988) to this list can be added:
- support for the end-user in navigation, manipulation and orientation.

It is clear that such systems do not yet exist. Nowadays, the coupling between the GIS and VR is usually modular and 'loose', using the GIS to create and process the geographical data and a transferable file format (most notably VRML) to pass this information to the VR package for representation (Berger *et al.*, 1996). Such VRGIS solutions are based on coupled systems, with a distinctive GIS module and a distinctive VR module. However, the coupling itself is improving and today it is possible to find VR elements that come as an integral part of the GIS package, or offer capabilities to access the GIS database from within the visualization package.

To summarise, it is possible to describe the main properties of VRGIS as:

- the system database is a traditional GIS. Furthermore, as the GIS is more suitable for performing analytical functions, the manipulation of the information is done within it.

- the VR functionality is used to augment the cartographic capabilities of the GIS. Even though there are some efforts to integrate higher level functionality within the VR environment, in most cases these functions are still performed in the GIS environment. Furthermore, they pose some interesting questions about human–computer interaction and the appropriateness of the VR environment for these tasks (Nilsen, 1998).
- more and more solutions use the VRML standard, despite its limitations. As a result, VRGIS comes with Internet functionality 'out of the box'.
- There is a trend towards PC-based systems, relying on desktop GIS. Even more interesting is the trend to combine high-end with PCs. By doing this, the overall costs of the implementation are reduced, while allowing the high-end systems to focus on creating and serving 3D models that can then be handled by the PCs.
- the VR and GIS software are loosely coupled. The graphic data are usually transferred through a common file format, and the synchronisation between the systems is based on communication protocols.

Future directions

VRGIS is in a fascinating technical state. After a single decade of development, VRGIS components have been integrated into commercial products, from desktop GIS such as 3D Analyst™ extension to ArcView™ (ESRI, 1997), to remote sensing oriented products such as the Imagine Virtual GIS™ (Erdas, 1997) and scientific visualization products such as AVS™. The latest PCs can handle reasonably small and medium-sized data sets but only the very high-end systems (such as a CAVE, with high performance computing such as IBM SP2) can support a real-time VRGIS.

Will we soon see a fully-fledged VRGIS? The survey reported in this chapter has exposed some problematical aspects of this combination. First, there is an inherently different world view in GIS and VR work that influences both the data structure and the algorithms used. These differences make the transition between the two models rather complex and computationally intensive (Gahegan, 1999). Furthermore, some of the problems with VR and its manipulation are not polynomial (therefore, can be described in the computer science jargon as 'hard problems'). This family of problems cannot be resolved by crude increases in CPU computing power. Second, there are serious questions of usability and HCI in the VR environment. It is impossible even to suggest that we have solved all the problems that are associated with user-interface to the traditional, 2D display GIS and, as mentioned earlier, there are voices who question the concept of 3D interface as the main method to access computers (Nilsen, 1998). It seems that at least some of the researchers who seek VRGIS solutions are aware of this issue (Gahegan, 1999; MacEachren, 1998; Neves *et al.*, 1997). Finally, on the more conceptual level there are issues such as the cognitive aspects of 3D representation or the meaning of adding one dimension of visualization to data space that might have hundreds of variables.

To conclude this chapter, it is worth listing some of the main research considerations in VRGIS. Among the most interesting are:

- the role of the raster model in VRGIS. Today, the main use of raster layers is to drape them over a polygonal representation of the terrain or the urban setting. What is the role of 3D raster GIS?
- a need to develop proper visualization techniques that can improve communications between the different users of VRGIS, and especially when these systems will be used for policy making and resource management (Gimblett, 1993).
- as Liggetti and her associates (1995) point out there might be an intrinsic difference in the database structure requirements between GIS and VR. There is room for research on the right method and into the data models that will be the most efficient for VRGIS.
- as Mennecke (1997) points out, the human interface part of any VRGIS should be studied in great detail in order to define and develop the proper environment for human interaction with these systems. This problem is enhanced by the growing number of 'GIS illiterate' users, who use GIS in their day-to-day work (sometimes without knowing that they use GIS).
- there are possibilities to combine 3D representations with 2D representations, as Neves and his colleagues (1997) and Verbee *et al.* (1998) have demonstrated. What are the tasks and variables that are better represented in each model?

References

Almond, P. 1997. British soldiers plug into virtual reality. *TechWeb News*: URL: http://www.techweb.com/ (accessed 8 December 1997).

Bell, G., Parisi, A. and Pesce, M. 1996. *The Virtual Reality Modelling Language – Version 1.0 Specification*: URL: http://vag.vrml.org/VRML1.0/vrml10c.html (accessed 8 December 1997).

Berger, P., Meysembourg, P., Sales, J. and Johnston, C. 1996. Toward a virtual reality interface for landscape visualization. *Proceedings of the Third International Conference/Workshop on Integrating GIS and Environmental Modeling, January 21–25, 1996, Santa Fe, New Mexico*, CD ROM.

Bishop, I.D. and Karadaglis, C. 1996. Combining GIS-based environmental modeling and visualization: another window on the modeling process. *Proceedings of the Third International Conference/Workshop on Integrating GIS and Environmental Modeling, January 21–25, 1996, Santa Fe, New Mexico*, CD ROM.

Buckley, D.J. and Berry, J.K. 1997. Integrating advanced visualization techniques with ARC/INFO for forest research and management. *Proceedings of the 1997 ESRI User Conference Proceedings, July 8–11, 1997, San Diego, CA*, CD ROM.

Campbell, D.A. and Davidson, J.N. 1996. Community and environmental design and Simulation – The CEDeS Lab at the University of Washington. In Bartol, D. and Foell, D. (eds) *Designing Digital Space: An Architect's Guide to Virtual Reality*. New York: Wiley, pp. 201–24.

Dodge, M., Smith, A. and Doyle, S. 1997. Urban science. *GIS Europe*, 6, 10, 26–9.

ERDAS. 1997. *IMAGINE Virtual GIS white paper*: URL: http://www.erdas.com/before/whitepapers/virtualgis_white_paper.html (accessed 8 December 1997).

ESRI. 1997. *Announcing ArcView 3D Analyst, ESRI – Press release 3 June*: URL: http://www.esri.com/base/news/releases/97_2qtr/3danlyst.html (accessed 8 December 1997).

Faust, N.L. 1995. The virtual reality of GIS. *Environment and Planning B: Planning and Design*, 22, 257–68.

Gahegan, M. 1999. Four barriers to the development of effective exploratory visualization tools for Geosciences. *International Journal of Geographical Information Science*, 13, 4, 289–309.

Gillings, M. and Goodrick, G. 1996. Sensuous and reflexive GIS exploring visualization and VRML. *Internet Archaeology*, 1: URL: http://intarch.ac.uk/journal/issue1/gillings_toc.html (accessed 8 December 1997).

Gimblett, H.R. 1993. *Virtual ecosystems*: URL: http://nexus.srnr.arizona.edu/~gimblett/virteco.html (accessed 8 December 1997).

Hughes, J.R. 1996. Technology trends mark multimedia advancements. *GIS World*, 9, 11, 40–4.

Jacobson, R. 1995. Virtual worlds: spatial interface for spatial technology. *Electronic Atlas*, 5, 4 (Electronic newsletter).

Liggetti, R., Friedman, S. and Jepson, W. 1995. Interactive design/decision making in virtual urban world: visual simulation and GIS. *Proceedings of the 1995 ESRI User Conference Proceedings, May 22–26, 1995, Palm Springs, CA*, CD ROM.

MacEachren, A.M. 1998. *Visualization – Cartography for the 21st century*: URL: http://www.geog.psu.edu/ica/icavis/poland1.html (accessed 22 December 1998).

Mennecke, B.E. 1997. Understanding the role of Geographic Information technologies in business: application and research directions. *Journal of Geographic Information and Decision Analysis*, 1, 1, 44–68.

Neves, N., Silva, J., Goncalves, P., Muchaxo, J., Silva, J.M. and Camara, A. 1997. Cognitive spaces and metaphors: a solution for interacting with spatial data. *Computers & Geosciences*, 23, 4, 483–8.

Nilsen, J. 1998. *2D is better than 3D*: URL: http://www.useit.com/alertbox/981115.html (accessed 22 December 1998).

Pesce, M.D., Kennard, P. and Parisi, A.S. 1994. Cyberspace. *Proceedings of the First International Conference on the World Wide Web, May 25–26, 1994, CERN, Geneva.*

Rhyne, T.M. and Fowler, T. 1998. *Geo-VRML Visualization: A Tool for Spatial Data Mining*: URL http://www.geog.psu.edu/icavis/rhyne98.html (accessed 22 December 1998).

Rhyne, T.M. and Taylor, D.A. 1997. *Carto project survey report March 1997*: URL http://siggraph.org/~rhyne/carto/cartosurv.html (accessed 11 March 1997).

SGI. 1996. Making A difference in Bosnia. *IRIS On line*, 4, 2 (Electronic newsletter).

Verbree, E., van Maren, G., Germs, R., Jansen, F. and Kraak, M.-J. 1999. Interaction in virtual world views – Linking 3D GIS with VR. *International Journal of Geographical Information Science* 13, 4, 385–96.

6 Visual exploration of virtual environments

Menno-Jan Kraak

Introduction

Consulting a map is the preferred method to get an overview of an area of interest. A map puts spatial relations into perspective and helps us understand our world. Today's maps can give the answers to many questions concerning the area depicted, since they are often used with a direct link to spatial databases. Not only do they visually reveal information such as the distances between objects, object locations in respect of each other, and the nature of distribution patterns, additional information, such as multimedia objects or database listings, is available at a single mouse click by way of programmed-in links.

Our view of maps has changed in at least five ways.

- First, of the large number of maps now being created using computers only a few are intended as a final product. Admittedly, these are the maps that most people use and they often serve a wide audience. Examples are topographic maps, newspaper maps, road maps and atlases. However, even these cartographic representations will change in their appearance, because in the near future it will be the user composing their content. However, the majority of today's maps are created for a single use only, to display alternative routes between A and B, to show the contents of a data set, or to view the intermediate results of a spatial analysis. These maps are used to explore the available data.
- Second, professional cartographers no longer create most maps. Map users have become their own cartographers. Because of their experiences in their own discipline, these new cartographers have different expectations of a mapping environment. This wider use has exposed mapping to positive external influences of which scientific visualization and the World Wide Web are examples (Hearnshaw and Unwin, 1994; Kraak, 2000; MacEachren and Taylor, 1994; Plewe, 1997). Both of these influences are characterised by both interaction and dynamics. Interaction refers both to navigating and manipulating the data.

New users want to be able to click on a map and expect an immediate, dynamic reaction.

- Third, these developments are also forcing cartographers to adopt a more demand-driven approach. Morrison (1997) referred to this trend as the democratisation of cartography, a result of which, he notes, is that 'using electronic technology, no longer does the map user depend on what the cartographer decides to put on a map. Today the user is the cartographer ... users are now able to produce analyses and visualizations at will to any accuracy standard that satisfies them...'
- Fourth, advances in technology have improved the whole spatial data handling process and, in turn, this has put new demands on visualization. This has resulted in both multiple and more complex data sets with more frequent updating and greater dimensionality of the data.
- Fifth, the World Wide Web has played an increasingly important role in these developments. In the framework of the spatial data infrastructure, it is used both to disseminate and to search for spatial data. A consequence of this has been that, in addition to their traditional presentation function, maps can have three major other functions (Kraak, 2000):

    - as an index to the available data. Where maps are used as indexes, they can guide users to other information. It is, for example, possible to click on a region, or any other geographic object, to lead to a list of links referring to other maps, spatial data sets, or other multimedia data.
    - as a preview of the available data. The quest for spatial data can be assisted by maps that visualize the data that can be obtained. Several organisations and data vendors offer the potential customer a preview from which to determine the data's suitability.
    - as part of a search engine, maps can represent the location component of the data. The attribute and/or temporal component are often complemented by textual search.

Figure 6.1 summarises the above trends from the point of view of visualization. It shows two different strategies being used. On the one hand, the visualization process is part of the traditional presentation realm of cartography, involving a well-structured approach to presenting well-known spatial data sets. On the other, the visualization process coincides with the new cartography involving the World Wide Web. Maps are created to explore spatial data sets, to support a geoscientist in solving a problem. They stimulate what has been called the visual thinking process. This type of cartography is called exploratory cartography (Kraak, 1998; MacEachren and Kraak, 1997).

Irrespective of the strategy applied, maps have their strengths and weaknesses. An important characteristic is their ability to offer an abstraction of reality. Maps simplify by selection, but at the same time, when well

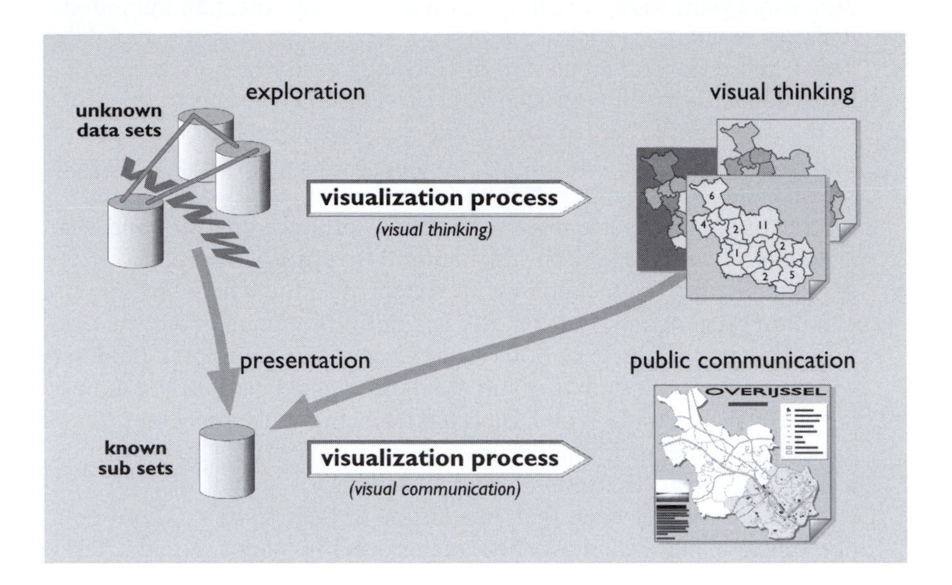

Figure 6.1 The cartographic visualization process.

designed, they put the remaining information into a clear perspective. This characteristic is well-illustrated when one compares a map with an aerial photograph of the same area, but it also shows that selection means interpretation.

Virtual reality

With their link to spatial databases, maps can be seen as a visual interface to spatial data. This abstract map view on the data has proved its usefulness and is currently going through a revival. Couclelis (1998) argues that a geographic view, represented by maps, will be helpful in dealing with large amounts of data, but she also highlights the problem people have in reading conventional maps. However, according to her, they seem to adapt to crude realistic environments of the sort currently provided by VR quickly, although Stanney (1995) claims that they get lost easily in larger environments.

VR creates a three-dimensional environment from the data and developments in it promise a much more involved way of interacting with data. Critically in VR, users become part of the data set, part of a digital world where they can explore and interact with the data. The experience of being part of the data, called 'immersion', is enhanced by the use of stereoscopic three-dimensional images, sound and real-time interaction. The VR environment gives the user a sense of the real environment. According to Neves and Camara (1999) this ability to 'feel' results of particular actions leads to a clearer and more natural understanding of these actions.

Immersion is realised because of the direct coupling of the viewing position and the image on the display. In advanced systems, the user's head position is constantly tracked and fed into the display algorithm to calculate at each moment a correct stereoscopic and perspective display of the scene, resulting in the immersive illusion. An example of full immersion is a CAVE system, for example a 3×3×3m cube with through-projection on three sides and a fourth projection on the floor offering its user a stereoscopic surround projection. At the so-called 'low-end', desktop VR offers just a three-dimensional view on a particular environment. Typically, these systems use VRML to create and manipulate these display environments (Rhyne, 1999).

Currently, as seen in other chapters in this book, many experiments involve some combination of VR and GIS. Earlier studies concentrated on the more technical issues, such as how to link both, but a major result of these early experiments was the realisation that VR can be seen as a natural interface to visualize and interact with the data through a very intuitive interface (Faust, 1995).

In their work Verbree *et al.* (1999) suggest combining both a two-dimensional GIS interface with a three-dimensional VR interface. This results in an interface to the data that includes a plan view, a model view and a world view. The user can switch between those views when needed, but the most suitable view depends on the type of user, the task at hand and the medium (Stanney, 1995). Figure 6.2 presents this multiple view approach.

The plan view is the conventional two-dimensional map interface. Here the user can create and manipulate the objects as symbols and can access standard GIS functionality through graphical interaction and alphanumeric queries. Navigation is possible through panning and scrolling. The

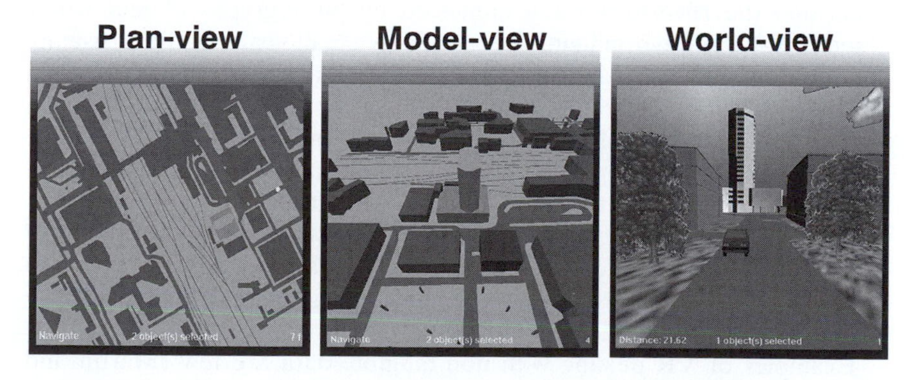

Figure 6.2 VR as interface to spatial data: a multiple view approach. Left, the plan view. Middle, the model view, and right, the world view.

level of detail can be adjusted by zooming in and out, by adapting the scale of the map, or changing the level of detail of the map representation.

In the model view the data are presented as a simple and symbolic perspective in 2.5 dimensions, a bird's eye view. The user looks down on the model as if it were a three-dimensional scale model but is not immersed within it. This form of representation is very suitable for manipulating the individual objects such as is needed for positioning and orientation. Objects can be grouped or organised in a hierarchy or layer and manipulated using these relations. Navigation is by 'fly-through'.

The world view is the immersive view of a virtual world. The user sees the model from a position within the model. The purpose of this view is to give a realistic impression of the environment and so real-time image update is needed to maintain the illusion of 'immersiveness' and to allow 'walk-through'.

Each of these types of view will have a specific way of interaction to, for example, select objects and specify queries. Each will also limit the possibilities of the user getting lost since they all will simultaneously show where the user is. Each is also suitable for presentation to a wide audience through the World Wide Web although, with current technology and bandwidth, this approach is limited to a desktop VR version with the world view limited in its use by the high data density required.

Working with geographical data in virtual environments

It is instructive to examine these three views of data in relationship to whether or not the reality to be visualized is tangible or non-tangible. The first deals with views of real topography whereas the second deals with views of, for example, climate or population density. Most current VR applications use data representing the tangible natural and built environment. These are environments that can also be traversed by users in reality as well as in the virtual world. Built environments are relatively large scale, and the realism is often enhanced by photographs of real world objects. Natural environments are relatively small scale, and are usually enhanced by devices such as draping satellite imagery over digital terrain models. For both scales, and in many applications, one often tries to reach as high a degree of realism as possible, but this is not always the case. Indeed, from a cartographic perspective, one can question if this realism actually increases insight into the data. The multiple view approach presented in the previous section offers both an abstract and a realistic view of the data. Research is currently being undertaken to judge the effectiveness of the realistic environments when using geographic data sets with the aim of getting better understanding of these data.

Examples of VR dealing with non-tangible data are few (Martin and Higgs, 1997). In this category one can distinguish between physical and socio-economic data. Examples of the first are temperature, rainfall and gravity, while population density, income and language fit into the second. In Chapter 18 of this book, Moore provides an intermediate example in which the 'real' urban environment of Leicester is represented by greatly simplified buildings coloured according to the land use at all floors of the buildings.

Figure 6.3 shows how these non-tangible data can be mapped into the multiple view approach outlined above. The plan view is straightforward. This is the familiar thematic map. Similarly, the model view is not new, but, because of technological constraints in the past it has not been often used. It presents the data in a 2.5 dimensional 'perspective' view and a good example is the so-called 'prism' version of the standard choropleth map. In the case of the world view and non-tangible data one can ask many questions. In Figure 6.3 examples are given for population density and rainfall. Photographs of many people together might give a feel for high population density and it is possible that photographs of people with umbrellas could give an idea of heavy rain. How we might incorporate the notions that can be derived from photographs and the like into an immersive world, and in ways that actually increase our understanding, is currently an open question. One can, of course, also think of introducing senses such as sound or smell in the immersive environment.

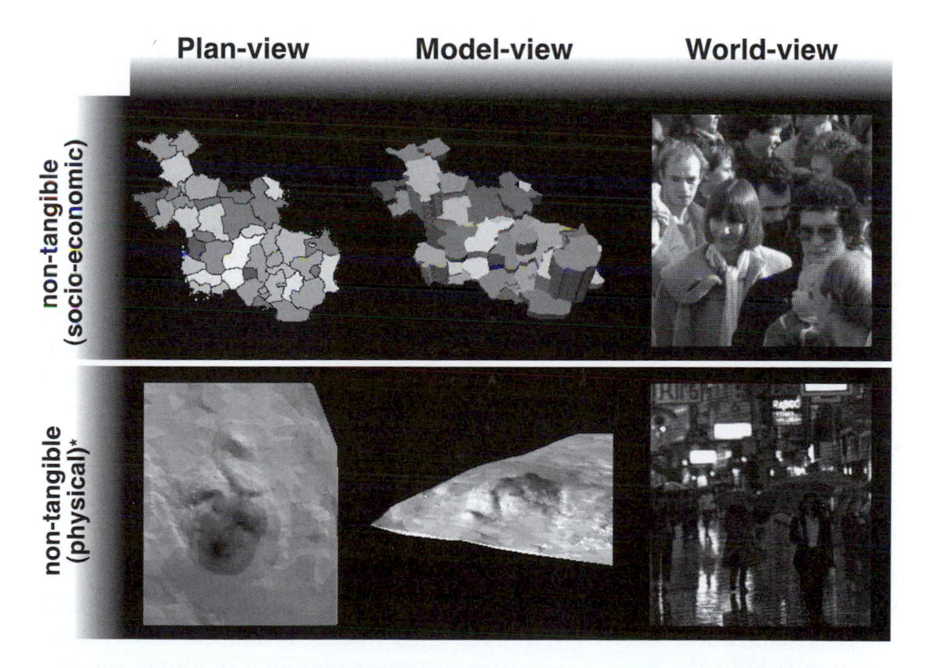

Figure 6.3 The multiple view approach applied to non-tangible data. Above, the example of population density; below, the example of rainfall.

Visual exploration of geographic data in virtual environments

Virtual environments are created with a specific purpose in mind. This could, for example, be to give an audience an idea of the past, present or future state of a particular area, and the virtual environment could also be used to investigate the area by visual means. The visual exploration of geographical data involves a discipline expert, dealing with mostly unknown data, creating maps. These generally have a single purpose, and function as expedients in the expert's attempt to solve a geographical problem. This kind of exploratory visualization will strengthen the shift from supply-driven cartography to one that is demand driven and the development of proper tools for it is one of the most pressing tasks in cartography.

How does the multiple view approach fit in? Functionality to explore large geographic data sets has been described in Kraak (1998), but how well can these functions be used when dealing with tangible and non-tangible data in virtual environments? At minimum, the following tools seem to be needed for data exploration.

- *Basic display*. Each view, whether plan, model or world, needs tools to allow the user to pan, zoom, scale, transform and rotate the image. These geometric tools should be view independent.
- *Navigation and orientation*. These involve the keys to the map. At any time, users should be able to know both their location and what the map symbols mean. Orientation functions are of particular importance in the model and world views, since it is relatively easy to get lost in them. However, working with multiple views on screen together should be of great help in maintaining the user's orientation. The meaning of the symbols can be derived from a pop-up legend or by clicking the objects themselves.
- *Data query*. At any moment, and from each of the views, the user should have query access to the spatial database from which the images were generated. The questions should not necessarily be limited to simple 'what?', 'where?' or 'when?' and might well be of a different nature in each view.
- *Multi-scale*. Combining different data sets is a common operation in an exploratory environment but it is unlikely that these sets will have the same data density or the same level of abstraction. In virtual environments, and especially in the world view, the level of detail problem needs attention.
- *Re-expression*. To stimulate visual thinking, an unorthodox approach to visualization is often required. This needs options to manipulate data behind the map, or to offer different mapping methods for displaying the data. In the map-like plan and model views these options can and have been realised, but in the more realistic world view it is not so obvious if such re-expression is needed.

- *Multiple dynamically-linked views.* The whole multiple view approach is based on dynamic linking of the views. The user is able to view and interact with the data in different windows, all representing related aspects of the data and additional windows can be provided to display extra maps, video, sound or text.
- *Animation.* Maps often represent complex processes that can be displayed using animation. Animation can be used for temporal as well as non-temporal changes in spatial data. Since exploratory users will also be the creators of the animations, they should be able to influence its flow.

There are no real differences in functionality for tangible or non-tangible data except that, as we have noted, we do not yet know how best to represent non-tangible data in the world view. Figure 6.3 suggests that this is rather different, and this might result in other needs for data exploration in these environments.

Discussion and conclusion

The combination of virtual reality and geographic data has promise. From a combined perspective, the multiple view approach offers realism via the world view, while the model and plan view provide abstract and selective map views on the data. From a mapping point of view, it is interesting to see how these developments can be combined with the dissemination of spatial information via the World Wide Web and which should lead to widespread use of advanced visual exploration.

What are the current possibilities? Looking at the combination VR and visual exploration, the upper table in Figure 6.4 shows what options are available at the time of writing. Working with tangible data, most of the functions described above can be realised using standard computing technology. For the non-tangible data, fewer options exist. Functions in the world view do not yet exist, and even for the other views experiments have to be conducted to evaluate this approach.

If we look at the possibilities of combining the multiple view approach with the World Wide Web, it can be seen from the lower table in Figure 6.4 that, at the moment, the options to use the world view hardly exist. For tangible data this is due to the need for a photo-realistic view of the geographic data that currently requires too much bandwidth. For non-tangible data it is due to the fact that we simply do not know how to represent these data in world view. From the diagram and earlier discussions in this chapter, it becomes clear that many questions remain and that experiments are needed to demonstrate how potential solutions might work in practice. However, any solution to this problem promises to provide an unorthodox new view for the advancement of visual exploration of geographic data to support users in their visual thinking.

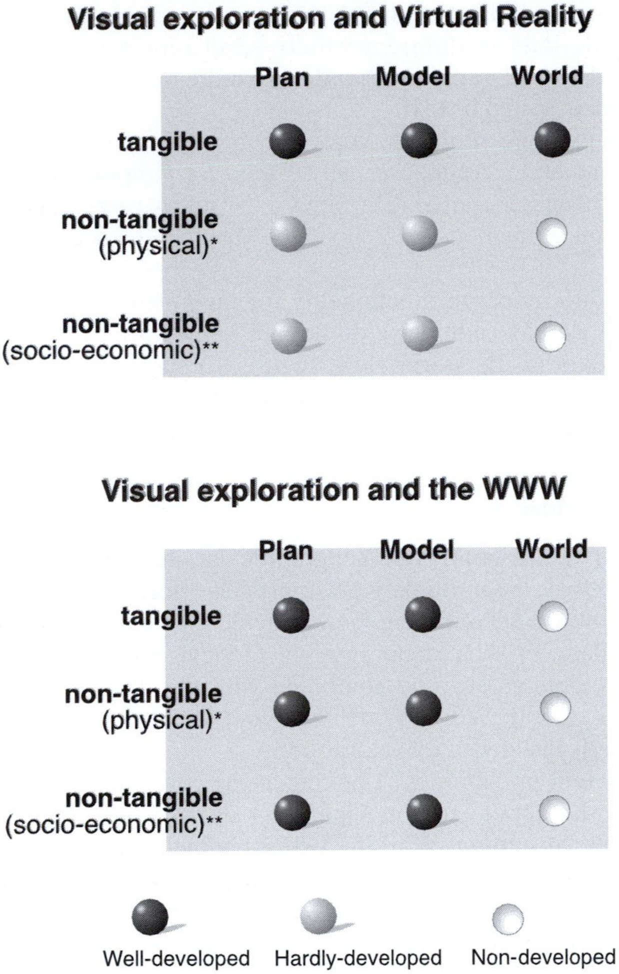

Visual exploration and Virtual Reality

Plan Model World

tangible

non-tangible
(physical)*

non-tangible
(socio-economic)**

Visual exploration and the WWW

Plan Model World

tangible

non-tangible
(physical)*

non-tangible
(socio-economic)**

Well-developed Hardly-developed Non-developed

Figure 6.4 Multiple views and visual exploration of geographic data. The upper diagram looks at options in relation to virtual reality, while the options to use the World Wide Web as a medium are given in the lower diagram. In current practice, non-tangible data of a physical nature are dealt with by scientific visualization software (see MacEachren *et al.*, 1999; Tufte, 1997), while the socio-economic non-tangible data are visualized using experimental software (Andrienko and Andrienko, 1999; Dykes, 1997). Tangible data are manipulated by GIS software and dedicated VR software.

References

Andrienko, G.L. and Andrienko, N.V. 1999. Interactive maps for visual data exploration. *International Journal for Geographic Information Science*, 13, 4, 355–74.

Couclelis, H. 1998. Worlds of information: the geographic metaphor in the visualization of complex information. *Cartography and Geographic Information Systems*, 25, 4, 209–20.

Dykes, J. 1997. Exploring spatial data representation with dynamic graphics. *Computers & Geosciences*, 23, 4, 345–70.

Faust, N.L. 1995. The virtual reality of GIS. *Environment and Planning: B, Planning and Design,* 22, 227–68.

Hearnshaw, H.M. and Unwin, D. (eds). 1994. *Visualization in Geographical Information Systems*. London: Wiley and Sons.

Kraak, M.J. 1998. Exploratory cartography, maps as tools for discovery. *ITC Journal,* 1, 46–54.

Kraak, M.J. 2000. *WebCartography Developments and Prospects*. London: Taylor & Francis.

MacEachren, A.M. and Fraser Taylor, D.R. (eds). 1994. *Visualization in Modern Cartography*. London: Pergamon Press.

MacEachren, A. and Kraak, M.J. 1997. Exploratory cartographic visualization: advancing the agenda. *Computers & Geosciences* 23, 4, 335–44.

MacEachren, A., Wachowicz, M., Edsall, R. and Haug, D. 1999. Constructing knowledge from multivariate spatiotemporal data: integrating geographic visualization with knowledge discovery in database methods. *International Journal of Geographic Information Science*, 13, 4, 311–34.

Martin, D. and Higgs, G. 1997. The visualization of socio-economic GIS data using virtual reality tools. *Transactions in GIS*, 1, 4, 255–66.

Morrison, J.L. 1997. Topographic mapping for the twenty-first century. In Rhind, D. (ed.) *Framework of the World.* Cambridge: Geoinformation International, pp. 14–27.

Neves, J.N. and Camara, A. 1999. Virtual environments and GIS. In Longley, P., Goodchild, M., Maguire, D. and Rhind, D. (eds) *Geographical Information Systems: Principles, Techniques, Management, and Applications*. New York: Wiley and Sons, pp. 557–65.

Plewe, B. 1997. *GIS Online: Information Retrieval, Mapping and the Internet*. Santa Fe: OnWord Press.

Rhyne, T.M. 1999. A commentary on GeoVRML: A tool for 3D representation of georeferenced data on the web. *International Journal of Geographic Information Science*, 13, 4, 439–43.

Stanney, K.M. 1995. Realizing the full potential of virtual reality: human factor issues that could stand in the way. *Proceedings of Virtual Reality Annual International Symposium '95*. Research Triangle Park, North Carolina: IEEE Computer Society Press, pp. 28–34.

Tufte, E.R. 1997. *Visual Explanations*. Cheshire, Conn: Graphics Press.

Verbree, E., van Maren, G., Germs, R., Jansen, F. and Kraak, M.-J. 1999. Interaction in virtual world views – linking 3D GIS with VR. *International Journal of Geographical Information Science*, 13, 4, 385–96.

7 Creating information-rich virtual environments with geo-referenced digital panoramic imagery

Jason Dykes

Introduction and objectives

The introduction to this section identified virtual reality (VR) as an interface between computer and user in which digital information is presented to the user graphically, in an immersive and engaging format, whereby the user's experience of operating in the real world can be used successfully to interact with the information. A whole series of techniques, devices and combinations of hardware and software can be used to present information in this way, with a wide range of levels of interaction, immersion and engagement that utilise the user's experiences of the real world to a greater or lesser extent. Within this scope, the VR representations created by computer programs can vary greatly in terms of their authenticity and realism.

This chapter presents an approach to VR that involves lower levels of engagement than most of the other offerings contained in this book. Whilst the interface to spatial information relies less directly on the relationships between the real and computer-derived environments than the majority of these, the imagery displayed on the screen is photo-realistic. The approach utilises a relatively new digital data type, the 360-degree panoramic image and describes how it can be used in combination with comparatively small amounts of metadata to create an information-rich and implicitly spatial virtual environment. A software implementation is presented and the chapter discusses experiences gained from dealing with panoramic imagery in this way. Circumstances in which this use of digital panoramic imagery is appropriate are identified and an assessment made of likely near-future developments within the field of multimedia cartography.

The chapter is presented in the context of a real application, that of using virtual reality to support fieldwork in undergraduate teaching. This is particularly appropriate in an assessment of VR since the use of fieldwork in the curriculum has parallels with the use of real-world metaphors in computer interfaces to represent information in a familiar format. However, when teaching fieldwork the relationship is in reverse. Selected experience of the real world is used by the teacher in order to examine and test familiar theoretical information presented to students during their education. Fieldwork is a much used and effective learning device (Gold *et al.*, 1991). The work presented here aims to provide insight into this new medium with examples and experiences of its use and examine the potential for exploiting some of the benefits of a 'virtual reality' interface to spatial information with digital panoramic imagery. Because these representations do not attempt to convey information in three dimensions, in the context of this book as a whole this is an unusual application. However, this restriction enables us to consider the essence of VR and so examine its bounds and review its roles and uses in geography.

Advances in cartography

Tobler's view of cartography as a series of transformations (Tobler, 1979) was introduced in Chapter 2. His model presents the cartographic process as a flow of information from the real world, into a formal recognised model, from which a particular cartographic representation is created. This results in a map image being produced by a map user. Although this approach provides a useful model for the cartographic process, through which map makers can ensure that their product is fit for the intended use, advances in computer technology have affected the manner in which maps are produced in a number of ways. Initially they drew attention to the fact that an infinite number of maps can be produced from the same information. Robinson *et al.* (1996) describe the cartographer's task as being 'to explore the ramifications of each mapping possibility and to select the most appropriate for the intended task'. Digital databases meant that successive maps could be created with far less effort and expense than their hand-drawn equivalents, but also meant that more maps were produced by those without the necessary skills. Digital methods were developed directly to address and resolve these issues. Higher levels of interaction between screen and digital map user have led to a well-documented shift in the roles of the cartographer and map user, whereby the latter has been able to change map parameters and undertake cartography on the fly to suit their requirements (Monmonier, 1991). An extreme case is that of VR, where the user changes the location from which successive oblique views of geographic information are calculated from a model by interacting directly with the representation. As this volume demonstrates, the levels of interaction and detail of the models themselves are increasing rapidly. These advances have resulted in a new role for interactive maps in the exploratory stages of the research process, known as 'visualization' (MacEachren, 1994).

An additional feature of the evolution of information technology has been the development of new data types and formats (Cartwright, 1997). The collection of many of these contrasts sharply with the traditional techniques of census, ground survey and remote sensing. The data themselves, however, are often associated with higher levels of currency, realism and availability than those collected through formal survey. They are often purchased from independent suppliers and so relieve the cartographer of any control over the first cartographic transformation. They can provide important ancillary information for geographical analysis and augment the cartographic toolbox with exciting new means of representing geographic information.

This chapter reports how a certain set of circumstances led to the development of cartographic software that incorporates and maps one such accessible data type in order to create a virtual environment that can be used for visualization. Uniquely in this book, the approach does not use three-dimensional graphics to create its virtual environment and the degree to which the user and representation are united is very low. Yet it is well suited, under particular constraints, to the specific purpose for which it was designed. It is highly interactive and engaging, utilises real-world affordances, and projects a sense of place and space. This encourages consideration of what we mean by, and want to achieve with, VR, and centres attention on producing a successful correspondence between VR application and purpose, requirements and resources. This alternative to generating VR models ‘because the technology exists’ corresponds with the focus that visualization has put upon map use (MacEachren and Kraak, 1997) and addresses an issue highlighted in Chapter 2 with a practical application.

A virtual reality application

All maps should be based upon a clearly-specified set of objectives, and VR interfaces to geographic data are no exception. Evidently, VR has a huge number of practical applications. That considered here is the support of teaching undergraduate students in the field. VR is an attractive way of enhancing fieldwork for a number of reasons. First, there is a parallel between the way in which students learn by operating on the real world when in the field and ways in which they use VR representations of digital spatial information. Laurillard (1997) identifies the important role that computer models can have in providing adaptive feedback during the process of academic learning. Second, it provides the potential to fulfil a number of objectives of fieldwork teaching. In a survey of all UK academic institutions running field courses the primary objectives of fieldwork stated by respondents included teaching students to observe and interpret the physical environment and encouraging students to make comparisons between different places, models and reality and over time (Williams *et al.*, 1997). Third, a computerised environment for fieldwork might improve a well-known deficiency in fieldwork teaching, that the amount of time that students spend analysing and interpreting field data does not justify the efforts put in to data collection (Haigh and Gold, 1993). Using VR to support fieldwork can ameliorate this by supplying analytical software in a relatively seamless manner with digital information that reminds students of their field experience during the debriefing and follow-up work that occurs on returning from the field trip.

Suitable software that provides a virtual field trip can be used in all four of the fieldwork teaching modes identified by Kent *et al.* (1997), namely preparation, engagement in the activity, analysis and debriefing. Students can prepare for their trip by using VR to navigate through data in a geographical way that mirrors the freedom and provides the sense of exploration that they will have in the field. They might, for example, record a route based upon data representations in VR and follow it in the field. Sufficiently information-rich virtual environments provide a suitable means for engaging in fieldwork by making observations and comparisons in the VR. Locations that cannot be visited or viewed for reasons of safety, accessibility, history or geography are particularly appropriate. Analytical functionality can be provided by the digital virtual environment, with locally-collected data being added to the representation for analysis during the field trip. This immediacy is extremely advantageous (Dykes *et al.*, 1999). In addition, these data can be made accessible in combination with further secondary information to aid analyses and to add context to the interpretation of results during debriefing on returning from the field (Williams *et al.*, 1997)

These considerations gave rise to the Virtual Field Course (VFC) project (Virtual Field Course, 2000), which was funded by the Joint Information Systems Committee (JISC) of the Higher Education Funding Councils of England, Scotland and Wales, to examine the potential for providing VR to support teaching in the field and to produce appropriate software for use in higher education institutions in the UK. In undertaking this remit VFC software was developed using a number of 3D environments that permitted a range of virtual realities. These included VRML/Java (Moore, Chapter 18), Java/OpenGL (Wood, Chapter 12) and ESRI/Sense8 (Raper *et al.*, 1998). Each of these alternatives renders a two-dimensional oblique image on the computer screen based upon a digital elevation model. A variety of data drapes can be loaded onto the models. The user has control of the position from which the view is calculated and can select any location or series of locations in turn. These are chosen either by clicking locations on a traditional planimetric map, loading a route, or interacting directly with the oblique view. Each piece of software is sufficiently fast and engaging whilst retaining adequate detail for the user to call on their cognitive experience and perceive that they are interacting with a three-dimensional virtual environment. This visual trickery results in a strong sense of immersion being felt despite the fact that the computer is merely presenting a flick-book of successive two-dimensional images on screen. Users frequently move with the surface, duck whilst navigating around obstacles, and express their desire to 'go' somewhere, or the fact that they are 'lost'.

These data surfaces are characterised by infinite flexibility in terms of the user's selection of locations and movement between them, but have finite detail at any particular locality. This results in very real limitations regarding the user's ability to undertake an important fieldwork objective that of performing detailed observation of the immediate locale. The actual limitations to resolution of this approach to VR can be determined by hardware (processing speed, memory or graphical capabilities), software (for a discussion of the limitations of VRML browsers see Moore *et al.*, 1999) or data. A number of significant efforts have been made to address hardware limitations and map larger, higher-resolution data sets less inefficiently so that greater detail is rendered at positions that are close to the viewpoint. Mip maps (Williams, 1983) vary drape resolution in this way. The geo-VRML group are using level of detail (LOD) successfully to overcome some of the limitations of the VRML 97 specification (Reddy, 1998). Both model and drape change resolution in real-time in response to the position of the viewer and features of the terrain with Muchaxo's wavelet-based LODs (Muchaxo *et al.*, 1999). View-dependent meshes (Hoppe, 1997) are particularly impressive in that the resolution of the elevation model varies in relation to the positions of the viewers so that parts of the model that are within the field of view, closer to the viewers and inclined towards them, contain higher levels of data density than those which are out of view, distant or inclined away. Certain technologies are able to stream relevant information across the Internet and use it in creating a new view as and when required by the user's interaction with the interface (Abadjev *et al.*, 1999).

These developing technologies are extremely impressive and promise to endow home PC users with more detailed and realistic VR in the near future. In the case of the VFC software, the limiting factors are related to the hardware specification of the standard higher education institution laboratory PC, the cost of digital elevation data for the UK, the resolution of data sets required as drapes for selected field areas, and the time and resources available to teaching staff in order to provide their students with relevant and useful VR. This final point is important. In order to be beneficial to the largest possible proportion of the community it was evident that software needed to be simple and intuitive to use with a rapid learning curve, flexible with respect to the locations and subject matter with which it could be used, and relatively straightforward for the teacher to set up. It was anticipated that the use of non-GI data types and formats, which could be found on the Internet or collected rapidly on the ground, would make the software accessible to the greatest number of teachers. Effectively, when designing interactive mapping applications for teaching, these considerations of usability, utility and efficiency are part of the assessment of map use undertaken by traditional cartographers. The delivery of a virtual environment that corresponded to these requirements and in which detailed observation could take place necessitated the investigation of an alternative approach to that of the surface model, which complimented other VFC software products.

Under these circumstances it was decided to explore the utility of a new medium that provides high levels of detail on screen to permit observation and comparison: the digital panoramic image. In effect, when you want to see specific details, which are necessary for observation in the field, imagery provides a solution. A large number of academic applications take advantage of imagery to add realism in VR (for an example, and strong support for the role of visual realism in understanding process, see Bishop, 1994). In the gaming arena, *Championship Rally*™, developed by Magnetic Fields Software Design Ltd and distributed by Europress, uses thousands of photographs taken from video that was collected along every yard of a 700 km route across the UK to provide extremely high levels of detail and realism. This results in 'a feeling of reality on these tracks that can't be achieved any other way' (IGN, 1999). Bamboo.com uses linked panoramic imagery for commercial partners to provide more seductive tours of cruise liners, TV studios and real estate than are achievable with 3D CAD models.

A number of authors have taken advantage of panoramic imagery within the spatial sciences. Shiffer (1995) reports the successful use of panoramic views in public planning meetings and uses an HTML image map to present Quicktime VR (QTVR) panoramas in NCPC's Virtual Streetscape (Ferreira *et al.*, 1996) and Doyle *et al.*'s (1998) Wired Whitehall presents a number of panoramic images through a viewer that permits spinning, zooming and hyperlinking in a similar way; Krygier (1999) identifies the utility of comparing map and image when examining nineteenth-century reports of the exploration of the American West and notes the benefits of viewing the maps in terms of the panoramic landscape views with his multimedia software. Each of these authors takes advantage of the fact that, when linked to a map, interactive panoramas present photo-realistic detail in combination with spatial information. However, in these cases the links between map and image are restricted to a static map displaying image locations that is used as a spatial menu for selecting images. In terms of the VFC software requirement, it was determined that additional interaction between maps and images could result in a virtual environment that was both sufficiently immersive to embrace the advantages of VR in teaching whilst incorporating sufficient photographic detail to support virtual 'fieldwork'. A novel form of virtual environment was thus proposed whereby the third dimension, that is so fundamental to evoking the senses of immersion and engagement in VR applications, was eliminated from the data model and interface. Instead, digital panoramic photography was employed to provide familiar, realistic views. High levels of interaction between view and map, and user and software, place strong emphases on location and direction in order to ensure the required impressions of place and navigation, affordances of real world, sense of immersion and high levels of engagement. This form of presentation possesses a number of characteristics that contrast with those of software used to produce surface models. Some of these are outlined in Figure 7.1

Such an interface could be used to present an array of multimedia data types in addition to panoramic imagery. These would enrich the environment, encourage the sense of immersion and aid in the fieldwork. In this way, readily collected geo-referenced data could be added to the virtual environment in whatever digital medium they were gathered. Additionally, the requirement for analysis can be readily supported in a 2D application through the provision of interactive graphics for visualization.

Collecting, stitching, geo-referencing and linking panoramic imagery

The production of the specified environment requires the collection of photographs, their combination into panoramic images, and their geo-referencing and linking in software that will display them appropriately so that navigation can take place. The data collection aspect of this process is relatively straightforward, hence the attraction of the digital data type for fieldwork teaching. A digital camera with the equivalent of a 35 mm lens (my experience is with the Olympus™ Camedia 800L) requires nine pictures in order to record the compete horizon through 360 degrees with adequate overlap. Alternatively an analogue camera can be used and the developed images scanned. A 28 mm lens uses approximately seven photographs in order to record the required information. Once these data are acquired, a number of software packages are available to stitch the pictures into a single image. These include Spin Panorama™ (PictureWorks Technology Inc., 1999), QTVR™ (Apple Computer Inc., 1999a) and PhotoVista™ (Live Picture Corporation, 1999). The process usually involves the selection of the images that make up a panorama, an automated preview and an element of tinkering by the user in order to ensure that the images are correctly matched. In the preview stage the software adjusts the image to take account of the distortions generated by the specified lens, detects common features in adjacent images, and overlays the images to create a single strip of imagery. The success of the initial operation depends upon the software, the degree of contrast between images, and the existence of recognisable shapes in the area of overlap between successive photographs. The software is sufficiently robust to take account of minor variations in the height at which adjacent images were taken and so eliminates the need for a tripod or panoramic tripod head, meaning that data collection is a quick and simple process that simply involves pointing and clicking (nine times!). The exact positioning of each image can be

	Virtual environment based upon 3D surface model	*Virtual environment based upon imagery and 2D map*
Viewer position	Infinite control over viewer position	Finite viewer positions
Screen detail density (data per pixel)	Low local, higher at distance	Constant across view
Spatial detail density (data per unit ground area)	Spatially constant detail density (excellent coverage)	Spatially variable detail density (information rich at photographed points, sparse elsewhere)
Data collection	Data acquisition requires survey	Data addition requires photography on ground

Figure 7.1 Comparing panoramic and surface-based virtual environments. The table highlights the differences between virtual environments based upon continuous surface models and that presented here using digital photography. The diagrams use grey scale to represent detail density, with darker shades denoting greater detail. Those associated with screen detail density represent a virtual view of the environment through the VR interface with a horizontal line depicting a nominal horizon. These are the equivalent of an oblique view. Those associated with spatial detail density show the resolution of the model and constitute planimetric views. This is constant for a continuous lattice or height field, but diminishing from points of high detail, and obscured by opaque objects close to the points at which photographs are taken in the panoramic example.

Figure 7.2 Creating panoramic images from digital photographs. A series of nine images are taken with small sections of overlap (top). These are blended and stitched by appropriate software. Here PhotoVista (Live Picture, 1999) is used. The sections shown demonstrate that adjacent images that contain features with distinct shapes, such as the section showing the island (centre left), tend to be blended together automatically with considerable accuracy. Those that do not, such as the images spanning the horizon across a stretch of sea (centre right), tend to be merged into a single image and require input from the user before a full and exact panoramic image can be created (bottom).

varied and confirmed manually before a final stitching process combines the photographs by blending the areas of overlap. Figure 7.2 shows that images can then be saved in any of the popular proprietary formats. If saved as standard images (in GIF or JPEG format, for example) the data form a rectangular strip with left and right ends that represent a break in the information, but which join precisely and seamlessly if connected together. An alternative technique for collecting suitable photographic data is to use digital video and suitable software. By panning a video camera steadily around the horizon, data are recorded from which a single panoramic image can be created with specialist software such as Videobrush Panorama™ (Dykes, 2000; Videobrush Corporation, 1998).

A number of options exist for displaying panoramic images. Whilst proprietary software and formats will permit spinning, zooming and linking to other images, the specification under consideration here required the flexibility to embed the images within specifically-programmed software functionality. A number of options exist for doing this, including the Java API for QuickTime VR™ (Apple Computer Inc., 1999b), which provides the ability to embed QTVR objects in Java programs. In this instance the data were presented using the Tcl/Tk scripting language, which provides a series of high-level programming commands, functions and objects that are suitable for highly-dynamic cartography (Dykes, 1996, 1999). These include support for the display and manipulation of digital imagery, the specification of vector graphics with particular observer-related behaviours, and the ability to issue external commands that run additional computer processes. The high-level nature of the environment makes it particularly suitable for the kinds of rapid prototyping associated with experimental software design and the investigation of new types of cartography and map use.

Geo-referencing panoramic images and linking them both to a planimetric map with symbols that identify the direction of view and each other are not new individually. Egloffstein's sweeping, 180 degree views produced during the Great Reconnaissance were marked with notations and symbols explicitly connecting panoramas and maps (Krygier, 1997). Landscape features are named along the bottom of the panoramas with more distant features being labelled closer to the bottom edge of the panoramic view. Each of these is lettered and the vantage points of the panoramas are noted on the expedition maps which Krygier (1997) describes as 'straining towards the abstraction of the map'. Imhof (1951) identified the utility of relating oblique and planimetric geographic information, using radial lines on a map view linked to vertical lines on an associated perspective landscape view to reinforce the comparison. The examples produced by Doyle *et al.* (1998) and Ferreira *et al.* (1996) contain static maps identifying image locations and permit linking between views. The software specification and implementation presented here is unique in that it combines these qualities and adds dynamic symbolism and a degree of interaction that force the application into the bounds of VR. The links between panorama and map are strong, direct and mutual, and in consequence the software is dubbed 'panoraMap'. In addition, it extends the dynamic mapping capabilities to provide visualization functionality and is sufficiently flexible to incorporate additional multimedia data and panoramic images at the click of a button at an appropriate location on the map.

This specification is achieved by supplying a few significant elements of metadata in association with each datum used by the software. Knowledge of the extents in each direction of any map using a rectangular projection enables the calculation of any position within it and so each map requires these metadata. Maps are loaded from CompuServe™ GIF images, which fulfils the requirement of avoiding GI-specific formats and using popular data types as GIFs are readily generated by software or screen dump and are used a great deal on the World Wide Web. Panoramic images read by panoraMap must also be in GIF format and require metadata denoting their location in the coordinate space of the map on which they are to be plotted, a bearing in degrees clockwise from north denoting the direction of the left-hand side of the image (the point at which a 360-degree panorama is split), and an angle indicating the extent of the horizontal view visible in the image strip (360 degrees for a full panorama). These metadata are equivalent to Egloffstein's noting of the vantage point and the left and right extreme bearings at the bottom of the panorama (Krygier, 1997). In combination these descriptions of locations and angles enable points to be plotted on the map at which any panorama is located, bearings to be made on the maps from a panorama location that relate to a position on the relevant panoramic image, horizontal locations along a panoramic image to be related to bearings on the map from the relevant panorama location and bearings between any number of panorama locations to be calculated (Dykes, 2000).

These calculations and the capabilities provided by Tcl/Tk to display images in graphical objects, copy sections of other images into these, add interactive symbols to graphical objects and program images and symbols with interactive behaviours, are the basis of panoraMap. A base map, generated from a GIF with appropriate metadata, forms the basis of the display in panoraMap. All panoramas presented to the software with suitable metadata are represented on this by circular symbols. Whenever a symbol is clicked a section of the panorama is displayed in a viewer along with a pair of arrows on the base map that represent the angular extents of the displayed section of image. Dragging the map symbols or the image with the mouse and cursor will pan the image and move the arrows appropriately. Additionally, any other panorama with a bearing that lies within the section of the panorama displayed in the viewer is symbolised with a vertical arrow on the horizon, the length of which indicates the proximity of this panorama. Touching one of these symbols with the cursor highlights the appropriate symbol on the map view (and vice versa) and clicking it opens the relevant distant panorama so that the view back to the original location is apparent (Figure 7.3).

A series of alternative base maps can be loaded, enabling realistic photographs to be compared with information derived from data surfaces or thematic information such as geology, ownership or real estate value. This method of navigation allows users of the software to teleport across spaces populated by panoramic imagery, whilst identifying features of interest on planimetric maps and viewing local details and larger scale topography from selected locations in an engaging and semi-immersive manner that takes advantage of many of the stated benefits of using VR to support fieldwork. The objective of helping students compare and contrast reality with digital cartographic representations of it is further enhanced by functionality that reads data logged by global positioning systems (GPS) used in the field. Routes taken and points recorded are displayed with interactive symbols that show the time at which a position on a route was visited and the name of a logged point when interrogated, aiding

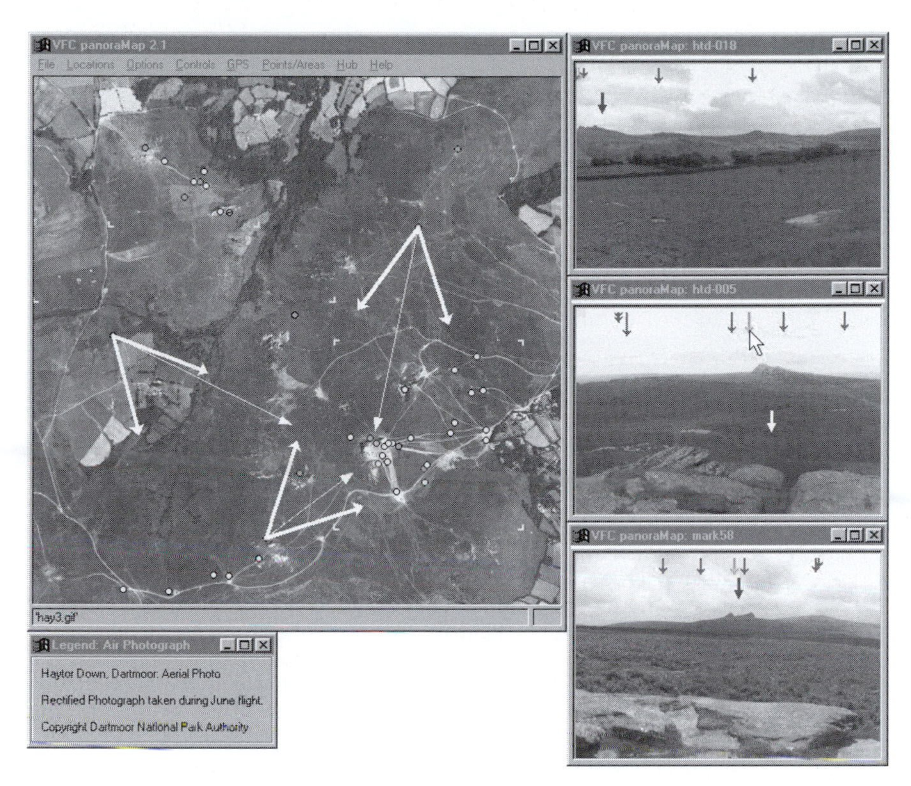

Figure 7.3 Navigating through discrete geographic spaces in panoraMap. Orange symbols on the map relate to panoramic images. Clicking a symbol reveals the panoramic view and arrows on the map showing the direction of view from the location at which the image was taken. Moving the cursor over an image as it spins produces an additional symbol and a thinner arrow on the map showing the direction on the ground. If other panoramas occur within this field of view, they are shown by arrows on the skyline at the appropriate bearing. When these are clicked the relevant panoramic image is displayed, permitting touring. Dynamic links mean that related symbols are highlighted in all views when touched with the cursor. The three displayed panoramas represent a tour from the symbol at the left of the map (top), to that at the bottom of the map (centre), and up to the symbol at the top of the map (bottom). The panoramic image represented by the arrow touched by the cursor in the middle view is visible in all three panoramas which converge on a feature of interest. (Aerial photography by ukperspectives.com). See CD ROM for colour version of this figure.

direct comparison of experience in the field with digital information during a field trip and prompting memory in the debriefing and feedback teaching mode. The functionality can be used in reverse, so that students are able to select points of interest from a digital database and plan and digitise a route between them. This is logged in a global positioning system receiver,

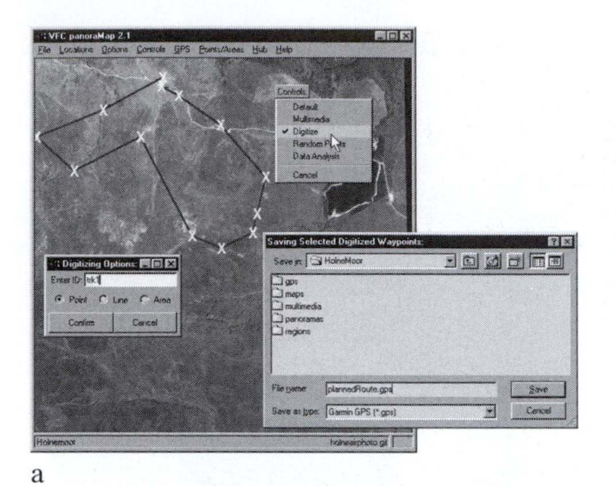
a

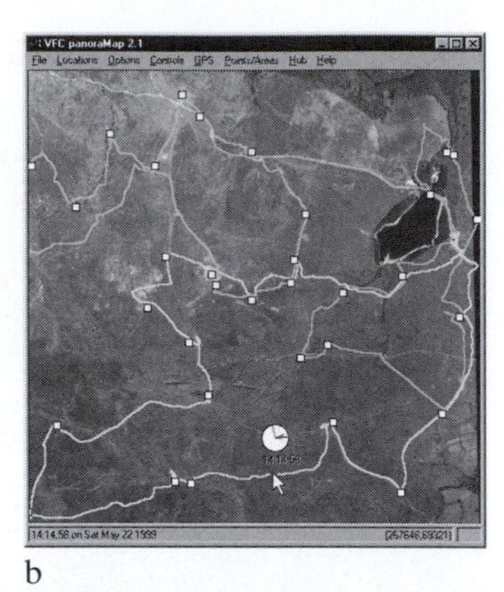
b

Figure 7.4 Linking the virtual environment and the real world with GPS. The left-hand image shows that, having selected 'Digitize' from the control menu, a series of locations can be selected in order, using the base maps available, collectively named, and saved into a suitable format for upload into GPS receivers. The GPS can then be used to follow the specified route on the ground. The image on the right shows a series of three routes taken in the field in order to collect images at thirty-four points of interest, identified and digitised through panoraMap. When a route is touched with the cursor, the time at which the section was covered is displayed. The map covers an area of 3.5 km by 3.5 km. The data were collected during a single day in the field on Holne Moore, Dartmoor. (Aerial photography by ukperspectives.com).

which can be used to navigate to the selected points in the field (Figure 7.4(a)). The degree of 'collapse' (see Chapter 2) between user and representation experienced by students utilising this functionality is surprisingly high for a two-dimensional application.

Additional specifications outlined above include low data-acquisition costs and the ability to collect and add data with ease. Data collection involves the processes of taking and stitching photographs as outlined on pages 74–77. The GPS route planning functionality can ease this process. By way of example, Figure 7.4(b) shows a series of positions selected using panoraMap. These identify hilltops and sites of specific interest in terms of the archaeology, vegetation and soil type. A route was planned between these points that included three image-logging sessions at suitable locations at which to park a vehicle and the data uploaded into the GPS receiver. This was then used to navigate to the correct positions in the field. The routes shown are for a day's data collection in a 3 km^2 area of heavy ground and steep undulation. Adding data to the software requires the use of the familiar windows 'File Selection' interface, and the population of a series of metadata fields for which templates can be created as shown in Figure 7.5. Any panoramic image added to the software will be

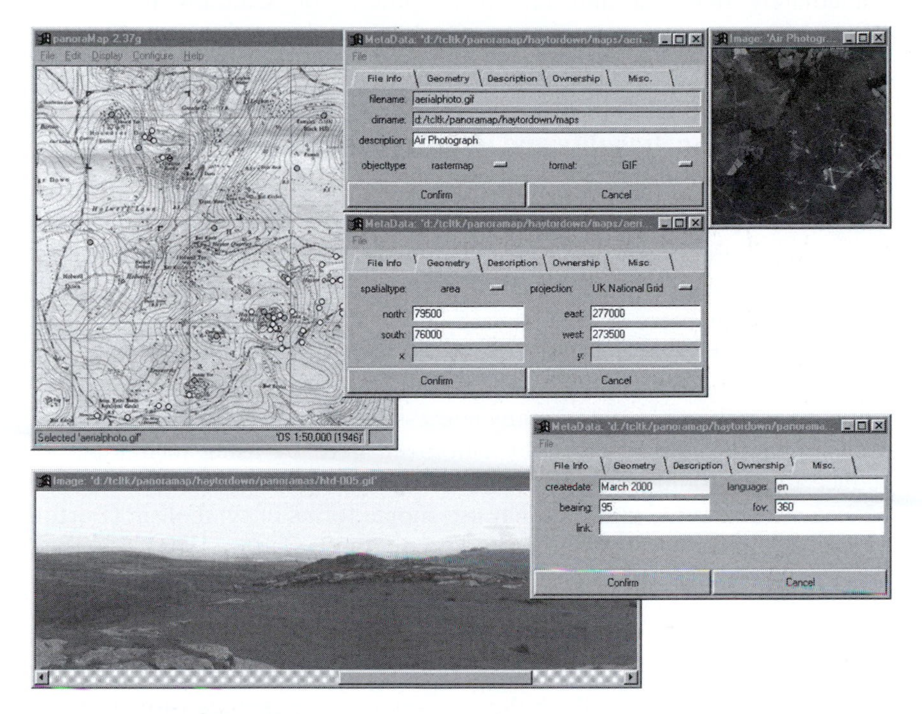

Figure 7.5 Adding data and metadata to panoraMap. Two items of data are being added in this figure. To the top left, the familiar planimetric interface shows the locations of panoramas and geo-referenced multimedia data items. Data can be added through the 'File' menu or by clicking the map. Each of these options produces a file selection dialogue. The windows at the top right show that a file containing a map has been chosen for addition into a panoraMap session. Preview and metadata entry windows are displayed. Metadata items organised into five categories are filled with defaults by the software and completed by the user via the keyboard for text and by clicking on the map for coordinates, or by loading a text file containing a metadata template. Here the 'File Info' and 'Geometry' categories are displayed. At the bottom of the figure a panoramic image has been selected by clicking on the map and choosing an appropriate file from the file selection interface. The 'Misc.' metadata category of the metadata GUI is shown, in which the field of view and bearing of the image are entered. On adding data items in this way, they become available for selection either through the map, the menus or by navigating between panoramic images, depending upon the data type. (Aerial photography by ukperspectives.com). See CD ROM for colour version of this figure.

immediately detected and linked to others for seamless navigation. PanoraMap sessions, containing specified data and metadata, can be saved and reloaded when required.

Visualization functionality and multimedia data in a VE

In order to achieve the aims of supporting data analysis, a number of features were incorporated that take advantage of the capabilities for interactive graphics and prototyping provided by Tcl/Tk. These include graphics for visualization and geo-referenced multimedia that extend the forms of data available to users of the software.

Whilst the three-dimensional approach does not preclude the employment of interactive visualization techniques (George Mason University, 1999; Symanzik *et al.*, 1997); many successful visualizers and visualization systems parallel the cartography employed here by using techniques to reduce the multidimensional nature of a data space to two dimensions. Manageable minimal graphics that use simple forms of symbolism (Bertin, 1983) and maximise the ratio of data to 'ink' (Tufte, 1983) are then plotted. These take advantage of interactive techniques for interrogation and graphical exploration. Examples include the Grand Tour (Cook *et al.*, 1995) which employs projection pursuit to generate successive two-dimensional views of a data space, parallel coordinate plots (Inselberg, 1995; Wegman, 1990) in which the geometry of hyper-dimensional spaces are characterised by linked dot and box plots, and the mosaic plots and extraordinary levels of flexibility and interactivity provided by MANET (Unwin *et al.*, 1996). PanoraMap adds visualization functionality to the virtual environment by providing an interface through which geometric and attribute data can be loaded, shading area or point symbols according to data values and allowing successive variables to be selected and suitable interactive graphics produced. Figure 7.6 demonstrates this using data collected during fieldwork around Haytor Rocks, Dartmoor, UK.

Multimedia data can add to the sense of realism in the virtual environment and provide important contextual information for analyses. This might include ambient sounds, interviews, or video that provides a realistic impression of quantitative values such as those representing traffic flows or tourist pressure (e.g. Shiffer, 1999). These data and metadata are added to panoraMap by simply clicking a desired location on the map and selecting the digital file that relates to the chosen geo-reference. The completion of

---

Figure 7.6 Visualization and multimedia in panoraMap. The top left of this figure demonstrates visualization functionality. Circular symbols on the map relate to the locations of a number of sampling points selected using the software and located with GPS receivers (top left). A menu shows the data that have been collected and loaded, and the selected variable ('% Grass' in this instance) is shown through the shading of the symbols. An

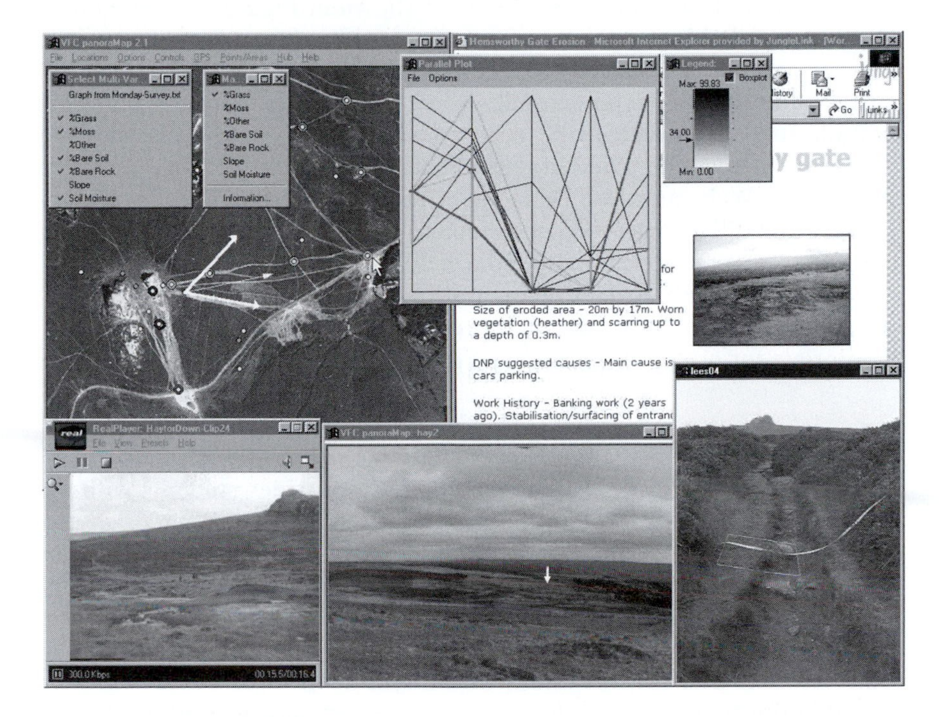

additional menu allows multiple variables to be selected and displayed in a parallel coordinates plot (top centre). Here five variables have been selected in the menu and are shown along parallel axes. Bi-directional linking means that touching a symbol with the cursor temporarily highlights the appropriate symbol in the map and parallel plot and reveals the data value in the box-plot legend. Clicking a symbol permanently highlights related symbols in all views. Here four locations along a single path have been permanently highlighted to assess the multivariate nature of the data collected along this path in comparison with the data collected at other locations in the region. Area data such as that collected in national surveys or census can also be visualized in this way. The bottom and right of the figure show qualitative data available through panoraMap. These include panoramic images (bottom centre) and their associated symbols on the map, and also digital video, imagery and hypertext that contains information about the area of interest. These are geo-referenced and represented by the symbols on the map (top left). The hypertext shown here contains secondary data, namely imagery and reports gathered and provided by the Dartmoor National Parks Authority outlining past and present levels of erosion and management policy. During fieldwork digital images were recorded at each data point in order to substantiate quantitative data with qualitative information. The image displayed at the bottom right of the figure relates to the point that is being clicked by the cursor on the map, and is highlighted in the parallel plot. The view up the path contrasts with the view down the path shown in the panoramic image (bottom centre). (Aerial photography by ukperspectives.com). See CD ROM for colour version of this figure.

a metadata GUI of similar form to those shown in Figure 7.5 results in a draggable symbol being produced, which plays the file in the appropriate piece of external software when clicked. In Figure 7.6 digital video and hypertext are shown. Hypertext allows one-to-many relationships between location and media data to be accommodated through the inclusion of URLs and means that external data resources, such as real-time imagery collected by Web cams, can be included in the representation. Locations logged with GPS receivers whilst collecting data can be plotted and clicked in order to geo-reference multimedia field data items or qualitative data values for visualization. This configuration provides a flexible and extensible 'interconnected barrage of representations' (Krygier, 1997) that is appropriate for prompting insight and gaining scientific knowledge.

Assessing the virtual environment

Theoretically speaking, the approach certainly provides students with an alternative route back through the series of transformations made in the cartographic process to the real-world experiences in the field. It permits observation and aids in their efforts at comparison and in relating models to their experience of the physical environment. The software has proved to be useful in terms of preparation, as students have assessed likely data collection sites and considerations that might need to be made whilst in the field prior to undertaking fieldwork. It can certainly be used to demonstrate the field environment in a particular season, under certain weather conditions, in the past or under pressure (during a public holiday, rush hour or fire event, for example).

In practical terms, whilst panoraMap underwent formative evaluation, no formal assessment of its utility and achievement of the specified goals has been undertaken. It has, however, been used in a number of fieldwork situations spanning a broad range of subject matter at a variety of locations. Certain goals have evidently been achieved, others can be assessed empirically by reactions to the software and its 2D virtual environment.

Data collected during fieldwork have been incorporated into the software and used in the field at locations including Haytor Down, Holne Moor and Fernworthy Reservoir, Dartmoor; Brancaster, North Norfolk; Exeter, South Devon; Coniston, the Lake District; Victoria Park, Leicester and a remote area of northern Greece. Both staff members and student groups have reacted positively to the way in which the software contributed to and enhanced the fieldwork by synthesising information from diverse sources in an interpretable manner, displaying field data in the context of additional secondary information so soon after returning from the field and providing a means for initial analyses of these data through visualization. Indeed, one member of staff regarded the day's fieldwork that was coordinated through the software as the best he had been involved with and described the analyses undertaken by the first-year geography undergraduates on data that they had collected in the field as more impressive than those of students considering the same topic at Masters level. An experience that was particularly testing for the author involved the member of staff who had coordinated two days of fieldwork using panoraMap making an impromptu enquiry to the tired and bedraggled group as to the value of using the software during their study. Their response was encouragingly positive, as was that reported by the teacher. Whilst experience demonstrates that panoraMap constitutes a useful learning device, these tests were undertaken during the development phase using software prototypes with the author providing a high level of support to those doing both the learning and the teaching. It would appear, however, that the techniques are accessible and straightforward enough to enhance fieldwork and make the representations useful. The usability of the software has not been assessed and careful programming in the final stages of software production has been required to achieve this by responding to users' reactions. The potential of the approach has, however, led to a number of academics collecting data and metadata in the required formats and using software prototypes in teaching.

Ultimately, as is the case with most maps and any computer-assisted learning device, the suitability of the identified data type and success of this kind of representation depends upon a range of external factors. These include the geography of the area under study, the specification of the machines used, the quality and quantity of data and metadata, the organisation of the fieldwork and the enthusiasm of the staff involved. Using panoraMap with an information-rich database is likely to be more beneficial than doing so with one that is less comprehensive, as long as the data quality is high. Locations with flat homogeneous areas work less well than those with undulation that display characteristics of the ground cover and easily identifiable features (such as Dartmoor, UK). Smaller areas are more suited to this form of representation than large ones (linked panorama locations need to be within view for the sense of navigation to be effective) and rural areas require less densely-spaced panoramas than urban ones. In certain homogeneous and urban areas, forms of VR that use continuous models might be more appropriate. This leads to an important conclusion. Developing techniques such as these need not be regarded as mutually exclusive or even competing. They add to the range of tools and representational methods available to map makers and map users as they select appropriate cartography for the tasks in hand, and their most effective use may well be complimentary (see Figure 7.7).

Multimedia can be used extremely effectively within VR, but it is currently hard to provide highly dynamic and effective symbolism for interactivity, and the addition of multimedia seems more appropriate from the kind of holistic view that 2D maps provide than those that rely exclusively 3D VR interfaces. Ultimately, the memory and processor configuration of the machine on which the software is mounted might determine the most

Figure 7.7 Combining panoramic and surface-based data in a virtual environment. This addendum to Figure 7.1 demonstrates that a combination of technologies provides both detail and continuous coverage, confirming that the synthesis of spatial data is desirable. In practice the suitability of the approaches presented here for particular tasks will be determined by a variety of factors such as scale, data cost and availability, hardware configuration, the geographic location under examination, the intended users and the use to which the VR is being put

	Virtual environment based upon 3D surface model	*Virtual environment based upon imagery and 2D map*
Combination	 Results in advantages of continuous information with pockets of additional detail that can be augmented by teacher in order to maintain currency and direct focus of representation	

appropriate form of representation. Those used in one of the examples cited above were equipped with 486 processors and were unable to render the 3D model satisfactorily. Finally, no teaching enhancement teaches on its own. It is the duty of the teacher or instructor to enthuse and motivate students through compelling teaching and imaginative assignments that involve the software (DiBiase, 1999). Good teachers can make poor maps, representations or software effective learning devices, whilst those without such skills might fail to enhance the learning experience with the most magical teaching aids. To achieve its goal, this VR application relies firmly on highly-skilled real people.

Developments and near-future issues

The future utility of both the software and the data type will be determined by developments in computing. Influences such as the Internet, real-time data collection, mobile computing, software integration and new media will impact greatly upon the technology under consideration here. Jo Wood's VFC Hub mechanism (Dykes *et al.*, 1999) allows panoraMap to communicate with a series of remote databases so that students and teachers can share information. Data and metadata entered during a panoraMap session can be uploaded to a database through the hub server–client mechanism which supports a series of user privileges through a password-protected security system. Additionally a remote hub can be chosen by entering the IP address, a username and appropriate password. Data and metadata can then be downloaded and added to a panoraMap session that mixes and matches local geo-referenced data with those derived from a series of remote databases. The user can seamlessly navigate around an area of interest using panoramic images and viewing multimedia data items obtained from each of these sources. Remote databases can also be searched for adjoining maps allowing the user to pan across geographic space to areas mapped by data accessible from remote hubs. Facilities for sharing and reusing data and transferring them across the Internet enhance the software significantly and make it far more usable for the specified purpose. Internet technologies such as those that permit streaming and the use of developing compression techniques such as enhanced compressed wavelet technology (GISL, 1999) will speed up this process and provide detailed geographic information at close proximity to the position occupied by the virtual viewer but at lower resolution at distance. Streaming software might make deductions about data that are likely to be required next by assessing the direction in which a user has been moving and the way in which they behave in a virtual world. Ultimately geographically-relevant elements of spatial databases will be streamed into the VR 'on the fly' as the user is able to see them, at a level of detail appropriate to their distance. This will increase as objects are approached, whilst detail will be dropped from memory as the user moves away.

Advances in real-time data acquisition and pattern recognition are likely to make imagery a more important source of quantitative data. For example, image processing and photogrammetric techniques are being used to provide continuous information on beach characteristics and nearshore processes and features from digital video at a number of locations across the globe. Furthermore, techniques are being developed to stitch and visualize panoramic imagery that forms a fully-enclosing sphere for more convincing immersion as opposed to the open cylindrical images utilised here (Szeliski and Schum, 1997; Wasabi Software, 1999).

Rapid improvements in the graphical and multimedia capabilities of mobile computing devices (Casio Inc., 1999; Palm Inc., 2000), and their incorporation with communications and positioning technology means that current efforts to provide location-specific information in the field are far in advance of their recent predecessors (e.g. Pascoe and Ryan, 1998). Indeed much of the functionality presented here might soon be available in hand-held devices that contain GPS receivers and digital cameras, provide interactive colour maps and link to the Internet (ArcPad, a GIS for hand-held computers, has recently been launched – Environmental

Systems Research Institute, 2000). Views could be selected on screen from software such as panoraMap by a user in the field and the GPS receiver contained within the device used to lead them to the location at which the view was taken in order to evaluate changes. Location-specific information is already provided in hand-held devices. Whenever the aerial of a Palm VII is flipped up a local weather forecast is supplied through its Web Clipping Application (WCA) and the Starbucks site offers directions to their nearest coffee house (Hill, 2000). Integration with GPS 'goto' functionality is an obvious and achievable next step. Libraries of panoramic images used in this way could have a range of applications that focus on environmental change or landscape assessment, including the identification of areas of erosion, the recognition of changes in species diversity, or public participation in planning decisions. Equally, views collected in the field could be geo-referenced and provided with metadata automatically and uploaded to a remote spatial database from the hand-held device. Such applications, with their basis in the real world, form perplexing examples when considering the 'degree of collapse' between user and representation in VR. The development of a fully-functional *Hitchhiker's Guide to the Galaxy* that provides spatially-specific information and navigation across the globe is an exciting prospect (The Digital Village Limited, 1999) and an investment by MapInfo in location-based mobile applications using wireless application protocol (WAP) suggests a promising and exciting future for field computing (MapInfo, 2000).

Increasing software compatibility will make the complimentary views presented by 2D image-based and 3D model-based environments more easily synchronised and used in conjunction. In the case of the VFC software, the development of closer links between Java and Tcl/Tk (Johnson, 1998) will facilitate the sharing of positional information between applications that use these technologies. This will benefit the fieldwork application, as a realistic view of vegetation will be able to be compared precisely with an abstract view of subsurface geology, soil-type, land-ownership or rainfall from the same viewing position and under the same viewing conditions.

These developments and others, such as the decreasing prices and increasing image qualities associated with digital cameras, are likely to make panoramic digital imagery, that is presented in interactive software such as panoraMap, an increasingly common and useful medium with application in planning, tourism, retail, and as a means for providing contextual information for more rigorous statistical and geographical analysis than those presented here (Dykes, 2000).

Conclusion

This chapter has attempted to provide some insight into the medium of digital panoramic imagery and shown how a virtual environment can be constructed from such data. It has specified a mapping requirement and examined the cartographic product that was produced in response to this demand. In doing so it has endeavoured to emphasise the transformations that take place in the production of VR interfaces and to demonstrate that this process is similar to more traditional forms of cartography. In each case a certain product is the result of a series of transformations from real world to user's image, in which the map designers use their skill and judgement to fulfil certain map-use objectives for a defined user set under a rigid series of criteria. Acknowledgement of these stages enables the creators of virtual worlds to ensure that their virtual realities are closely tied to the applications and users for which they are being designed at all stages. Examples of the data type in use are presented throughout and experiences recounted in an assessment both of the fulfilment of the outlined requirements and of the potential for exploiting some of the benefits of a 'virtual reality' in an interface to geographic information that relies upon digital panoramic imagery. The approach presented here is explicitly spatial, provides a level of engagement and immersion, and uses realistic views as an interface to lead users into a dataset that represents a geographic area. Empirical evidence suggests that the panoraMap software utilises the medium usefully and achieves the specific set of identified objectives. Linked panoramic imagery provides an especially useful means of adding detailed local information with relative ease for comparatively small geographic areas in a format that supports exploration and navigation and is suitable for undertaking observation. This corresponds well with the model of fieldwork that involves short but concentrated visits to a number of constrained locations within a larger region of study, and the requirement of the fieldwork teacher to supplement their students' learning with the efficient and effective data collection programmes and resources.

The two-dimensional and image-based nature of the virtual environment is unique in this book. It may result in a reconsideration of the nature of VR, and at the very least should compel those who use VR to consider the use to which their interfaces and data are put in priority to a focus on the technology with which they are built. The contribution draws attention to the combination of VR in geography with multimedia data and visualization functionality and an acceleration in the trend towards the integration of technologies. The software presented here was designed to be used by those without experience of VR or geographical computing. Users' responses suggest that the product achieved its goals and high levels of interest in VR exist amongst geographers who do not regularly use advanced graphical technologies, suggesting that the technologies outlined in this book will continue to flourish as developments continue apace.

Note: Demonstration software can be downloaded from the World Wide Web: www.soi.city.ac.uk/~jad7/VRinGeography/.

References

Abadjev, V., del Rosario, M., Lebedev, A., Migdal, A. and Paskhaver, V. 1999. MetaStream. *Proceedings VRML 99: Fourth Symposium on the Virtual Reality Modelling Language.* New York: ACM, pp. 53–62.

Apple Computer Inc. 1999a. QuickTime VR, www.apple.com/quicktime/qtvr/.

Apple Computer Inc. 1999b. QuickTime 4 API documentation: Summary of QuickTime for Java, developer.apple.com/techpubs/quicktime/qtdevdocs/Java/ tp_java_digest.htm.

bamboo.com Inc. 1999. Virtual tours, www.bamboo.com/index.shtml.

Bertin, J. 1983. *Semiology of Graphics.* Madison: University of Wisconsin Press.

Bishop, I. 1994. The role of visual realism in communicating and understanding spatial change and process. In Hearnshaw, H.M. and Unwin, D.J. (eds) *Visualization in Geographical Information Systems.* Chichester: Wiley, pp. 60–4.

Butts, S. 1999. Rally Championship 99 Preview, pc.ign.com/previews/11031.html.

Cartwright, W. 1997. New media and their application to the production of map products. *Computers and Geosciences*, 23, 447–56.

Casio Inc. 1999. Casio – Mobile Information Devices. Cassiopeia E105, www.casio.com/mobileinformation/.

Cook, D., Buja, A., Cabrera, J. and Hurley, C. 1995. Grand tour and projection pursuit. *Journal of Computational and Graphical Statistics*, 4, 155–72.

DiBiase, D. 1999. Evoking the visualization experience in computer-assisted geographic education. In Camara, A.S. and Raper, J.F. (eds) *Spatial Multimedia and Virtual Reality.* London: Taylor & Francis, pp. 89–101.

The Digital Village Limited. 1999. Hitchhiker's Guide to the Galaxy: Earth Edition. On the Move – The Vision, www.h2g2.com/A281486.

Doyle, S., Dodge, M. and Smith, A. 1998. The potential of web-based mapping and virtual reality technologies for modelling urban environments. *Computers, Environment and Urban Systems*, 22, 137–55.

Dykes, J.A. 1996. Dynamic maps for spatial science: A unified approach to cartographic visualization. In Parker, D. (ed.) *Innovations in Geographical Information Systems 3.* London: Taylor & Francis, pp. 171–81.

Dykes, J.A. 1999. Scripting dynamic maps: Some examples and experiences with Tcl/Tk. In Cartwright, W., Peterson, M.P. and Gartner, G. (eds) *Multimedia Cartography.* Berlin: Springer-Verlag, pp. 195–204.

Dykes, J.A. 2000. An approach to virtual environments for fieldwork using linked geo-referenced panoramic imagery. *Computers, Environment and Urban Systems*, 24, 127–52.

Dykes, J.A., Moore, K.E. and Wood, J.D. 1999. Virtual environments for student fieldwork using networked components. *International Journal of Geographical Information Science*, 13, 397–416.

Environmental Systems Research Institute Inc. 2000. Press Release – ArcPad 5 Unveiled at CeBIT, www.esri.com/news/releases/00_1qtr/arcpad.html.

Ferreira, J., Schiffer, M., Singh, R. and Chandonnet, J. 1996. NCPC Virtual Streetscape, yerkes.mit.edu/ncpc96/home.html.

George Mason University. 1999. Project ScienceSpace, www.virtual.gmu.edu/.

GISL Limited. 1999. Image Web Server, www.imagewebserver.co.uk/iws_info.htm.

Gold, J.R., Jenkins, A.J., Lee, R., Monk, J., Riley, J., Shepherd, I. and Unwin, D.J. 1991. *Teaching Geography in Higher Education: A Manual of Good Practice.* Oxford: Blackwell.

Haigh, M.J. and Gold, J.R. 1993. The problems of fieldwork: a group-based approach to integrating fieldwork into the undergraduate curriculum. *Journal of Geography in Higher Education*, 17, 21–32.

Hill, R. 2000. Wireless Wonder. *Mobile Computer User*, March–April, 21–4.

Hoppe, H. 1997. View-dependent refinement of progressive meshes. *Computer Graphics, Proceedings SIGGRAPH 97*. New York: ACM, pp. 189–98.

Imhof, E. 1951. *Terrain et Carte*. Eugen Rentsch Verlag: Erlenbach-Zurich, 261 pp.

Inselberg, A. 1995. Parallel coordinates for visualizing multidimensional geometry. In Kloesgen, W. (ed.) *EuroStat: New Techniques and Technologies for Statistics II*. Amsterdam: IOS Press, pp. 279–88.

Johnson, R. 1998. Tcl and Java Integration, www.scriptics.com/java/tcljava.pdf.

Kent, M., Gilbertson, D.D. and Hunt, C.O. 1997. Fieldwork in geography teaching: a critical review of the literature and approaches. *Journal of Geography in Higher Education*, 21, 313–30.

Krygier, J.B. 1997. Envisioning the American west: Maps, the representational barrage of 19th century expedition reports, and the production of scientific knowledge. *Cartography and GIS*, 24, 27–50.

Krygier, J.B. 1999. Cartographic multimedia and praxis in human geography and the social sciences. In Cartwright, W., Peterson, M.P. and Gartner, G. (eds) *Multimedia Cartography*. Berlin: Springer Verlag, pp. 245–56.

Laurillard, D.M. 1997. Learning formal representations through multimedia. In Hounsell, D., Marton, F. and Entwhistle, N. (eds) *The Experience of Learning*. Edinburgh: Scottish Academic Press, pp. 172–83.

Live Picture Corporation. 1999. PhotoVista, www.livepicture.com/products/photovista/content.html.

MacEachren, A.M. 1994. Visualization in modern cartography: Setting the agenda. In MacEachren, A.M. and Taylor, D.R.F. (eds) *Visualization in Modern Cartography*. Oxford: Pergamon, pp. 1–12.

MacEachren, A.M. and Kraak, M.-J. 1997. Exploratory cartographic visualization. *Computers & Geosciences*, 23, 335–43.

MapInfo Corporation. 2000. Press Release – MapInfo invests in wireless application protocol (WAP) technology, www.mapinfo.com/corporate_info/press/ releases/382.html.

Monmonier, M. 1991. *How to Lie with Maps*. Chicago: University of Chicago Press.

Muchaxo, J., Neves, J.N. and Câmara, A.S. 1999. A real-time, level of detail editable representation for phototextured terrains with cartographic coherence. In Câmara, A.S. and Raper, J.F. (eds) *Spatial Multimedia and Virtual Reality*. London: Taylor & Francis, pp. 137–46.

Palm Inc. 2000. Palm VII Connected Organizer Web Clipping, www.palm.com/ products/palmvii/webclipping.html.

Pascoe, J. and Ryan, N. 1998. Mobile computing in a fieldwork environment, www.cs.ukc.ac.uk/research/infosys/mobicomp/fieldwork/index.html.

PictureWorks Technology Inc. 1999. Spin Panorama, www.videobrush.com/spinpano/.

Raper, J.F., McCarthy, T. and Williams, N. 1998. Georeferenced four-dimensional virtual environments: Principles and applications. *Computers, Environment and Urban Systems*, 22, 1–11.

Reddy, M. 1998. The QuadLOD Node for VRML, www.ai.sri.com/~reddy/ geovrml/new_lod/.

Robinson, A., Morrison, J., Muehrcke, P., Kimmerling, A. and Guptill, S. 1996. *Elements of Cartography*, 6th Edition. New York: Wiley, 674 pp.

Shiffer, M.J. 1995. Environmental review with hypermedia systems. *Environment and Planning B: Planning & Design*, 22, 359–72.

Shiffer, M.J. 1999. Augmenting transportation-related environmental review activities using distributed multimedia. In Camara, A.S. and Raper, J.F. (eds) *Spatial Multimedia and Virtual Reality*, London: Taylor & Francis, 35–45.

Symanzik, J., Cook, D., Kohlmeye, D.B., Lechner, U. and Cruz-Neira, C. 1997. Dynamic statistical graphics in the C2 virtual environment. *Computing Science and Statistics*, 29, 35–40.

Szeliski, R. and Shum, H.-Y. 1997. Creating full view panoramic mosaics and environment maps. *Computer Graphics: Proceedings SIGGRAPH 97.* New York: ACM, pp. 251–8.

Tobler, W.R. 1979. A transformational view of cartography. *The American Cartographer*, 6, 101–6.

Tufte, E.R. 1983. *The Visual Display of Quantitative Information.* Cheshire CT: Graphics Press.

Unwin, A.R., Hawkins, G., Hofmann, H. and Siegl, B. 1996. Interactive graphics for data sets with missing values – MANET. *Journal of Computational and Graphical Statistics*, 5, 113–22.

Videobrush Corporation. 1998. Panorama V2.0, www.videobrush.com/videobrush/products/panorama/.

The Virtual Field Course. 2000. Welcome to the Virtual Field Course, www.geog.le.ac.uk/vfc/.

Wasabi Software. 1999. SkyPaint: 3D Panorama Paint Tool, www.wasabisoft.com/.

Wegman, E.J. 1990. Hyperdimensional data analysis using parallel coordinates. *Journal of the American Statistical Association*, 85, 664–75.

Williams, I. 1983. Pyramidal parametrics. *Computer Graphics: Proceedings SIGGRAPH 83.* New York: ACM, pp. 1–11.

Williams, N., Jenkins, A., Unwin, D.J., Raper, J.F., McCarthy, T., Fisher, P.F., Wood, J.D., Moore, K.E. and Dykes, J.A. 1997. What should be the educational functions of a virtual field course CAL '97, University of Exeter School of Education: Exeter, 351–4.

Part II

Virtual landscapes

8 Introduction

Iain M. Brown, David B. Kidner, Andrew Lovett, William Mackaness, David R. Miller, Ross Purves, Jonathan Raper, J. Mark Ware and Jo Wood

Introduction

This chapter, on virtual landscapes, sets the context for subsequent chapters about the development and role of virtual reality (VR) in issues relating to non-urban environments. In that context, the chapters address key topics, including those of believability of imagery, spatio-temporal modelling, resolution and scale, and navigation of the VR model, with examples presented on the use of VR and discussion of its potential.

The applications of virtual reality to non-urban landscapes have, in large part, been motivated by broad societal objectives of environmental protection of terrestrial and coastal areas, recreation, education and, in particular, issues regarding change in the landscape, and the acceptability of such change. International agreements and protocols, such as those on sustainable development or conservation of biodiversity (UN, 1993), as implemented through legislation or guidelines on best practice, have led to increased demand for tools for the management of landscapes. The planning process is an example of where legislative requirements and public interest have driven improvements in tools for landscape visualization, and the analysis of the impacts of change on the rural landscape, often through scenario exploration. However, more rigorous environmental controls, at both national and international levels, and increased expectations from some authorities or agencies have created a demonstrable need, and opportunity, for the development of more robust methods to explore landscapes. This, in turn, has contributed to an improved awareness of the issues associated with processes of change in rural and coastal environments.

It is stakeholders in the rural environment who are essentially the end-users of the VR tools developed to improve and inform decision making, and these stakeholders include the general public, informed 'lay people' and those with a legal, professional or private interest in landscape issues. The nature and purpose of their stake varies by geographic location and extent, land use sector, degree of authority and expertise, but together they form a constituency for whom the landscape may be a home, work or leisure environment.

The demand for information has escalated alongside both the capacity for the collection of data on landscapes, and the requirements of the end-user, potentially placing VR tools in a key position to aid in the description, explanation and communication of often sophisticated concepts to audiences of differing reference levels and capabilities with regard to landscape-related issues. In principle, accessibility to the technology, suitable data, and a level of competence to utilise them in combination, provides a powerful facility for advocacy in relation to management practices and pressures driving landscape change. It also has the potential to enhance public education and participatory decision making across the stakeholder community, especially when coupled with a World Wide Web infrastructure (Carver *et al.*, 1997).

However, a number of issues govern the successful utilization of virtual landscapes; these include virtuality, data structures, contextualization, integration and visualization. The following sections summarize the key points relating to these issues in the context of the use of VR in non-urban areas and are brought together in a discussion of applications in visualizing landscape change.

Virtuality, contextualization and shape

The authenticity, belief and comprehension of the virtual landscape comes from immersion within it, which in turn depends on effective methods of navigation. A framework underpinning navigation is provided through a variety of cues that provide information on location, orientation, rate of movement and field of view (Figure 12.4), as described by Wood (Chapter 12). These orienting mechanisms provide the basis for the implementation of a range of metaphors of interaction (flying 'through' the landscape, alighting and walking, or teleporting between way-marked locations on the landscape; Purves *et al.*, Chapter 13).

System constraints invariably lead to the need for compromise between the provision of the context to support navigation, and interpretation of the landscape (the degree of realism), the method of interaction, and the speed of response. Activities such as the interpretation of a surface, the perception of pattern, and the assessment of change often require the user to travel between coarse and fine levels of information, and continuity between these scales is an important prerequisite to comprehension of the landscape. Ideally the components act collectively, both to immerse the user in the landscape, and to support a range of tasks, with continuity of movement providing a context which binds the detailed to the general, an issue which is dealt with in greater detail by Purves *et al.* (Chapter 13) and Moore and Gerrard (Chapter 14).

Intriguingly, conveying a credible view of a landscape is often achieved by means of 'white lies' (something that is equally true of conventional 2D maps). For example, viewing terrain from above using the commonly applied 'fly-through' metaphor tends to 'flatten' vertical relief. This can be countered by exaggerating the projection of space in a vertical direction, and thus geometric realism is sacrificed in order to enforce an impression of surface variation. The shading of relief, often used in traditional cartography, is one well-accepted method of enhancing the quality of virtual worlds.

Other visualization methods (such as shadowing and height colouring) are superimposed on the surface in order to reinforce such differences, and a balance must be struck between the salient information pertinent to the task and the contextual information required to support its interpretation. Depth-cueing methods can also be used to reinforce the, often large, viewing distances implied by the landscape vista, while reducing the colour saturation of features with distance, or occluding the landscape with fog, are also techniques that can be used to limit the depth of view and improve rendering speed. Many VR models also impose an artificial horizon (the curvature of the earth does not constrain VR models) by a process known as Yon Clipping. Such artificial horizons can be directionally dependent or viewer-height dependent; in either case the developer must 'design in' a horizon appropriate to the use of the expected model.

The visualization of virtual landscapes is critically dependent on the purpose for which they are intended, and in representing a landscape, the shape of our constructed reality is highly significant. A number of issues should be considered when designing the representation, including: appearance (i.e. is the virtual landscape credible?); does the representation of the landscape accurately reflect the features in which we are interested?; and, in producing the virtual landscape, are the above constraints met whilst the VR model remains usable? These topics are discussed by Lovett *et al.* (Chapter 9), Miller *et al.* (Chapter 10), Purves *et al.* (Chapter 13) and Moore and Gerrard (Chapter 14).

Integration and data characteristics

The suitability of a particular VR model, with respect to its intended application, has become more of an issue as larger-scale data become more widely available, and the costs of appropriate software and hardware configurations are reduced in real terms. Users are now faced with the dilemma of determining which data to use on the basis of a range of conflicting performance criteria (e.g. what is the most appropriate balance between the requirements of the accuracy and purpose of the representation, and acceptable user interaction performance?).

The end-user needs to be made aware of the data resolution or generalization inherent in the visualization, and its likely effect on accuracy (although, to date, most implementations of virtual landscapes give no indication of error). For example, the resolution of the Digital Elevation Models most commonly used in virtual landscapes means that even moderately steep slopes are smoothed, or in a 2.5D representation, where any one point in horizontal space (*xy*) has a unique value (*z*), overhangs (a relatively uncommon feature in nature) are not represented. In many cases the loss of such features is unimportant to the application, but in some instances, for example where virtual landscapes are used in some form of route planning, the smoothing between the real landscape and the virtual one can have significant implications.

An alternative to choosing a discrete data resolution is to integrate data from multiple scales, and sources into the VR model. This leads on to the question of choosing a suitable data structure for VR modelling. The traditional terrain modelling debate of TINs versus grids may become obsolete as the next generation of multi-resolution data structures evolves (Hoppe *et al.,* 1992). These data structures provide efficient methods for storing and accessing data at variable levels of detail. Current fixed-scale data models limit performance because they are unable to handle the large volume of data efficiently. Real-time visualization and rendering of large-scale VR landscapes imply a 'paradigm shift' in the application of spatial data models. In this context, many innovative developments, previously considered only of academic interest, now deserve reconsideration. With the limited bandwidth currently available on the Internet, multi-resolution models, which allow for the transmission of data at progressively fine resolution, represent the most likely solution. This is especially true when considering the in-built propensity for data-compression inherent in multi-resolution data structures, a topic discussed further by Brown *et al.* (Chapter 11) and Wood (Chapter 12).

These technical developments need to be considered alongside the dramatically increased performance of computer chips and 3D graphic cards, and also low-cost memory. GIS data structures also need to be integrated within the scene graph model (e.g. VRML, JAVA3D) which represent the 'state-of-play' with regard to the development of VR landscapes. Methods of conversion between TIN data structures and VR scene graphs have recently become more ubiquitous, making GIS-derived 3D dynamic viewing of non-urban landscapes in VR environments much more achievable (Raper *et al.*, 1998).

The development of VR models makes it possible to integrate different sources of data with that represented by, or with, the terrain. Since terrains can be regarded as a framework for many kinds of human activity and natural processes, their properties correlate with a wide variety of other phenomena. Carrying out integration with virtual terrains implies the availability of appropriate datasets. Those with the same extent as the terrain should be synoptic with it if the terrain changes, and should be resolution matched, if in raster form. Datasets consisting of static or moving objects should stand on/follow the terrain within specified tolerances and should change consistently with the terrain if the terrain is re-scaled or its vertical exaggeration altered.

Two basic techniques are available for enhancing virtual landscapes – the use of drapes and 'building' virtual objects in the landscape. Numerous sources are available for the production of drapes including aerial photography, satellite imagery, digital maps, land-use maps and data extracted from vector products to produce raster images. However, draping imagery over a terrain requires the computation of surface 'normals' orthogonal to the landsurface for each triangle or cell necessitating specialized tools. Adding 3D elements to virtual landscapes is also complex, generally involving vector data and computation of the intersection of 2D vector data with the 2.5D surface. Often ground survey is also required to indicate the nature of what may be represented in a dataset as, for example, a polygon coded as woodland may provide no information on tree type, height or density.

Integration of data on coverages (drapes) or objects (three-dimensional in extent) with the terrain brings the opportunity to add value to both datasets. It may be possible to identify objects in the draped coverage (e.g. air photo) or to interpret the pattern/location of objects (e.g. settlements) positioned on the terrain. Conversely, it may be possible to identify terrain positions (e.g. aspect) or shapes (slope angle) which provide an explanation for properties of the draped coverage or positioned objects. It may also be feasible to analyse the properties of the terrain or to difference two terrains and then drape the derived properties back over that terrain.

Visualization of landscape change

The ability to visualize landscape change has been an important factor encouraging the use of virtual reality techniques in geography. Motivations for visualizing changes can be conveniently grouped under two main headings. The first of these is to improve the understanding of natural processes such as the development of coastal spits, movement of river channels within floodplains, changes in glaciers or ecosystems, the diffusion of plant species, or plate tectonics. Examples of how the use of virtual reality techniques can inform the understanding of processes are provided by Mitas *et al.* (1997), Neves and Camara (1999) and Raper (2000).

A second category of applications is in assessing the impacts of human interventions on landscapes. In many environmental management contexts the ability to visualize a range of 'what if' scenarios is a valuable planning tool, particularly when it facilitates consultation or discussion across a variety of interested parties and perhaps enables agreement on a preferred course of action. Interventions may occur at a range of geographical scales, the most local involving the construction of individual buildings, features such as wind farms, or modifications to field boundaries, trees or other elements of land cover. Examples of these types of applications include Lovett *et al.* (Chapter 9) and Miller *et al.* (Chapter 10). At the regional scale, situations where virtual reality techniques may be beneficial include the evaluation of proposals for new residential developments, forestry plans for either new planting or compartment felling (Buckley *et al.*, 1998), and measures for coastal defence or 'managed retreat' in the face of anticipated sea level rise.

There are already a variety of methods for creating virtual reality representations of landscape change. One of the simplest approaches involves the use of image editing software to create photographs of a scene before and after a proposed development or policy change. This technique has been used effectively by Simpson *et al.* (1997) in their study of the Environmentally Sensitive Areas policy in Scotland, but it only provides the observer with isolated views and little scope for interaction with the landscape. More immersive, though still comparatively static, approaches involve the use of drape images on DEMs, CAD software, or languages such as VRML to present models of a landscape at several points in time. These visualizations can be explored by an observer in a more flexible manner and change can be conveyed in both the form of the terrain surface and the characteristics (such as colours, textures and objects) placed upon it. Change may also be represented by differencing of surfaces and then symbolizing the result (e.g. alterations in elevation) as colours on a drape.

If landscapes are modelled in VRML then it is straightforward to create animations of change as transformations from one model to another. Sequences of images such as GIFs or JPEGs can also be linked to provide visualizations of change. Usually the speed of change in such representations is faster than would occur in reality and, similarly, observers typically move through such landscapes in a more rapid manner than would occur in practice. Animating change in complex landscapes also tends to require more computing power than is commonly available in a desktop or portable PC at present, though this constraint may well become less significant in the future.

Conclusions

The following chapters discuss each of the broad issues described above in depth, and with examples of the actual use of VR. Naturally, some of the issues are similar to those relevant to VR in urban contexts, which are dealt with in Chapters 15 to 19. However, the following chapters provide examples of how and why non-urban environments can be, profitably, represented in VR, enabling the user to explore places not easily accessed in person, at periods in time, either in the past or in the future, which would be at best problematic, and probably not possible. In a more sophisticated manner, the user can be exposed to evidence of change due to man or nature, hopefully to enhance understanding of processes and the dynamics of the rural environment. The chapters and accompanying materials on CD provide a snapshot of progress in this sector of VR developments and some visions of how it may evolve into the future.

References

Buckley, D.J., Ulbricht, C. and Berry, J. 1998. The virtual forest: advanced 3-D visualization techniques for forest management and research. Paper presented at the ESRI 1998 User Conference, July 27–31, 1998, San Diego, California. The paper is available at http://www.innovativegis.com/products/vforest/contents/vfoverpaper.htm/.

Carver, S., Blake, M., Turton, I. and Duke-Williams, I. 1997. Open spatial decision making: evaluating the potential of the World Wide Web. In Kemp, Z. (ed.) *Innovations in GIS 4*. London: Taylor & Francis, pp. 267–78.

Hoppe, H., Derose, T., Duchamp, T., McDonald, J. and Stuetzle, W. 1992. Surface reconstruction from unorganized points. *Computer Graphics,* 26, 2.

Mitas, L., Brown, W.M. and Mitasova, H. 1997. Role of dynamic cartography in simulations of landscape processes based on multivariate fields. *Computers and Geosciences*, 23, 4, 437–76.

Neves, J.N. and Camara, A. 1999. Virtual environments and GIS. In Longley, P.A., Goodchild, M.F., Maguire, D.J. and Rhind, D.W. (eds) *Geographical Information Systems: Volume 1, Principles and Technical Issues*. New York: John Wiley, pp. 557–65.

Raper, J.F. 2000. *Multidimensional Geographic Information Science.* London: Taylor and Francis.

Raper, J.F., McCarthy, T. and Williams, N. 1998. Georeferenced 4D virtual environments: principles and applications. *Computers, Environment and Urban Systems*, 22, 4, 1–11.

Simpson, I.A., Parsisson, D., Hanley, N. and Bullock, C.H. 1997. Envisioning future landscapes in the Environmentally Sensitive Areas of Scotland. *Transactions, Institute of British Geographers*, New Series, 22, 307–20.

UN, 1993. *Earth Summit '92*. Proceedings of the United Nations Conference on Environment and Development, Department of Public Information, Information Programme on Sustainable Development, United Nations, New York.

9 Visualizing sustainable agricultural landscapes

Andrew Lovett, Richard Kennaway, Gilla Sünnenberg, Dick Cobb, Paul Dolman, Tim O'Riordan and David Arnold

Introduction

Computer technologies have been used to visualize current or potential future landscapes in many different ways (Sheppard, 1989). Examples include approaches based on the digital manipulation of photographs (e.g. Simpson *et al.*, 1997) and applications of a variety of stand-alone visualization packages (e.g. Berry *et al.*, 1998; Bishop, 1994a). The ability to display landscape characteristics by draping a colour-coded image, such as a satellite-derived classification of land cover, on a digital terrain model has been an established feature of many commercial GIS for at least a decade, but until a few years ago the more sophisticated rendering capabilities of 3D visualization programs were not readily available to the GIS user (Sheppard, 1999). More recently, however, there has been a convergence of the two types of software, with published studies describing the export of feature outlines from GIS databases to visualization packages, as well as more closely coupled integrations (e.g. Bishop and Karadaglis, 1997; Mason *et al.*, 1997). There have also been improvements in the 3D visualization capabilities directly embedded within GIS, prominent examples being the 3D Analyst module of ArcView (http://www.esri.com/software/arcview/extensions/3dext.html) and the VirtualGIS facility of ERDAS Imagine (http://www.erdas.com).

Another important recent development with implications for visualization has been the rapid growth in the use of the World Wide Web (WWW). In particular, the development of Virtual Reality Modelling Language (VRML) since 1994 has provided an open standard for creating 3D 'worlds' that can be readily viewed and disseminated across the Internet (Hartman and Wernecke, 1996). This innovation makes 3D landscape visualizations much more widely accessible (a VRML model can be viewed on any computer with an appropriately configured WWW browser) and greatly increases the scope for using virtual reality techniques within participative decision-making contexts. Facilities to generate VRML models of features now exist within several GIS packages (e.g. ArcView 3-D Analyst or ERDAS VirtualGIS) and there are a number of published applications using a combination of GIS and VRML, particularly to visualize characteristics of urban areas (e.g. Batty *et al.*, 1998; Doyle *et al.*, 1998; Dykes *et al.*, 1999; Martin and Higgs, 1997).

This chapter provides another example of the integration of GIS and VRML for visualization purposes, though the context is one of potential changes to a rural landscape. The key issue under consideration is the scope for achieving more 'sustainable agricultural landscapes', so the next section briefly examines how this term might be defined. This is followed by descriptions of the study area investigated, the initial fieldwork undertaken, and the landscape change scenarios that were generated through interviews with a range of interested parties. The sources and methods used to compile a GIS database for the study area are then outlined, leading on to a discussion of the software employed to create VRML representations of current and possible future landscapes. Several examples of the visualizations are presented, followed by an appraisal of their effectiveness in terms of the practical work required and reactions to them from representatives of different organisations. The chapter concludes with a more general evaluation of the current strengths and weaknesses of VRML as a means of visualizing landscapes and other geographical phenomena.

Sustainable agricultural landscapes

The concept of a sustainable agricultural landscape is discussed in detail by Cobb *et al.* (1999). A review of the literature soon reveals many interpretations of sustainable agriculture, with approaches emphasising enhanced environmental management (e.g. reduced use of chemicals and increased biodiversity), protection of traditional rural economies or more efficient food production (CEC, 1992; CAP Review Group, 1995). Furthermore, many scientific uncertainties remain, especially regarding matters of soil health and biodiversity. The issue of property rights is also important, since a greater degree of co-operation between farms appears necessary if there is to be much scope for creating more ecologically-diverse or culturally-pleasing landscapes. Such a principle of 'whole landscape management' is recognised in several of the agri-environmental schemes already operational in the UK (e.g. Environmentally Sensitive Areas) and is also a feature of the countryside character initiative being promoted by the Countryside Agency (http://www.countryside.gov.uk/activities/special/character_01.htm). The idea of whole landscape management is therefore not new, but the practicalities of implementing such a concept still require considerable thought and investigation.

Cobb *et al.* (1999) conclude that there are several components inherent in the idea of a sustainable agricultural landscape. First, the landscape has to be perceived as an integrated unit, irrespective of current and previous ownership or management. Second, sustainability has to be defined on both a physical and a cultural basis. Physically this means maintaining and enhancing flora and fauna diversity, minimising environmental damage from fertilisers or pesticides, and farming in accord with soil characteristics. Culturally, in addition to the strong prospect of greater public enjoyment arising from enhanced ecological diversity, the concept entails managing the landscape in the interests of all stakeholders (i.e. not just farmers or landowners), ensuring both access and appreciation, and linking aesthetic considerations to the food and leisure needs of the local community. In this way, agricultural practices may also become connected to more sustainable local economies and lifestyles.

Inherent in the above is a recognition that there is no single blueprint for a sustainable agricultural landscape. There may be a range of acceptable choices, as the final results will be socially and culturally determined. Decisions could well focus on different mixes of priorities in terms of agricultural productivity, biodiversity conservation, aesthetic impact, economic opportunity or least environmental damage. Ideally the compromise between these choices, in the context of the various interests arguing their case, would increase the 'education' of all parties on what constitutes an acceptable outcome. Whatever the theoretical benefits, however, the viability of such a procedure in practice is far from certain. As a consequence, a research project was devised to assess the scope for achieving sustainability objectives through more coordinated landscape management in one particular region, the overall aims of the study being:

- to assess impacts of existing or proposed farm management on biodiversity,
- to identify the management practices that need to be adopted on a whole landscape basis to meet objectives for maintaining and enhancing biodiversity and landscape character,
- to evaluate the potential of GIS and virtual reality techniques as means of visualizing alternative future landscapes for stakeholders,
- to establish whether farmers will be prepared to act on a whole landscape basis, and if so, under what conditions or incentives.

The study area

The research site was selected to fulfil certain criteria. It needed to be a landscape unit with components that were recognisable, but which, when combined, formed an integrated whole with distinctive character. Another consideration was that the region should include some protected or designated areas, but also be a part of the wider countryside. To examine the prospects for inter-farm co-operation, the farms all needed to be adjacent, ideally with a mix of ownership, specialisations and operating constraints.

After some preliminary investigation, a research site centred on the National Trust (NT)-owned Buscot & Coleshill Estate was chosen. This estate is situated on the boundary of Oxfordshire, Gloucestershire and Wiltshire (see Figure 9.1) and consists of eleven farms, all of which are

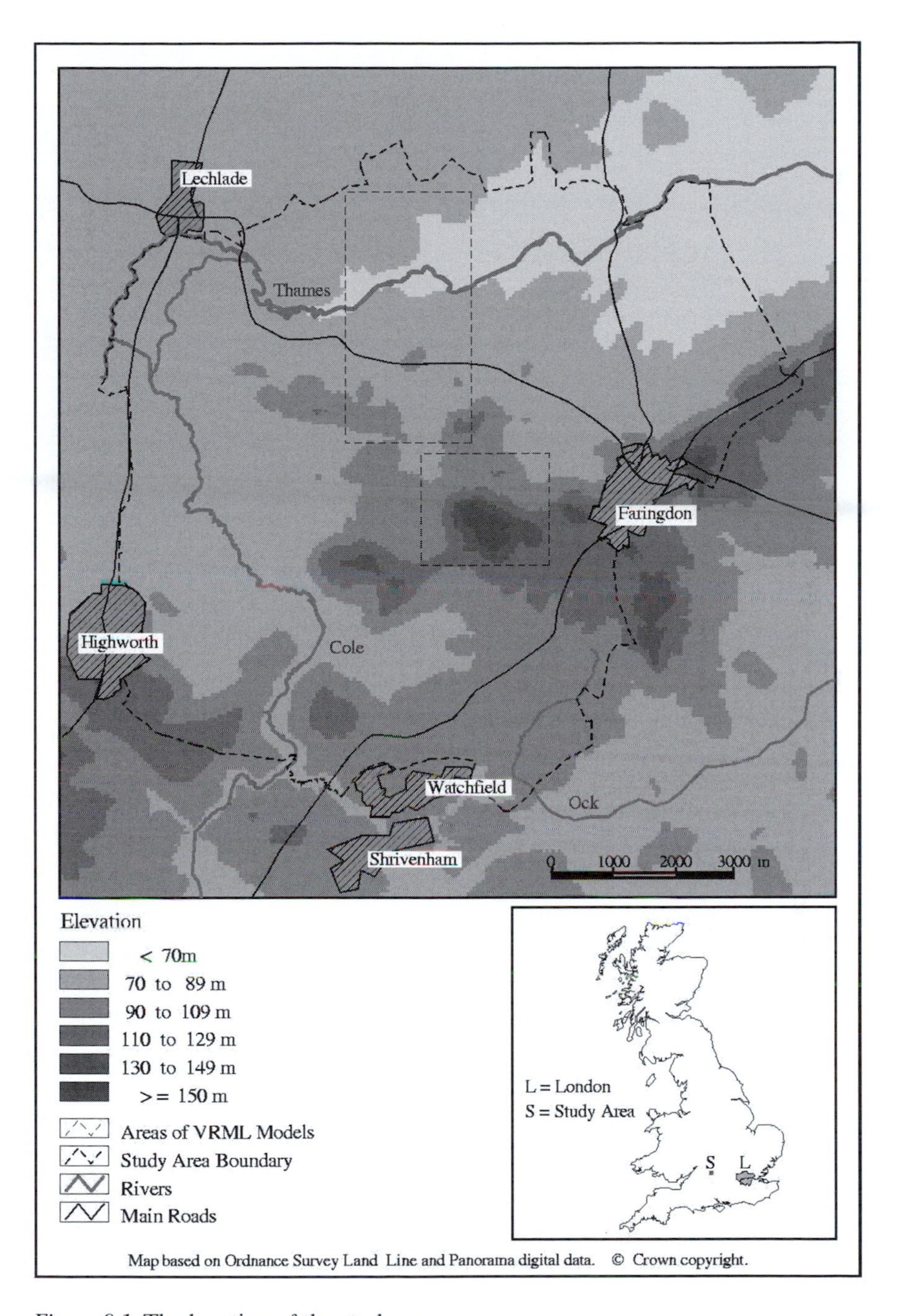

Figure 9.1 The location of the study area.

Figure 9.2 A view across part of the study area from south of the River Thames.

participating in the study. A further twenty farms in the area around the NT estate also agreed to co-operate, making a total study area of some 8,200 hectares (approximately 3,000 of which constitute the NT estate).

The northern part of the study area consists of a zone of clay vale either side of the upper River Thames. A photograph taken from south of the river is included as Figure 9.2 and gives a sense of the relatively flat and open landscape in this area. More undulating topography occurs along the Midvale Ridge which crosses the region from Highworth to Faringdon (see Figure 9.1), while to the south there is another zone of clay vale overlooked by White Horse Hill. Several areas of high landscape value are included (e.g. views from Midvale Ridge across the river flood plains) and the region also encompasses SSSIs, County Wildlife Sites and part of the Upper Thames Environmentally Sensitive Area (ESA). The majority of the study area, however, does not have any protected status.

Initial research

Ecological fieldwork

A primary requirement was to establish the state of the existing environment. An ecological survey was therefore carried out across the study area. Some 3,400 field boundaries were identified, coded in terms of feature characteristics (e.g. hedges were classified in terms of height, width and completeness), and recorded on a database. Seven hundred hedges were examined in more detail to collect information on management practices. A subset of ninety-eight hedges was then chosen (covering contrasting farm regimes) and studied further to collect data on hedgerow species composition, field margin flora, and the birds, butterflies and small mammals present. In addition to providing a fine resolution ecological baseline, statistical analysis of these data, in conjunction with an extensive literature review, enabled the consequences of various hedge and field margin management strategies for different 'emblematic' species to be assessed. This information, in turn, provided empirical support during subsequent scenario generation and validation.

Survey of farmers

A structured interview was undertaken with each of the thirty-one farmers in the study area. This covered subjects such as the farmers' agricultural practices, financial situation, management intentions, and likely responses to possible changes in European Community policy.

The results of the survey confirmed the diverse nature of farms in the study area. Two farms were as small as 150 acres, while a farm manager was responsible for some 8,500 acres (mostly outside the study area). Some farms had a mixture of land ownership under their control: thirteen farms had both owned and tenanted land; nine had only rented land; and the remaining nine were 100 per cent owned. One farmer had operated on an organic basis for over ten years and two others were converting some of their acreage. Other farmers in the area were intent on yield maximisation, often with the added strategy of enlarging their farmed area (sometimes by increased ownership, but also through contract farming). Most of the sample operated mixed farming methods (twenty-seven had livestock and arable enterprises); three concentrated almost exclusively on arable, while a single farmer had no arable acreage. Fifteen farmers had dairy herds and twenty beef cattle, while four had sheep and four kept pigs. One of the farms had a thriving equestrian centre in addition to running a large arable-only enterprise (including an expanding contract farming business).

Farming strategies within the landscape were therefore very mixed. This had obviously influenced the present appearance of the area and was generally expected to do so into the future. There was some interest in the Upper Thames ESA (of those with eligible land, eight farmers had entered part of their farm and eight had not), and participation in the Countryside Stewardship scheme (eight farmers had applied and three more were seriously considering the option). Many farmers reliant on dairying or intensive arable practices, however, expressed reluctance to participate in either scheme.

Alternative landscape scenarios

Drawing upon information obtained during the ecological fieldwork and the survey of farmers, as well as discussions with a range of stakeholder

Table 9.1 The principal landscape changes in each scenario

	Scenario	*Main landscape changes*
1	Business as usual	Limited conversion of arable to grassland on several farms. Creation of buffer strips and enhanced field margins on one farm.
2	Landscape character	Conversion of selected riverside fields to improved grassland. New deciduous woodlands to screen urban areas. Planting of trees as linear features alongside roads, rivers and streams. Hedgerow restoration and replacement of some existing fences by hedges. Creation of an area of open space around the top of Badbury Hill.
3a	Biodiversity conservation	Floodplain reversion to extensive grass or marshland along main rivers. New riparian woodlands. Hedgerow restoration and replacement of some existing fences by hedges. Buffer strips around all streams and ditches (10 m on arable land, 5 m on improved grassland). Uncropped margins around all remaining arable fields (width dependent on presence of buffer strip).
3b	Supplemented biodiversity conservation	All elements of Scenario 3a. Creation of 50 m buffer zones around designated wildlife sites. Conversion of fields around springs on Midvale Ridge to rough grassland.

organisations (e.g. representatives of voluntary bodies, statutory organisations and local authorities), four scenarios for the future landscape of the study area were devised. The sources and thinking underpinning the scenarios are described below, while Table 9.1 summarises the main types of landscape change associated with each alternative.

- *Scenario 1 – Business as usual* was based on each farmer's own plans for future land management. The main source of information was the survey of farmers, as part of which each farmer was asked about current and past farming strategies, as well as reactions to the proposals included in Agenda 2000 (CEC, 1997). Farmers were also asked to respond to a series of statements on farming, conservation and rural economy issues, ranking these on a 'strongly agree' to 'strongly disagree' scale. Few farmers indicated that they were planning major changes to arable operations, none intended to cease dairying, and only five stated that they would abandon beef production (more as a result of difficulties arising from the BSE crisis than general changes in agricultural policy). As a consequence, the landscape changes in this scenario were relatively muted, involving a small shift from arable to grassland and a few habitat protection measures (e.g. buffer strips around streams).

- *Scenario 2 – Landscape character* focused on maximising visible amenity. The scenario was compiled based on discussions with stakeholder organisations and information in a variety of documents including local authority plans, Countryside Commission landscape character assessments, NT landscape plans, and ESA provisions. As biodiversity provides important amenity value, the proposals included some conservation management, but with a focus on popular species of flora and fauna. The main types of landscape changes therefore included hedgerow restoration, conversion of some fields near rivers to grassland, and an increase in deciduous woodland (mainly to screen urban areas or to act as linear features along roads, rivers and streams).
- *Scenario 3a – Biodiversity conservation* was designed to deliver substantial nature conservation and biodiversity benefits. It was based on detailed discussions with statutory and non-statutory organisations, including English Nature, local County Wildlife Trusts and the RSPB, supplemented by the scientific literature and expertise of the research team. The proposals also reflected provisions in the Midvale Ridge and Thames and Avon Vales Natural Area Profiles that applied to the study area as a whole, relevant prescriptions in the Countryside Stewardship scheme, and updated recommendations for the Upper Thames ESA. Blanket compliance was assumed across all habitats, the main landscape changes being a reversion of floodplain farmland to extensive grass and marshland (with hay and grazing in the summer, but possible flooding in winter), increased riparian woodland, hedgerow restoration, the provision of buffer strips around all streams or ditches, and uncropped margins for all arable fields.
- *Scenario 3b – Supplemented biodiversity conservation* incorporated all the components of Scenario 3a, together with a number of measures for specific locations in the study area. Examples included the conversion of fields around springs on the Midvale Ridge to rough grassland, and the creation of large scrub or grass buffer zones around designated wildlife sites. The landscape implications of this option were therefore the most substantial of the four scenarios.

Database construction

Parallel with the definition of landscape scenarios, a digital map database was created for the study area. Two main sources were used, Land-Line® vector mapping and Land-Form Panorama™ elevation data (see http://www.ordsvy.gov.uk). Both were generously supplied free of charge by the Ordnance Survey (OS) for the purposes of the research project.

Land-Line™ data for rural areas are digitised from 1:2500 scale maps. The data have a high degree of positional accuracy, but consist of only point, line and annotation features. Polygon topology has to be explicitly created within a GIS and is a straightforward operation for some types of features (e.g. buildings and areas of water which have seed points defined), but rather more awkward for others (e.g. fields tend to need some manual editing). Point symbols representing some types of land cover (e.g. marsh, scrub or woodland) are included, but they are primarily a cartographic device and cannot be used reliably to code polygons (e.g. several symbols may occur within the same boundary). The classification of line attributes is also restricted, one category described as 'general line detail' encompassing hedges, fences, walls and many other boundaries. A final complication is the principle of hierarchical feature coding, whereby if a line represents multiple features (e.g. an administrative boundary and a riverbank) only one attribute (in this case the former) is recorded (Ordnance Survey, 1997).

Due to these complexities, considerable effort was required to convert the Land-Line™ data to more relevant and structured formats. Manipulation and editing of the data were undertaken using the Arc/Info GIS (see http://www.esri.com/software/arcinfo/index.html), with information from paper maps, classified Landsat imagery, the ecological survey and farmer interviews being used to help distinguish and code different types of polygon or line features. Ultimately three main vector map layers were created:

1 a land cover map containing some 30,000 polygons classified into thirty-five categories,
2 a line coverage distinguishing ten types of field boundary and with additional information for hedges (e.g. width, height and completeness),
3 a polygon coverage defining farm and estate boundaries within the study area.

An excerpt from the land cover data for one of the VRML model areas marked on Figure 9.1 is shown in Figure 9.3. This map includes the National Trust property of Buscot House in the south western corner and extends to cover part of the clay vale around the River Thames and the village of Kelmscott further north. Several land cover categories have been amalgamated for purposes of display, but the level of plan detail provided by Land-Line™ is still apparent.

The Land-Form Panorama™ data provided a 50 m resolution lattice of elevation estimates and were used to model the topography of the study area (see Figure 9.1). In conjunction with details from Land-Line™, the elevation model was also employed to define floodplains around the River Thames and Cole. The outline generated was subsequently validated

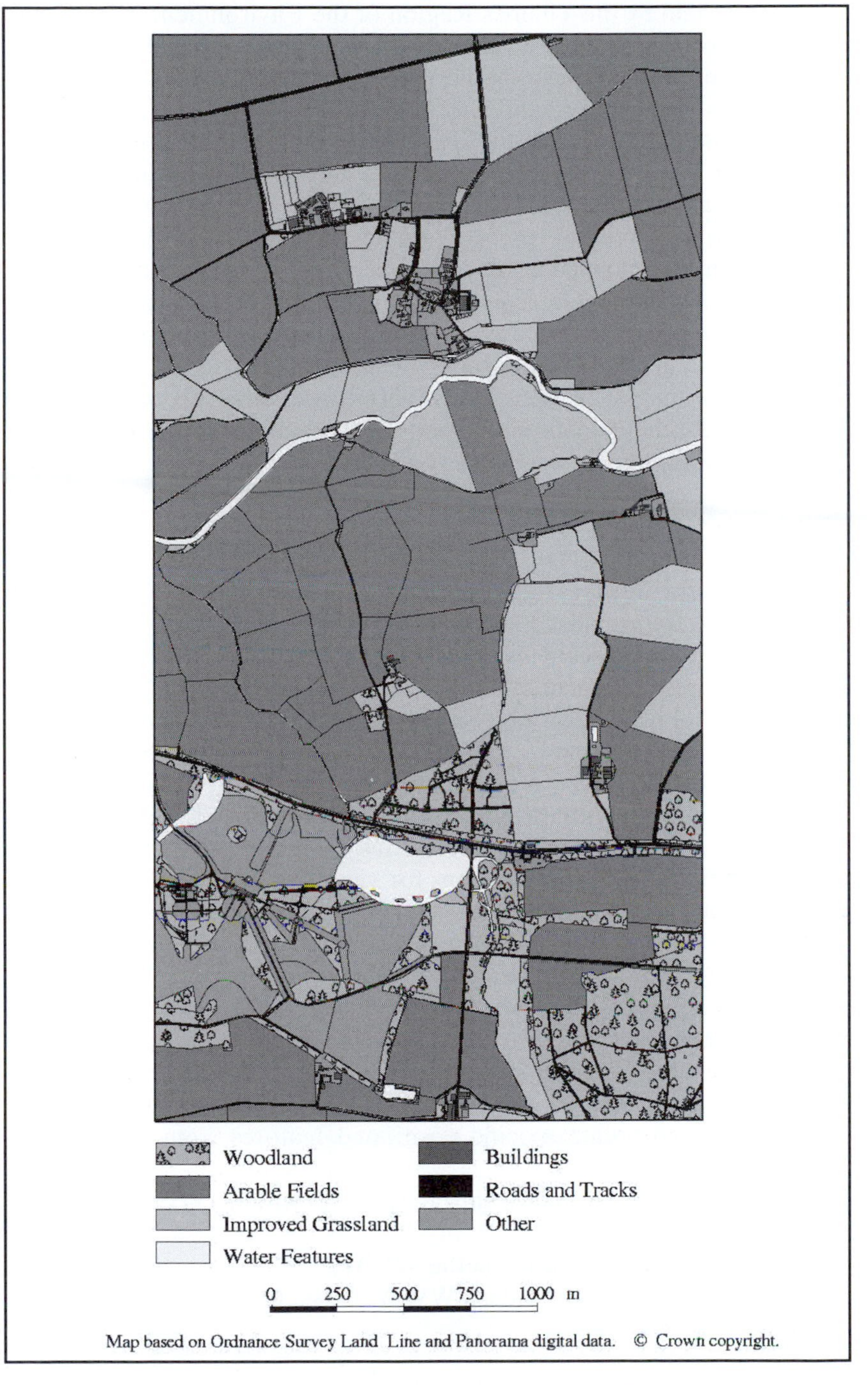

Figure 9.3 Current land cover in the Buscot House and Thames Valley VRML model area.

against maps held by the Thames Region of the Environment Agency and then used to code relevant polygons on the land cover layer. This step was necessary in order to identify the floodplain reversion areas that were key components of Scenarios 3a and 3b.

Once a database depicting the current situation had been constructed, the next stage was to create equivalent map layers for the different scenarios. In some instances (e.g. Scenario 1) the degree of change was so minor that it was straightforward to copy existing files and make the necessary alterations through manual editing and recoding. In more complex cases, however, Arc Macro Language (AML) code was written to implement sets of operations and then merge together a number of intermediate coverages. Such an approach was used to generate the different sized buffer strips and field margins in the Biodiversity conservation scenarios. It was also employed to implement an element of randomness in the selection of field boundaries for hedge replanting or restoration.

Visualization methods

Changes in land cover and field boundary characteristics under the different scenarios were visualized either as a series of large (A0) colour maps or as 3D models for selected areas of interest.

Figure 9.4 presents a map of land cover under Scenario 3a (Biodiversity conservation) for an area around Kelmscott village and the River Thames. This map has been designed specifically for purposes of greyscale display, and so the shadings and categories are different from the larger colour version. Comparisons with the corresponding section of Figure 9.3 indicate substantial alterations in land cover, but 3D visualizations are really necessary to get a full sense of implications for views across the landscape. This is primarily due to topographic influences, but also because of screening and texture effects from features such as buildings, trees and hedges. It was therefore decided to supplement the scenario maps with 3D visualizations for key regions within the study area. Several types of visualization software were investigated, but ultimately an approach based on VRML was selected for the following reasons:

1. ability to incorporate specific positioned features from a GIS database,
2. practicality of generating visualizations from within a GIS,
3. platform independence,
4. potential for dissemination via the WWW,
5. ease of viewing with a standard WWW browser,
6. limited software costs.

The following section provides a brief introduction to VRML, after which the specific software and procedures used in the research are discussed.

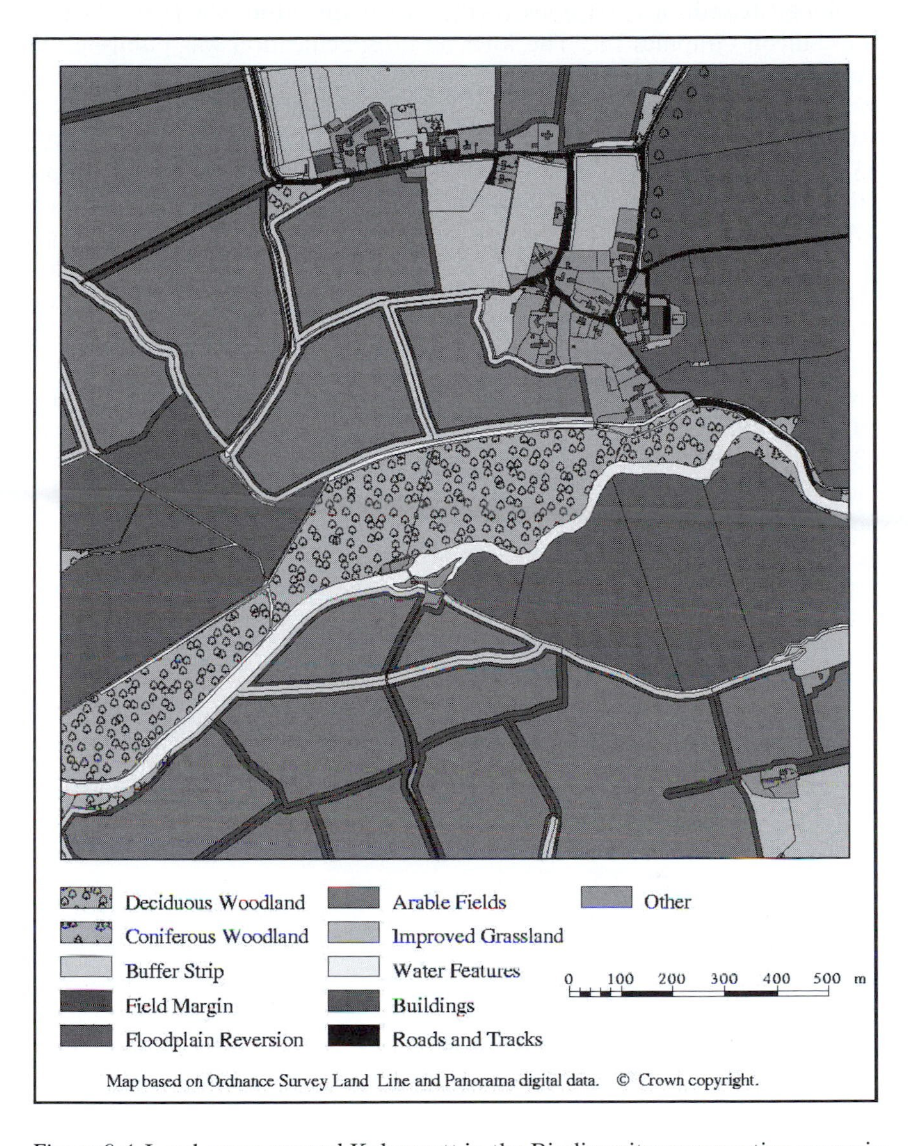

Figure 9.4 Land cover around Kelmscott in the Biodiversity conservation scenario.

Virtual Reality Modelling Language

VRML is an open standard for 3D multimedia and shared virtual worlds on the Internet (http://www.vrml.org/fs_specifications.htm). The concept originated from discussions at the First International Conference on the World Wide Web held in Geneva during spring 1994, and (stimulated via debate on an electronic mailing list) a specification was subsequently developed based on extensions to the Open Inventor ASCII file format from Silicon Graphics Inc. The VRML 1.0 specification was published in May 1995 and a working group then began to consider proposals to incorporate facilities for animations, directional sound and generally greater interactivity. These, and other enhancements, were included in the VRML 2.0 specification released in August 1996. Since then, the development of VRML has been the responsibility of a formally constituted consortium with an elected board of directors (http://www.web3d.org/fs_aboutus.htm). In December 1997 VRML was recognised as an ISO international standard (VRML, 97) and in early 1999 the VRML Consortium changed its name to the Web3D Consortium (see http://www.web3d.org or http://www.vrml.org).

A VRML 'world' consists of one or more files (conventionally with a wrl suffix) that together describe the geometry and attributes of objects in a 3D scene. The fundamental building blocks of VRML files are nodes. These can be of several types and are used to define shapes, properties such as textures, relationships between entities (e.g. hyperlinks or groupings), or special features such as viewpoints, spotlights and sound sources (Carey and Bell, 1997; Hartman and Wernecke, 1996). The following is a simple VRML file which defines a solid yellow cylinder.

```
#VRML V2.0 utf8
Shape {
appearance Appearance {
  material Material {
    diffuseColor 1 1 0
    shininess 1
    }
}
geometry Cylinder {
  radius 2
  height 4
  side TRUE
top       TRUE
bottom    TRUE
}
}
```

In this example the first line is a header and the remainder constitutes a Shape Node with parameters to define the appearance and geometry of the cylinder. Like all VRML, the file consists of standard ASCII characters and in this instance was created using an ordinary text editor. When viewed in a suitably configured WWW browser, however, the text commands are processed to draw the cylinder. Recent releases of Netscape Navigator (http://www.netscape.com/browsers/index.html) and Internet

Explorer (http://www.microsoft.com/windows/ie) have VRML interpreters pre-installed, but it is possible to download suitable software as a plug-in for almost any browser. The two most widely used VRML interpreters are CosmoPlayer (http://www.cai.com/cosmo) and WorldView (also known as Microsoft VRML 2.0 Viewer).

Figure 9.5 shows the example VRML file viewed using Netscape Navigator and CosmoPlayer. The buttons on the control panel at the bottom of the Netscape window provide facilities for using the mouse to move around the VRML world in various ways (e.g. zoom in/out, pan, slide, rotate and tilt), so it is possible to explore the virtual environment in a flexible and interactive manner. To produce the scene shown in Figure 9.5 several movement options were used to shift from the initial view of a vertical cylinder to a position above one end. It is also possible to define specific viewpoints during the creation of a VRML world and these are then readily selectable from a box (here labelled 'Entry') on the control panel. Another visualization option is to link several viewpoints in a sequence and generate a simple animated tour through a VRML scene.

Although it is perfectly possible to produce VRML files with just a text editor, as the objects involved become more complex so the practicality of this approach declines. For more sophisticated development work it is common to use a variety of authoring tools, usually either 'world builders'

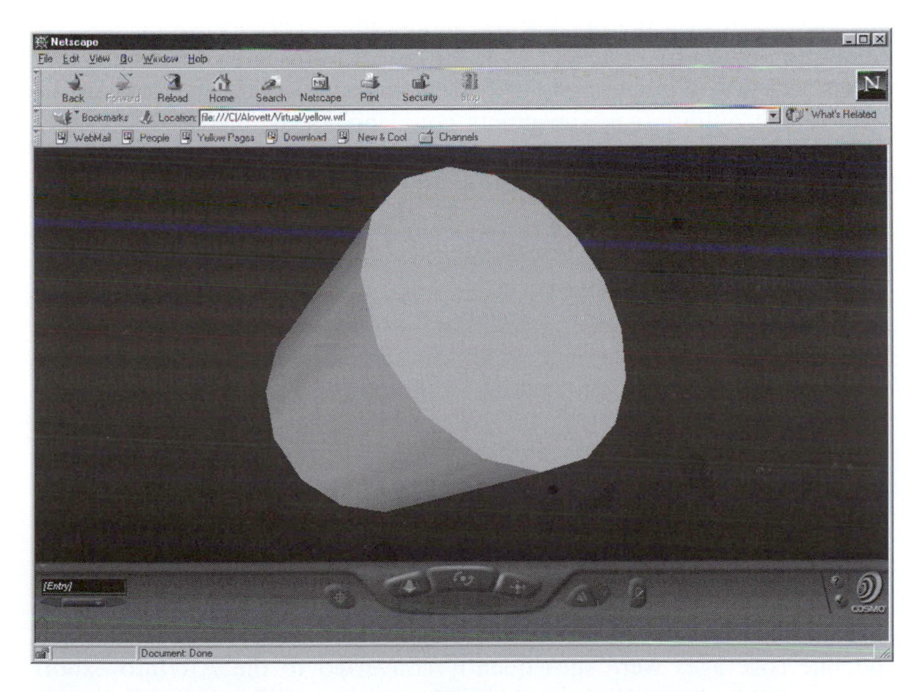

Figure 9.5 A view of the example VRML file using Netscape Navigator and CosmoPlayer.

(3D drawing applications which create VRML worlds) or conversion programs which generate VRML from other data formats. Many of these utilities are freely available via the Internet (see listings at http://www.web3d.org/vrml/3dauth.htm). For further information on aspects of VRML the following are all good starting points, with many downloadable resources and links to other WWW sites.

- The Web3D Consortium http://www.web3d.org
- VRML Repository http://www.web3d.org/vrml/vrml.htm
- ACOCG VRML/ Java3D Centre http://www.scs.leeds.ac.uk/vrmljava3d/
- VRMLWorks http://home.hiwaay.net/~crispen/vrmlworks

The Pavan™ Virtual Reality Toolkit

A number of authoring tools were investigated for purposes of creating VRML visualizations of the landscape scenarios. Several of those freely available had the limitation that they could only utilise information from the Arc/Info database if it was first converted to CAD formats (such as DXF) and this, in turn, made it difficult to transfer sets of polygon and line attributes. Other software was either too expensive or only available for operating systems (e.g. IRIX) that we did not have access to. Consideration was given to using the VRML generation capability within the 3-D Analyst module of ArcView, but at the time the research was being conducted this was a single command with restricted ability to control the form of the output (e.g. viewpoints could not be defined). Much more comprehensive VRML authoring tools were available in the Pavan™ software (http://www.pavan.co.uk; Smith, 1997) but, unfortunately, this only operated within the MapInfo™ GIS. Nevertheless, it was eventually decided that Pavan™ represented the most efficient means of generating the 3D visualizations required.

Figure 9.6 summarises the main stages in the production of the VRML landscape models. Early in the research it became apparent that it would not be feasible to create realistic 3D visualizations of the entire study area, and so attention was focused on two smaller zones (outlined on Figure 9.1). One of these selections represented a transect across part of the clay vale (see Figure 9.3 for further detail) and the other was centred around Badbury Hill on the Midvale Ridge. These areas were chosen as examples of contrasting landscapes where different types of changes were anticipated in the scenarios.

The first step in the modelling process was to clip out the land cover, field boundary and elevation data for the selected subset areas. Polygon and line coverages were subsequently converted to the Arc/Info export format (e00) and imported into MapInfo as tables with associated graphic objects. Individual land cover categories were then selected in MapInfo and

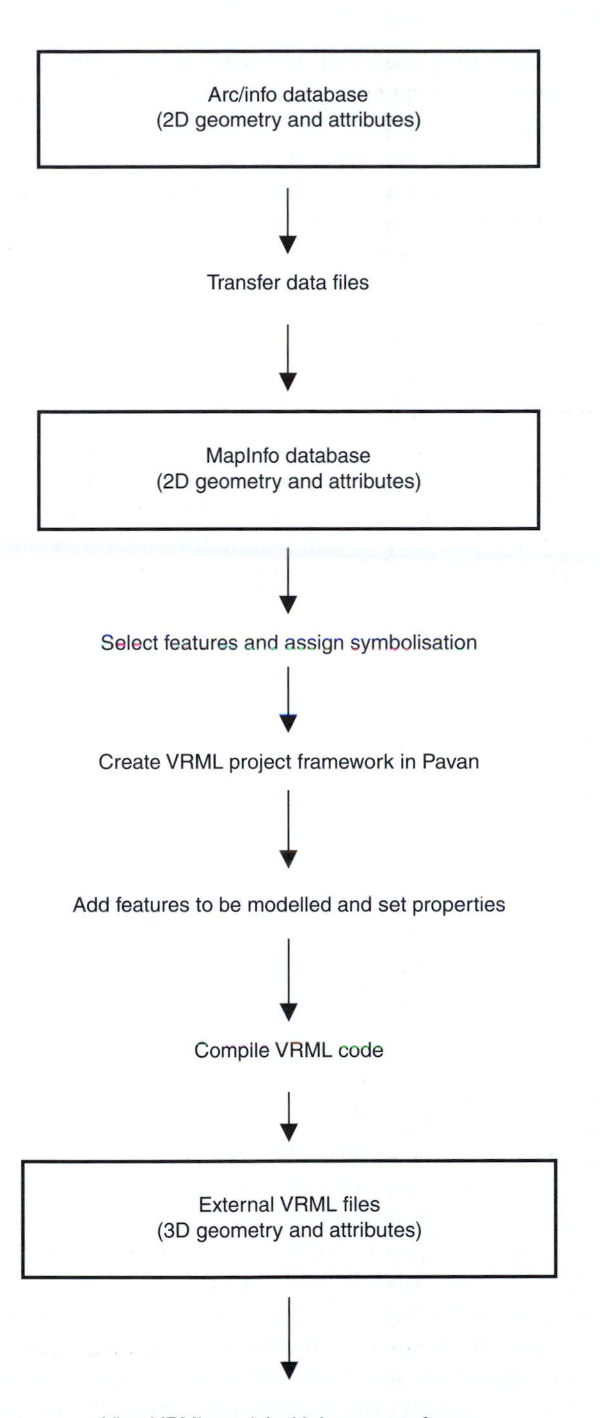

Figure 9.6 The main stages in the production of the VRML landscape models.

the polygons assigned colours which matched as closely as possible those used on the large format maps. A similar process was used to symbolise particular field boundary types as lines of different style and thickness.

Each new VRML model is created in Pavan as a separate *project*. Initiating a new project involves defining a series of parameters including the study area limits, projection type, elevation grid interval and the number of *cells* that the model should be subdivided into. The cells constitute the basic building blocks of a VRML world and, in Pavan, each cell consists of several layers representing the different types of features present. One advantage of this structure (discussed further on page 119) is that it is possible to create or modify a VRML model in a cumulative and very flexible manner.

For the selected area shown in Figure 9.3 (2 km by 4 km in extent) a cell resolution of 500 m was eventually decided upon. This size was chosen so as to be appreciably larger than the elevation grid interval (50 m), but not to involve too much complex line or polygon detail for the processing capacity of the Pavan software and principal computer used (a 233 MHz Pentium II with 64 Mb RAM). Once cell framework had been established the following steps were undertaken:

- MapInfo tables representing the land cover polygons and field boundary lines were added to the project,
- the lattice of elevation values was imported as an ASCII file in *x,y,z* format and the vertical exaggeration was doubled by applying a scale factor of 2.0 (the terrain otherwise appeared artificially flat when viewed from a height),
- building outlines were extracted from the land cover polygons, saved as a separate layer, and then defined as simple 3D block features,
- trees and bushes were added to the model using Pavan vegetation modelling tools,
- point lights were positioned to illuminate surface features,
- viewpoints were defined and, in some instances, linked into sequences to provide simple animated tours.

Several aspects of the above require additional comment. Most of the buildings were deliberately kept simple with a default height of 6 m and a flat roof. A few of the most prominent buildings (e.g. Buscot House) were made taller and given a ridged roof. Pavan includes facilities to vary individual building heights, but there were no suitable data to allow this to be done in a comprehensive manner. It would also have been feasible to place bitmap images or vector drawings on the sides of buildings, but a pilot investigation soon revealed that this substantially increased the processing and rendering requirements. As the focus of the research was on landscape change, it was therefore decided to keep the appearance of buildings at a basic level.

Vegetation was added to the landscapes by first specifying the characteristics of the required feature (e.g. type, height and colour), and subsequently positioning it using a mouse click. Unfortunately there were no facilities to cover selected polygons with set densities of trees and so defining areas of woodland was a rather laborious operation involving a multiplicity of mouse clicks. Trees were also positioned to represent known avenues or shelter belts, and scattered along the most complete sections of hedge in accordance with data from the ecological survey. No attempt was made, however, accurately to mimic the location of specific trees and consequently the end result was a somewhat symbolic representation.

Other difficulties arose in the visualization of field boundaries. The original intention had been to show hedges or fences as 3D extrusions above the terrain surface, but in Pavan it was only possible to digitise the spine of such objects rather than utilising lines from an existing GIS database. Given the number of field boundaries in the selected areas, extensive on-screen digitising was not really practical and there were also some concerns about the processing load associated with so many extrusions. A different method was therefore adopted with thick lines draped over the terrain surface to represent hedges and thinner ones to depict other boundary types. In addition, bushes were placed along the lines symbolising hedges, the density of these 3D objects reflecting (in a relative sense) the completeness of the vegetation as classified in the ecological survey. Complementing the line thicknesses in this manner helped with the visualization of changes such as hedgerow restoration, but only partially compensated for the visual barriers provided by hedges and fences at ground level. The modelled representations were consequently stylised and more visually transparent than the real field boundaries.

Compiling the landscape features as VRML code was very much a gradual and cumulative operation. In most instances an initial model was created based on the land-use polygons, field boundaries and elevation data. If this appeared satisfactory when viewed with Netscape then further work was undertaken to add elements such as vegetation, buildings and viewpoints. Defining the latter was particularly time consuming as it was necessary to set the view parameters (e.g. position, height and angle) within Pavan, compile the VRML, and then examine the result obtained in Netscape. Many iterations were required to get some viewpoints exactly as desired, but the exercise was assisted by the manner in which Pavan allows specific cells or feature layers to be selected for VRML compilation. Figure 9.7 shows the software in operation, with the main display window focused on four cells around the village of Kelmscott. Locations where deciduous trees have been placed are shown by symbols and two viewpoints are represented by lines with arrow heads. Features in three of the cells have been converted to VRML previously and the upper right one has just been selected for compilation (hence the cross-hatch rather than

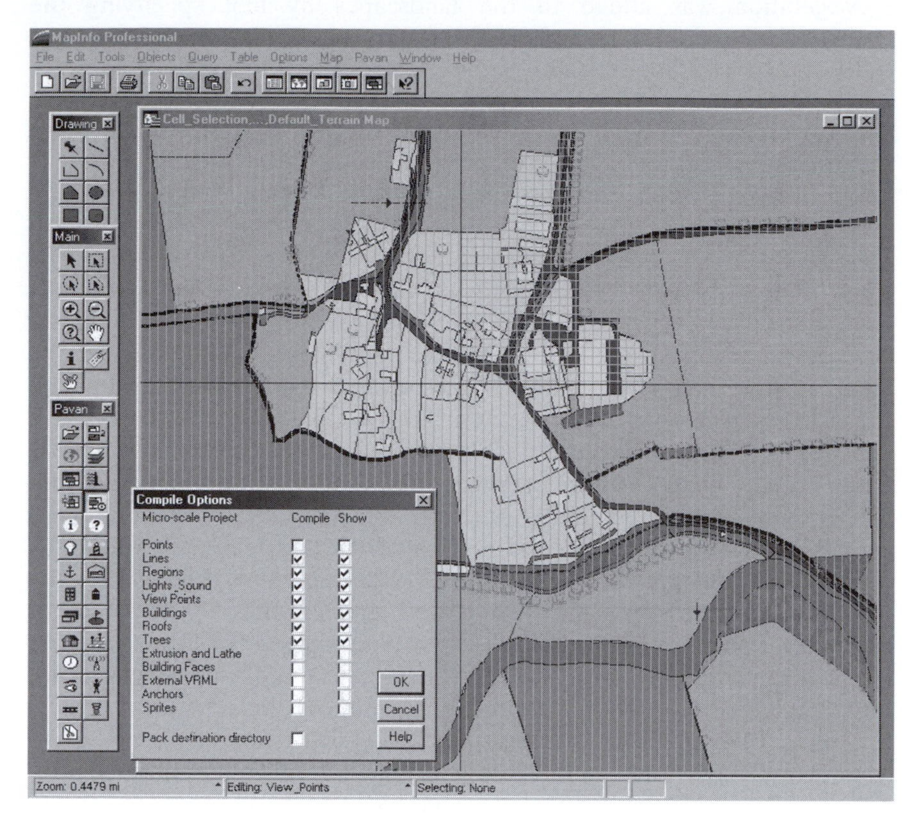

Figure 9.7 VRML compilation using Pavan.

vertical shading). In this example, several layers are being converted (see the checked boxes in the *compile options* window), but it would also be possible to select a single layer in one or two cells. This ability to make small refinements whilst retaining other code unaltered proved particularly useful in the final stages of landscape generation.

The production of the VRML model for the area shown in Figure 9.3 required a total of approximately forty minutes' processing time on the Pentium II used. For the second area (roughly half the size) around Badbury Hill the overall processing time was some twenty-two minutes. Once the current land-use models had been created it proved straightforward to copy and modify them to create versions for the various policy scenarios. This had the advantages of reducing the amount of preparatory work required and also keeping features like viewpoints consistent across different models for the same area.

One final point which should be made is that the 3D modelling work did not incorporate all the facilities available within VRML and Pavan. No attempt was made to include sound sources or hyperlinks to photographs and other external databases (see Smith, 1998). Other options would have been to use external VRML models to depict a greater variety of vegetation types or add satellite imagery to provide textures draped on the elevation surface. Demonstrations of several such possibilities are available at the Pavan WWW site (http://www.pavan.co.uk).

Landscape visualizations

The main products from the visualization work were sets of large colour maps and VRML models. These were supplemented with tables of statistics generated from the underlying digital data. Table 9.2 provides an example and indicates the extent of potential changes in land cover under the different scenarios. Similar outputs were produced to show possible changes in field boundaries across the study area and other statistics were compiled to summarise the implications of scenarios at the individual farm level.

Several illustrations of the map format used in the research have been presented in this chapter already (e.g. Figures 9.3 and 9.4). Two small excerpts from the VRML models are included on the CD accompanying this book and instructions on how to view them are provided in an appendix to this chapter. Illustrations of the larger VRML landscapes have been captured using the Paint Shop Pro software (http://www.jasc.com) and will be discussed in the following paragraphs. Colour JPEG images of these views are also included on the accompanying CD.

Table 9.2 Changes in land cover between scenarios

	Area in hectares				
Land cover type	*Current use*	*Scenario 1*	*Scenario 2*	*Scenario 3a*	*Scenario 3b*
Arable field	4,147	3,955	3,844	3,092	3,031
Improved grassland	2,803	2,964	2,976	1,952	1,902
Rough grassland	124	142	124	95	149
Parkland	169	169	169	169	169
Woodland	474	474	609	528	528
Buffer strip	0	5	0	146	143
Field margin	0	9	0	421	404
Marshland or floodplain reversion	5	5	5	1,330	1,330
Scrub	27	26	27	26	103
Water or riverbank	113	113	113	112	112
Road, track or verge	157	157	157	157	157
Building	27	27	27	27	27
Other	178	178	173	169	169
Total	8,224	8,224	8,224	8,224	8,224

Figure 9.8 A view of the VRML model for Buscot House and the surrounding parkland.

Figure 9.8 shows a view across the current land-use model for the area mapped in Figure 9.3. The viewpoint is in the south-west corner of the area and looks across the parkland surrounding Buscot House onto the clay vale south of the River Thames. Comparison of Figures 9.3 and 9.8 provides a sense of the complementarity between the two types of visualization, though it should be noted that the greyscale shadings on the map and VRML view are not completely identical. The greyscale on the image is a conversion of the colour view, while the map was deliberately designed as part of a set with a reduced number of land cover categories. In particular, buildings have a lighter shading on the image than the map, while the reverse is true for lakes or rivers.

Buscot House is prominent in the foreground of Figure 9.8 and the block representation of buildings is very evident. The simplified form of individual trees is also apparent, although as they are viewed in groups and at longer distances the degree of realism becomes greater. Several examples of lines representing hedges can be seen north of the parkland. Some of these appear quite apt, but many lines are more broken than intended due to the combined influence of topography and the limited level of detail that can be displayed.

Figure 9.9(a) provides another visualization of the area mapped in Figure 9.3, this time looking east from above the clay vale. Trees on the south bank of the River Thames dominate the foreground and part of Kelmscott village can be seen on the left side in the middle distance. Lines depicting the boundaries around the arable fields are evident on the clay vale, though individual bushes are difficult to detect against the dark background. Figure 9.9(b) shows a view from the same point across the Scenario 3a (Biodiversity conservation) landscape, and here the introduction

Figure 9.9 (a) A view of the VRML model depicting current land cover near the River Thames.

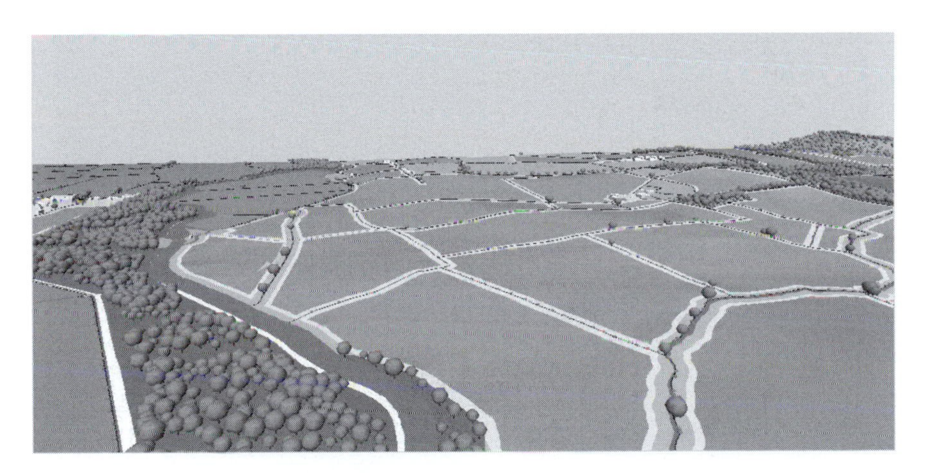

Figure 9.9 (b) A view of the VRML model showing land cover in the Biodiversity conservation scenario.

of features such as buffer strips, field margins and enhanced hedgerows results in a substantial visual contrast. Other alterations include new areas of woodland beside the River Thames and the reversion of floodplain zones to rough grazing or marshland. Part of the view can also be compared with the map in Figure 9.4, though again there are some differences in greyscale shading between the two. Note particularly that the field margins are darker on the map than in Figure 9.9(b).

A third example of landscape visualization comes from the Badbury Hill area. Figure 9.10 presents a map of the current land use in this region and can be compared with the view from the south-east corner shown in

Figure 9.10 Current land cover in the Badbury Hill VRML model area.

Figure 9.11(a). Badbury Hill itself is very prominent towards the centre of the image and other features such as areas of woodland and the main road are also readily identifiable. Changes in the landscape under Scenario 3b (Supplemented biodiversity conservation) are displayed in Figure 9.11(b) and essentially consist of some new field margins and scrub buffer zones around designated wildlife sites. The latter are most obvious near the top of Badbury Hill, but also occur adjacent to a few other blocks of woodland.

Several aspects of the VRML models are difficult to convey using static greyscale images. One distinctive characteristic of VRML is the ability interactively to navigate through a 'world' using the browser controls and,

Figure 9.11 (a) A view of the VRML model depicting current land cover around Badbury Hill.

Figure 9.11 (b) A view of the VRML model showing land cover in the Supplemented biodiversity conservation scenario.

indeed, this feature proved valuable in several demonstrations of the research. It is fair to note, however, that even with the VRML files stored on the hard disc of a computer (rather than being accessed over the Internet) the speed of response to the controls was often rather slow. Initial loading of the VRML model for the 2 km by 4 km area shown in Figure 9.3 typically took around 130 seconds on a 233 MHz Pentium II, with a further 25 seconds to shift from one viewpoint to another. For the smaller Badbury Hill model the corresponding figures were 100 and 15 seconds. Similar performance limitations have been recorded by Gahegan (1999), who notes that they reflect the interpreted (rather than compiled) nature of VRML. Moving at will around all the Oxfordshire VRML worlds was quite possible, but for purposes of audience display it was found easiest to define at least six viewpoints and an animated tour in each model. The latter was especially useful as it could be left to run for several minutes while members of the research team provided a commentary on what was being shown. Examples of viewpoints and tours are included in the VRML models on the CD accompanying this book.

Another characteristic of the VRML visualizations was the simplified and symbolic manner in which some landscape components were depicted. This was certainly true of buildings and field boundaries, but it also applied to certain types of land cover. Narrow or small features such as field margins and buffer zones were deliberately given less realistic, but bright, colours so that they would be clearly visible. This was important for emphasising the changes associated with the policy scenarios and also allowed elements of the VRML models (where there was no legend) to be readily identified on the colour maps. The artificial nature of some colours is not especially obvious on the greyscale prints included in this chapter, but should be immediately apparent if the JPEG images on the CD are examined.

Reactions from stakeholder organisations

A meeting attended by representatives from thirteen stakeholder organisations (including local authorities, wildlife trusts, English Nature, the National Farmers' Union, and the National Trust) was held in May 1999. The large colour maps and VRML models were presented to the participants and they were asked for their comments on these, and other aspects of the research. Overall reaction to the maps was very positive, though it was noted that they required careful study to compare the present situation in a particular locality with the different scenario outcomes. Opinions on the VRML models were more mixed. It was accepted that they had the potential to provide effective overviews of areas, but it was also thought that additional details and textures were necessary to make specific locations immediately recognisable by local residents. This latter point was considered especially relevant as far as the acceptance of the VRML landscapes by farmers was concerned. Other comments included the need to improve response speeds so that navigation around the landscapes could occur in a smooth and seamless fashion. It was also suggested that the models would be easier to interpret with hyperlinks to individual photographs, 360° panorama views, or legend information. In general, the two approaches to landscape visualization were regarded as complementary, one response being:

> The novelty of the VRML may be attention grabbing; but just as one part of the presentation package.

It is fair to say that some of the responses from the stakeholder meeting were not unexpected. Several of the perceived limitations of the VRML models are matters that were discussed earlier in this chapter and could be rectified with additional data, staff time or computing resources. The use of VRML in the research was experimental and further investment would certainly result in enhanced landscape models. At the time of writing, the farmers in the study area are being re-interviewed to obtain their reactions to the various policy scenarios. As part of this process they are being shown the maps, relevant statistical data for their farm, and a selection of the VRML models running on a laptop computer. This exercise should provide another perspective on the merits of the visualizations and ultimately enable conclusions to be drawn regarding the scope for achieving sustainable landscape management in the study area.

Conclusions

This chapter has described the use of GIS and VRML to visualize potential landscape changes in a rural district of England. The wider research project incorporating this work is still in progress, so it is not possible to present definitive findings on the viability of a sustainable agricultural landscape. It is clear, nevertheless, that the conversion of features from GIS databases to VRML represents an increasingly feasible and quite efficient means of generating 3D landscape visualizations. Major strengths of VRML include the open and flexible nature of the scene description code, platform independence, limited costs, the ability to incorporate hyperlinks to other data, and the manner in which models can be readily viewed and interacted with over the Internet. It is also relevant to note that VRML is still being actively developed and, for instance, there is a GeoVRML Working Group (http://www.geovrml.org; Rhyne, 1999) aiming to further enhance means of representing geo-referenced data within the standard.

There are several improvements in modelling tools which would make the process of converting GIS databases to VRML more straightforward. In the context of this particular project, facilities within Pavan™ automatically to cover polygons with trees or utilise existing digitised lines as the base of extrusions would have been of substantial benefit. More generally, a greater ability to define the parameters of VRML objects from feature attributes in a GIS database would save considerable time. Recent releases of visualization software such as World Construction Set™ (http://www.3dnature.com) and GenesisII™ (http://www.geomantics.com) have much enhanced capabilities to import features from GIS databases and, if access or dissemination via the WWW is not important, may represent an easier means of creating some types of 3D landscapes. VRML, however, still has an obvious role and the Pavan™ toolkit is being actively extended at present (see http://www.pavan.co.uk).

The provision of suitable digital data will also be vital for further advances in 3D landscape visualization. One of the most time-consuming aspects of the research was editing the OS Land-Line™ data to define polygons and add land cover codes. The need for such efforts may well be reduced in future as research is in progress to automatically convert Land-Line™ into a polygon-based structure (Murphy, 1998). Other projects are investigating the scope for combining OS digital data and high-resolution satellite imagery to produce land-cover classifications on a per-parcel basis (e.g. Aplin *et al.*, 1998; Murfitt, 1998). Less positively, however, current licence terms greatly restrict the use of OS data on the Internet. This means that the best sources of large-scale digital data in the United Kingdom cannot be legally employed to create VRML models accessible via the WWW. Such a constraint essentially negates one of the main strengths of VRML, and it is to be hoped that a more satisfactory solution to the problem of copyright protection will be found in the future.

Continued advances in computer technology can be expected to reduce some of the restrictions encountered in this project in terms of the size of area and level of detail that it was practical to include in a VRML model. Improvements in visual simulation are therefore almost inevitable, but it is perhaps questionable whether the most realistic landscape is always the best one for purposes of public presentation and policy formulation. Much still remains to be discovered about the most appropriate means of symbolising features in virtual landscapes and the ability of users to navigate through, and interpret, such worlds. Researchers have begun to investigate such issues (e.g. Bishop, 1997; Doyle *et al.*, 1998), but in some instances it may be that rather simplified and stylised landscapes are better communication devices than those with a very high degree of realism. Such questions remain for future examination, but for the moment this chapter has demonstrated the potential of a GIS and VRML combination for landscape visualization and provided some indications of what is currently feasible.

Appendix

Excerpts of the VRML models discussed in this chapter, together with additional documentation are included on the accompanying CD ROM.

Acknowledgement

This research has been primarily funded by the ESRC Global Environmental Change Programme (Grant number L320253243). Additional support has been provided by the Arkleton Trust, the Ernest Cook Trust, the Esmee Fairbairn Charitable Trust and the European Union ETHOS project (Telematics Applications Programme Project N1105). Thanks are due to the National Trust for giving us the opportunity to include their

Buscot and Coleshill Estate within the study area. We are also very grateful to the Ordnance Survey for the provision of digital map data and for permitting two VRML models to be included on the CD accompanying this book. These digital map data are reproduced by kind permission of OS Crown Copyright. Parts of the visualization research were completed while Andrew Lovett and Gilla Sünnenberg were visiting fellows at the Centre for GIS and Modelling (CGISM), University of Melbourne. We would like to thank Professor Ian Bishop for providing facilities in Melbourne and his interest in our research.

References

Aplin, P., Atkinson, P.M. and Curran, P.J. 1998. Combining fuzzy and per-parcel classification techniques to map urban land cover from fine spatial resolution satellite imagery. In *Proceedings RSS 98: 24th Annual Conference of the Remote Sensing Society*, University of Greenwich, September 1998, pp. 561–7.

Batty, M., Dodge, M., Doyle, S. and Smith, A. 1998. Modelling virtual environments. In Longley, P.A., Brooks, S.M., McDonnell, R. and Macmillan, B. (eds) *Geocomputation: A Primer*. London: John Wiley, pp. 139–61.

Berry, J.K., Buckley, D.J. and Ulbricht, C. 1998. Visualize realistic landscapes: 3-D modeling helps GIS users envision natural resources. *GIS World*, 11, 8 (August), 42–7.

Bishop, I.D. 1994. Using Wavefront Technology's advanced visualizer software to visualize environmental change and other data. In MacEachren, A.M. and Taylor, D.F.R. (eds) *Visualization in Modern Cartography*. Oxford: Pergamon, pp. 101–3.

Bishop, I.D. 1997. Testing perceived landscape colour difference using the Internet. *Landscape and Urban Planning*, 37, 187–96.

Bishop, I.D. and Karadaglis, C. 1997. Linking modelling and visualization for natural resources management. *Environment and Planning B: Planning and Design*, 24, 345–58.

CAP Review Group. 1995. *European Agriculture: The Case for Radical Reform*. Ministry of Agriculture, Fisheries and Food, London.

Carey, R. and Bell, G. 1997. *The Annotated VRML 2.0 Reference Manual.* New York: Addison-Wesley.

Cobb, D., Dolman, P. and O'Riordan, T. 1999. Interpretations of sustainable agriculture in the UK. *Progress in Human Geography*, 23, 2, 209–35.

Commission of the European Communities. 1992. *Towards Sustainability: A European Community Programme of Policy and Action in Relation to the Environment and Sustainable Development*. COM (92) final, Commission of the European Communities, Brussels.

Commission of the European Communities. 1997. *Agenda 2000: Volume 1 – Towards a Stronger and Wider Union*. DOC 97, Commission of the European Communities, Brussels.

Doyle, S., Dodge, M. and Smith, A. 1998. The potential of web-based mapping and virtual reality technologies for modelling urban environments. *Computers, Environment and Urban Systems*, 22, 2, 137–55.

Dykes, J., Moore, K. and Wood, J. 1999. Virtual environments for student fieldwork using networked components. *International Journal of Geographical Information Science*, 13, 4, 397–416.

Gahegan, M. 1999. Four barriers to the development of effective exploratory visualization tools for the geosciences. *International Journal of Geographical Information Science*, 13, 4, 289–309.

Hartman, J. and Wernecke, J. 1996. *The VRML 2.0 Handbook: Building Moving Worlds on the Web*. New York: Addison-Wesley.

Martin, D.J. and Higgs, G. 1997. The visualization of socio-economic GIS data using virtual reality tools. *Transactions in GIS*, 1, 4, 255–65.

Mason, S.O., Baltsavias, E.P. and Bishop, I.D. 1997. Spatial decision support systems for the management of informal settlements. *Computers, Environment and Urban Systems*, 21, 3/4, 189–208.

Murfitt, P. 1998. Imagery in mapping. In *Proceedings RSS 98: 24th Annual Conference of the Remote Sensing Society*. University of Greenwich, September 1998, pp. 9–14.

Murphy, D. 1998. Ordnance Survey geospatial data management systems: the next generation. Session 14–5 in *Proceedings of GIS Research UK, 6th National Conference*, Department of Geography, University of Edinburgh.

Ordnance Survey. 1997. *Land-Line User Guide: Reference Section*, Ordnance Survey, Southampton.

Rhyne, T.-M. 1999. A commentary on GeoVRML: a tool for 3D representation of georeferenced data on the web. *International Journal of Geographical Information Science*, 13, 4, 439–43.

Sheppard, S.R.J. 1989. *Visual Simulation: A Users' Guide for Architects, Engineers and Planners*. New York: Van Nostrand Reinhold.

Sheppard, S.R.J. 1999. Visualization software brings GIS applications to life. *GeoWorld*, 12, 3 (March), 36–7.

Simpson, I.A., Parsisson, D., Hanley, N. and Bullock, C.H. 1997. Envisioning future landscapes in the Environmentally Sensitive Areas of Scotland. *Transactions, Institute of British Geographers*, New Series, 22, 307–20.

Smith, S. 1997. A dimension of sight and sound. *Mapping Awareness*, October 1997, 18–21.

Smith, S. 1998. More than meets the eye. *Mapping Awareness*, July 1998, 24–7.

10 The application of VR modelling in assessing potential visual impacts of rural development

David R. Miller, Roger A. Dunham and Weiso Chen

Introduction

Three-dimensional modelling of the rural environment is being carried out in response to identifiable needs for greater understanding, and the communication, of processes of change in land use or form (Kraak, 1999). These needs operate at different levels of detail, including broad impressions of gross change, such as climate variables at a continental scale (Kesteven and Hutchison, 1996) or nitrate concentrations at the scale of the water catchment or estuary (e.g. Mitasova *et al.*, 1996) or the siting of individual objects such as wind turbines, and can represent a complex set of interactions.

Modelling and communicating change in rural environments using virtual reality (VR) techniques provides an opportunity to explore places at a period in time, or at a scale that would be otherwise impossible (e.g. in the past or the future). The visualization of a scene can also offer a means of communicating more abstract concepts, such as the level of visibility of a landscape (Bishop and Karadaglis, 1996; Miller and Law, 1997; Neves and Camara, 1999; Zewe and Koglin, 1995), as part of a visual decision support capability (Kraak *et al.*, 1995). However, while conventional maps can contain valuable data on the landscape, many people find it difficult to associate the two-dimensional map with the structure and activities in a three-dimensional landscape (Berger *et al.*, 1996; Lange, 1994). Presenting imagery and data in a VR environment is with the objectives of aiding communication of, and widening access to, information pertinent to a process of decision making (Daniel, 1992).

This chapter describes the use of virtual reality modelling in the development of techniques for presenting potential visual impacts of windfarm developments, with reference to a case study in the United Kingdom. The issues associated with the creation of the base data, the construction and refinement of a prototype model, and its further development and use, are discussed in the context of the provision of information to a public audience.

Case study

A major obstacle in realizing the potential for wind energy in some countries, particularly the United Kingdom, is concern about the visual impact of individual, or groups, of wind turbines (CCW, 1999; Dumfries and Galloway Regional Council, 1996; Kidner, 1998; Tan, 1997). Such impact assessments provide an important input to the process of planning the development of a windfarm. The role of modelling in the visual impact assessment process is to enable the exploration of the possible consequences on the landscape of building the windfarm and assess where, and to what extent, the changes would have an effect. That exploration can consider the implications of where the farm is to be located, the spatial distribution of the turbines, and the size and number of turbines with respect to viewer location and viewer perception (Miller *et al.*, 1999). In order to illustrate the potential of VR modelling in relation to assessing the visual impact of a change in land use, a hypothetical case study is evaluated below.

The study area selected to test and illustrate a role for VR is in the north-east of Scotland at the Glens of Foudland, 50 km northwest of Aberdeen. It is an area in which the wind climate has been monitored in recent years, and it is a likely candidate for future development. The study area is 3.25 km × 5.3 km, and in total, the locations of thirteen wind turbines were selected and used in the geographic analysis and modelling, with turbine heights of 78 m to the top of the rotor arc, and located approximately 100 m apart, across the hilltop.

There were three objectives of the modelling:

1. To test the ease and means of creating a Virtual Reality Modelling Language (VRML) model for a windfarm.
2. To provide demonstrator examples for assessing the value of dynamic models in the landscape.
3. To explore methods for presenting multiple (overlaid) datasets for the same geographic area.

The VR model was created as a VRML dataset, which used four different types of input data: high-resolution DEMs, orthophotographs, CAD models of landscape features, and derived datasets showing a quantitative estimate of the visibility of the wind turbines and the landscape. The development and content of the VR model is described in the following section.

Model creation

Input data

Two sources of input data are used to provide terrain and texture surfaces: aerial photography and CAD models. This combination takes advantage of digital photogrammetric techniques for creating the input DEM, including as it does certain terrain model features that will obscure the view of turbines from an observer (such as a block of woodland), and the detail of specific features designed in CAD software, such as individual trees.

1 Aerial photography
   Near vertical, panchromatic, aerial photographs, at a scale of 1:24,000, were scanned at 400 dpi, and processed using ERDAS OrthoMax software (ERDAS, 1997) to produce a DEM and the surface texture.
   i Terrain surface
      A high-resolution DEM was derived, with a horizontal resolution of 2 m × 2 m (with a RMS of 1.4 m) and a vertical resolution of 0.25 m (with a measured RMS of 1.2 m).
   ii Texture surface
      An orthophotograph was derived for use in the provision of textural data, draped across the elevation data, in the VR model. The orthophotograph was produced with an output resolution of 0.5 m, using a nearest neighbour re-sampling algorithm in the image transformation, utilizing the ERDAS Imagine Virtual GIS software package, VGIS (ERDAS, 1999).
2 CAD models of surface features
   The models of surface features have been derived from two sources:
   i A CAD package (MultiGen) was used to construct a model of a three-blade wind turbine, following the specification of a Westinghouse 600 Kw turbine and tower. This model was then converted into DXF format for use in ERDAS VGIS (ERDAS, 1999).
   ii Modelling software was used for the creation of 3D models of tree and shrub vegetation (Artifice, 1999), which enables a number of parameters that describe the shape, size and structure of a tree, to be set by the user (Table 10.1). The creation of these models used Internet access to the Artifice website, to produce a 3D file for subsequent conversion into DXF format.

Derived data

A visibility census has been derived from the DEM by calculating the number of 2 m × 2 m cells that were visible from each other, to produce a score of relative visibility. That is, the greater the number of cells that are visible from any one cell, the higher the visibility score of that cell. The

Table 10.1 Settings for tree model

Variable	*Setting*
Canopy shape	Spherical
Canopy width	3 m
Tree height	10 m
Leaf orientation	'Random' and 'grouped'
Leaf size	10 cm
Leaf number	300
Trunk choice	Deciduous

visibility of the locations of individual turbines can then be compared, and sites chosen according to their lower, or higher, levels of visibility.

The location of each turbine has been used to derive the Zone of Visual Influence (ZVI) from the top of the rotor arc. The output of the calculation is a dataset that records the number of turbines visible from each location. The turbine visibility surface has been draped across the DEM and orthophotograph to allow an interpretation of the number of turbines that would be visible from any point on the landscape.

VR model construction

The VRML models have been created in two steps.

Landscape model

The landscape model is produced by draping the orthophotograph across the DEM and then placing, scaling and orienting the wind turbines in the model at locations specified in the draft planning proposal. This step used the ERDAS VGIS software from which the model was exported into VRML, which produced a view of the landscape after development of wind turbines (Figure 10.1).

The derived datasets have also been draped across the landscape model to produce virtual landscapes in which the view is of different visual characteristics of the landscape, related to its visibility, or the visibility of the wind turbines. Figure 10.2 illustrates three views of the same scene. Figure 10.2(a) shows the wind turbines, with a backdrop of the orthophotograph to provide a context for the observer. Figure 10.2(b) shows the relative visibility of the scene, from the same location, from which open hillsides can be seen as being more visible than the enclosed land in a valley bottom. Figure 10.2(c) shows the derived visibility of the wind turbines from the viewpoint. In this model, the shading represents the variable of interest (i.e. the shade indicates the number of turbines that would be visible from any location), and the draping of the imagery across the orthophotograph provides the user with a context against which to assess

Figure 10.1 View of wind turbines, with model trees added, on an orthophotograph providing the textural backdrop.

the turbine visibility from different features. Figure 10.2(c) also gives an impression of the screening effect provided by the blocks of woodland, reducing the visibility of the wind turbines from the opposite side of the woodland.

Another means of using the VR model to illustrate where a greater or lesser number of turbines would be visible is presented in Figure 10.3. This model used the turbine visibility dataset as the surface over which the orthophotograph was draped, and turbines added. This model is an example of a virtual landscape within which the shape of the surface represents the variable of interest to the user (i.e. the higher the surface, the greater the number of turbines visible; and the lower the surface, the fewer turbines are visible). In Figure 10.3, the height of the surface on the opposite side of the woodland block to the turbines is lower than the turbine side as a consequence of the screening effect of the trees which are embedded in the DEM used in the analysis of visibility.

VR refinement

VGIS uses a rudimentary approach to creating VRML files resulting in robust, but large files. These files can be modified to improve the efficiency and functionality of the model. For example, the VR model derived directly from VGIS includes multiple copies of the same object (e.g. wind

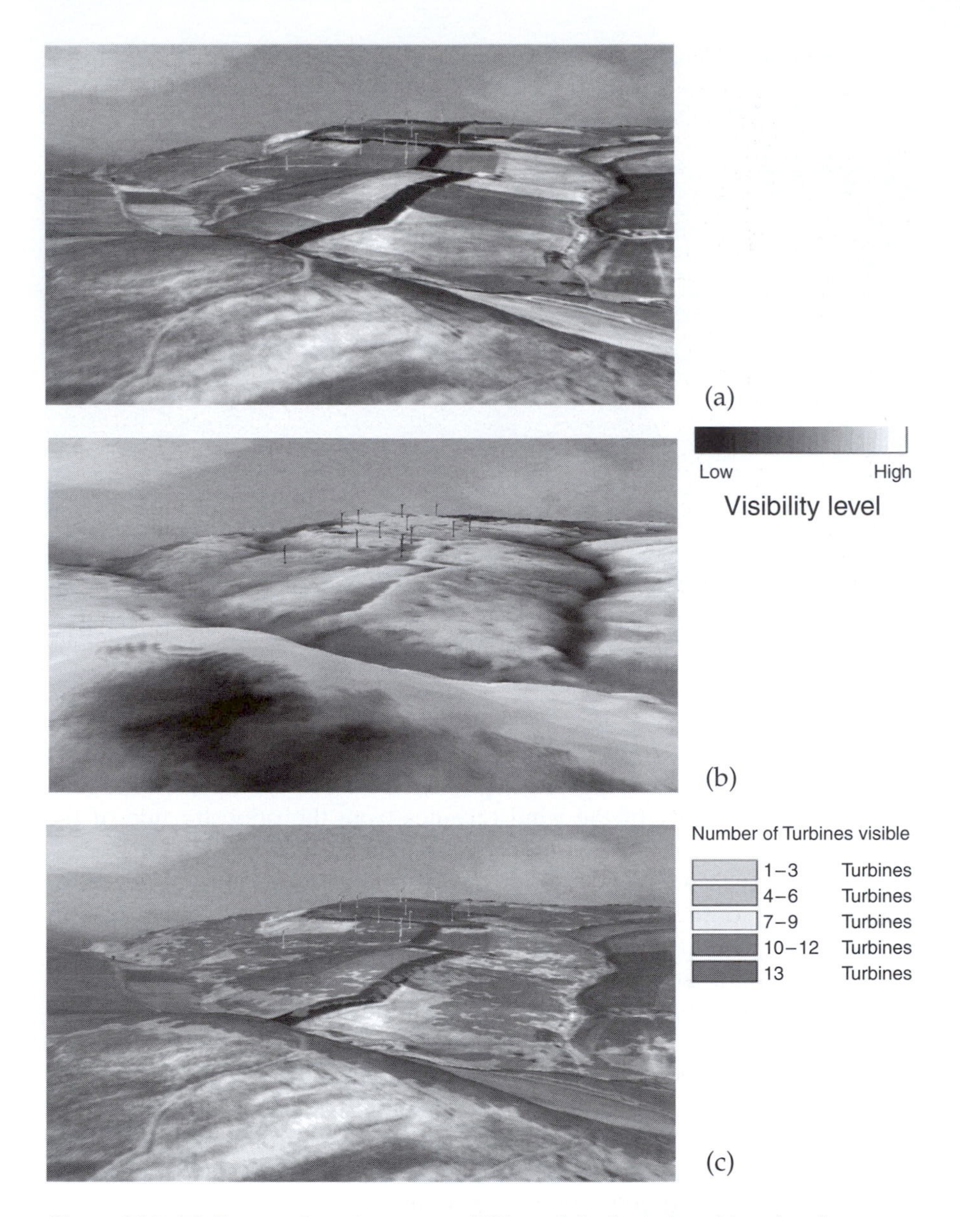

Figure 10.2 (a) Perspective view across VR model of wind turbine development. (b) Visibility census surface draped across DEM, with wind turbines located. (c) Combined Zones of Visual Influence (ZVIs) of wind turbines, draped across orthophotograph and DEM, with wind turbines located.

Figure 10.3 Turbine visibility surface used as a base over which the orthophotograph has been draped, and turbines added.

turbines), with every detail of each instance described separately within the exported VRML. By defining one instance of the object within the VRML file, further instances can be created and allocated locations based on this definition. Further efficiency gains can be achieved by redesigning models to be created from primitives (such as cones, boxes and cylinders), rather than importing them from DXF files. This can substantially reduce file sizes, and thereby improve rendering times. The models in Figure 10.4 were created using this approach.

A further modification to the original VR model is the addition of functionality to the virtual landscape. For example, simple program scripts have been added to the VRML file to enable the blades of the turbines to rotate, or to enable users to move the location of a turbine, and to change the height of stands of trees. These scripts are based on a subset of 'javascript' which is a high-level programming language that is widely used in the creation of 'pages' for the World Wide Web.

The VR model of wind turbines in Figure 10.4 is one in which the turbine blades rotate with the positions of blades at different times shown in shading. The objective was to introduce variation into the scene and avoid the appearance of blades rotating in a synchronized fashion, and to emulate the variation that occurs naturally due to differences in start-up times of the turbines or the impact of local variations in wind on blade

Figure 10.4 VRML model of wind turbines, with different orientations of the blades represented in different grey shades. See CD ROM for colour version of this figure.

pitch and speed. To achieve this effect, the rotation speeds and initial angles of orientation of the turbine's blades were set differently for subgroups of turbines.

Four viewpoints have been pre-set for the user, two to provide roadside views, one overlooking the site from an adjacent hilltop and a fourth in a close-up view of the wind turbines. An example of the VR model is accessible on the accompanying CD.

The role of the pre-set viewpoints is to help the user navigate the VR model for two different reasons:

1 The visual impact of developments such as wind turbines traditionally employs photomontages, with images of the view from locations of importance agreed between the planning authorities and the developer. The pre-set viewpoints in the VR model are indicative of the nature of those that would be selected for photomontages. That is, points of local interest (e.g. hill top viewpoint), laybys and road junctions. Within the VR model, the objective of including pre-set viewpoints would be to aid the users to evaluate for themselves the view from each point, and use that as a starting point for further movement around the model.
2 The second reason for identifying pre-set points is to provide the user with reference points to which they may return if 'freehand' navigation of the VR model leads them to a location, or viewing direction, within the model that is not of interest.

Discussion

Currently, the use of the VR models is limited by two factors:

1 The level of detail, and geographic extent, judged to be fit for the purposes of assessing the potential visual impact of the introduction of new wind turbines is currently too great for operational use by the specification of the PC that is available to the majority of users, as of the first half of 1999.
2 The only dynamic feature in the model is that of the moving turbine blades, whereas in reality many other features of the landscape are also dynamic, particularly lighting conditions and sky backdrops.

Nonetheless, VRML allows the user to gain an impression of the effects of blade movement, which is not possible in visualizations that use photomontage techniques. In the example presented in the accompanying CD ROM the speed of rotation of the blades is constant for each turbine, and the pitch and direction of the blades cannot be altered. However, variations in the behaviour of the turbine blades are simple to incorporate. Developments that will be more challenging are those that incorporate a greater sense of the dynamics of the landscape, such as trees that sway, water that flows and the movement of people and wildlife.

The addition of dynamic aspects to VR models does, however, represent an important part of their evolution in aiding in the communication of information between author and user. For example, in a potentially significant issue in the assessment of the visual impact of wind turbines is the effect of 'flicker' (Hetherington *et al.*, 1993), that is, the periodic appearance of parts of the rotating blades, while much of the tower and turbine structure are hidden from view by intervening features such as hill-crests. Examples of this could be created within the VR model by the selection of suitable pre-set viewing points where such an effect will be visible.

There are a limited number of options to facilitate the dissemination of information about the landscape to a public audience, and thus to enable discussion and evaluation of the consequences of a change in land use, such as wind farm development. That audience includes both members of the public and officials involved in the conductance of the planning process (Bishop and Leahy, 1989). Each option has its own advantages and disadvantages. Current approaches include:

1 paper maps, showing zones of visual influence (estimates of the land area from which features may be visible),
2 photomontages of current and future landscape views (Institute of Environmental Assessment, 1995),
3 videomontage or fly-throughs using pre-set routes, also showing the landscape pre- and post-development,

4 presentations at public meetings, using software that requires expert knowledge.

VRML, however, provides one alternative to these approaches, offering 'free' technology, generally suitable for use within a World Wide Web browser that is available on home computers. Therefore, there is increasing scope for members of the public to download models of development proposals in their own home, with the potential of selecting viewpoints, and routes, from which the proposal may be viewed. This has, potentially, greatly significant implications for the extent and effectiveness of consultation in the planning process.

In addition to the improvements in functionality of the model, the modifications made to the source code have resulted in more rapid download times when accessing the data over the Internet, and increased the speed of movement across the model. However, these gains in efficiency have had an associated cost, which can be broadly described as a loss of resolution. Specifically, the VR model available on the accompanying CD now contains a simplified representation of the specific make of wind turbine proposed for the site. The underlying elevation model is at a reduced resolution to the original, and no longer contains the three-dimensional structure of the stand of trees embedded in the DEM. Therefore, the balance between efficiency, operation, accuracy and authenticity is one which requires greater attention and there is an argument that VR models should include an equivalent of the metadata standards that are being developed, and accepted, in other areas of geographic information science.

It is likely that VR models will be formally included alongside other media in planning proposals in the foreseeable future. There are, however, several development stages that will be required before this can be considered to be a standard approach, and these are summarized below.

1 A measure, or other representation, of the level of accuracy of the VR model is required, since the validity of individual models will be questioned, just as wire-frame and photomontage imagery is queried at present.
2 An assessment of the level of detail of the simulation that the VR model should aim to achieve. It has been suggested that visualization of a scene need not be photorealistic for the observer to form an opinion as to their preference of the content of the scene (Daniel and Orland, 1997). If correct, this will allow attention to be focused on providing greater dynamism within the model.
3 Currently navigation of the VR model requires some practice. Further development is required so that the mechanics of navigation do not inhibit the interpretation of the model and thus reduce the value of VR within the application domain. At minimum, this can utilize a range of pre-set viewpoints (labelled by easily understood reference

sites, such as a farmhouse name or road junctions) and routeways. However, greater value will be realized by the non-expert user if they are able to navigate their own routes across the model.

4 Further additional factors need to be built-in to the models to aid in the objective of greater realism in the presentation of landscapes. These include:
   - i the relationship between detail and distance (variable object representation),
   - ii lighting models to describe illumination conditions,
   - iii 'conditional modifiers', e.g. a wet or snowy landscape,
   - iv 'environmental modifiers', e.g. atmospheric conditions such as haze,
   - v sky and cloud models that enable the user to represent either prevailing conditions, or conditions that occur less frequently.

5 Developments in hardware, and supporting infrastructure for VR (e.g. a 'cave', or a 'dome'), enable the potential for 'immersion' in the landscape. Headset technology for VR use will become a practical option in the near future, although still at a cost that may prohibit its operational use. The infrastructure for a fully-immersive environment is not portable, thus this currently restricts its use in the testing of VR technology for application in the planning process.

The use of this technology by companies and research agencies in the planning of current development proposals in the United Kingdom is still at an early stage, but VR modelling will find an increased role in research of the rural environment, and in the process of decision making that can significantly affect its content and appearance.

Acknowledgements

The authors would like to acknowledge the Scottish Executive Rural Affairs Department, the Forestry Commission, the Countryside Council for Wales and the Robert Gordon University for funding the work reported in this chapter.

Appendix

The VR model presented in this chapter, the source code and additional documentation are included in the accompanying CD ROM.

References

Anon. VRML, http://www.vrml.org/.

Artifice Inc. 1999. http://www.artifice.com/.

Berger, P., Meysembourg, P., Sales, J. and Johnston, C. 1996. Towards a virtual reality interface for landscape visualization. *Third International Conference/ Workshop on Integrating GIS and Environmental Modeling*. Santa Fe, New Mexico, January 1996.

Bishop, I.D. and Karadaglis, C. 1996. Combining GIS-based environmental modeling and visualization: another window on the modeling process. *Third International Conference/Workshop on Integrating GIS and Environmental Modeling*. Santa Fe, New Mexico, January 1996.

Bishop, I.D. and Leahy, P.N.A. 1989. Assessing the visual impact of development proposals: the validity of computer simulations. *Landscape Journal*, 8, 92–100.

CCW. 1999. *A consultation paper: CCW policy on wind turbines*. Countryside Council for Wales.

Daniel, T.C. 1992. Data visualization for decision support in environmental management. *Landscape and Urban Planning*, 21, 261–3.

Daniel, T.C. and Orland, B. 1997. Perceptual evaluation of SmartForest-II visualization system: comparison of analytic and landscape representations. *Data Visualization 97: Previewing the Future*. St Louis, Missouri, USA, October 1997.

Dumfries and Galloway Regional Council. 1996. *Wind Farm Strategy, Structure Plan Report of Survey. Background Paper*, No. 5. Dumfries and Galloway Regional Council, Dumfries.

ERDAS Inc. 1997. IMAGINE Users' Manual, ERDAS Inc., Atlanta.

ERDAS 1999 – http://www.erdas.com/products/imagine_virtualgis.html.

Hetherington, J., Daniel, T.C. and Brown, T.C. 1993. Is motion more important than it sounds?: the medium of presentation in environmental perception research. *Journal of Environmental Psychology*, 13, 283–91.

Institute of Environmental Assessment. 1995. *Guidelines for Landscape and Visual Impact Assessment*. London: E and FN Spon, p. 126.

Kesteven, J. and Hutchison, M. 1996. Spatial modelling of climatic variables on a continental scale. In Goodchild, M.F., Parks, B.O. and Steyaert, L.T. (eds) *GIS and Environmental Modelling*, CD ROM.

Kidner, D.B. 1998. Geographical Information Systems in wind farm planning. In Geertman, S., Openshaw, S. and Stillwell, J. (eds) *Geographical Information and Planning: European Perspectives*. Chapter 9, Springer-Verlag.

Kraak, M.-J. 1999. Visualizing spatial distributions. In Longley, P.A., Goodchild, M.F., Maguire, D.J. and Rhind, D.W. (eds) *Geographical Information Systems*, 1, 11, pp. 157–73.

Kraak, M.-J., Muller, J.-C. and Ormeling, F. 1995. GIS-Cartography: visual decision support for spatio-temporal data handling. *International Journal of Geographical Information Systems*, 9, 637–47.

Lange, E. 1994. Integration of computerized visual simulation and visual assessment in environmental planning. *Landscape and Urban Planning*, 30, 99–112.

Miller, D.R. and Law, A.N.R. 1997. The mapping of terrain visibility. *The Cartographic Journal*, 34(2) 87–91.

Miller, D.R., Wherrett, J.A., Fisher, P.F. and Tan, H.B. 1999. *Cumulative Visual Impacts of Wind Turbines*. Report to Countryside Council for Wales.

Mitasova, H., Mitas, L., Brown, W.M., Gerdes, D.P., Kosinovsky, I. and Baker, T. 1996. Modeling spatial and temporal distributed phenomena: new methods and tools for Open GIS. In Goodchild, M.F., Steyaert, L.T., Parks, B.O., Johnston, C., Maidment, D., Crane, M. and Glendinning, S. (eds) *GIS and Environmental Modeling: Progress and Research Issues*. GIS World Books, 345–53.

Neves, J.N. and Camara, A. 1999. Virtual environments and GIS. In Longley, P.A., Goodchild, M.F., Maguire, D.J. and Rhind, D.W. (eds) *Geographical Information Systems*, 1, 39, pp. 557–65.

Tan, B.H. 1997. *Predicting Visual Impact of Man-made Structures in the Scottish Countryside*. PhD Thesis, Robert Gordon University, Aberdeen.

Zewe, R. and Koglin, H.-J. 1995. A method for the visual assessment of overhead lines. *Computers and Graphics*, 19, 97–108.

11 Multi-resolution virtual environments as a visualization front-end to GIS

Iain M. Brown, David B. Kidner and J. Mark Ware

Introduction

In recent years, there has been a significant shift in 3D Geographical Information Systems (GIS) research, from presentation visualization to exploration visualization. This corresponds with evolving techniques in computer graphics, generally referred to as Visualization in Scientific Computing (ViSC), which have facilitated enhanced exploration of large multidimensional datasets (Brodlie, 1994). Hence, visualizing digital terrain models (DTMs) and other associated datasets in 3D is now considered a valuable technique not just for displaying data but also for evaluating its morphological or thematic context and validity; for instance, localised inaccuracies in digital surface models can be readily detected by comparing simulated views with reality (Dorey *et al.*, 1999; Figure 11.1).

Better tools for data exploration are becoming increasingly important due to the vast amounts of data that are now available, with respect to improved spatial, temporal and spectral resolutions. The growth in remote sensing means that it has become possible to develop terrain models in a continuum from the continental scale down to a very local level. Globally, topographic data from sources such as GTOPO (USGS, 1999) can be combined with the world-wide availability of thematic sensor data (e.g. AVHRR, LANDSAT, SPOT) to develop models covering very large areas. By contrast, at the local scale, technological developments, such as airborne laser-scanning by LiDAR (Flood and Gutelius, 1997), now mean we can model topography with centimetre precision. Furthermore, a new generation of remote sensors (e.g. IKONOS, OrbView, QuickBird) can provide data at resolutions of 1–2 m/pixel (Li, 1998), and such datasets are also being updated at shorter temporal intervals and often come with increasingly fine spectral bands.

Although high-quality interaction and fast rendering of graphics are desirable during presentation visualization, for data exploration they become essential. Virtual reality (VR) provides us with the scope for enhanced exploration because, by general definition, it includes motion and interaction within the visualization rather than just static scenes. A

Figure 11.1 Top: 'Real world' view to target location on the horizon. Bottom: VRML representation of buildings with roof structures generated from a triangulated irregular network (TIN) digital surface model: see CD ROM for colour version of this figure.

VR user-interface (VRUI) therefore provides the potential to 'zoom-in' from a large area to smaller areas of interest which have higher-resolution data, as well as to 'fly' over the landscape at the chosen resolution searching for anomalies or other features of interest. This is a much more powerful tool for data exploration than simply the ability to change viewpoint in 3D as found in most GIS user-interfaces.

Utilising the Internet

Interactive 3D visualization has been provided for several years in specialist software (e.g. AVS Express, Iris Explorer), and is now available, usually in a more limited context, in some GIS and remote-sensing packages. However, the remote visualization of multidimensional information via the Internet can bring considerable advantages, notably the non-requirement to have specialist local software beyond a low-cost browser. There is also the potential to search and filter out relevant datasets from databases much larger than could be stored locally (Abel, 1998). To fully enable this, however, requires enhancing the capability of the browser to provide a more flexible and intuitive interface, especially for spatial data exploration. As an example, the TerraServer project (Barclay *et al.*, 1999) has demonstrated how a well-designed user-interface can provide access to a global resource of high-resolution remote sensing imagery stored in a very large remote database. The principle of the VRUI is to extend this concept into 3D for a wide variety of spatial data sources (Figure 11.2).

Significant conflicts arise with the development of visualization systems for exploring large-terrain databases and associated datasets. Paramount amongst these issues is the need to maintain 3D interaction whilst keeping accuracy at an acceptable level. This becomes particularly critical when developing a user-interface to datasets across the Internet, because generally the network connection only permits a limited transfer of data (Figure 11.3; Goodchild, 1997).

Our current VRUI is based on the Internet standard language of the Virtual Reality Modelling Language (VRML) (Carey *et al.*, 1997). This has the advantages that:

- it is an open standard linked to the development of the Web and other Internet tools,
- it is an inexpensive method with which to develop 3D models,
- it requires no software for the end-user other than a freely-available browser.

On the other hand, VRML was not designed specifically to handle large-scale spatial datasets. In particular, the standard VRML data structures are certainly not optimal for our purposes, and require some additional programming. This has hindered the development of automatic conversion

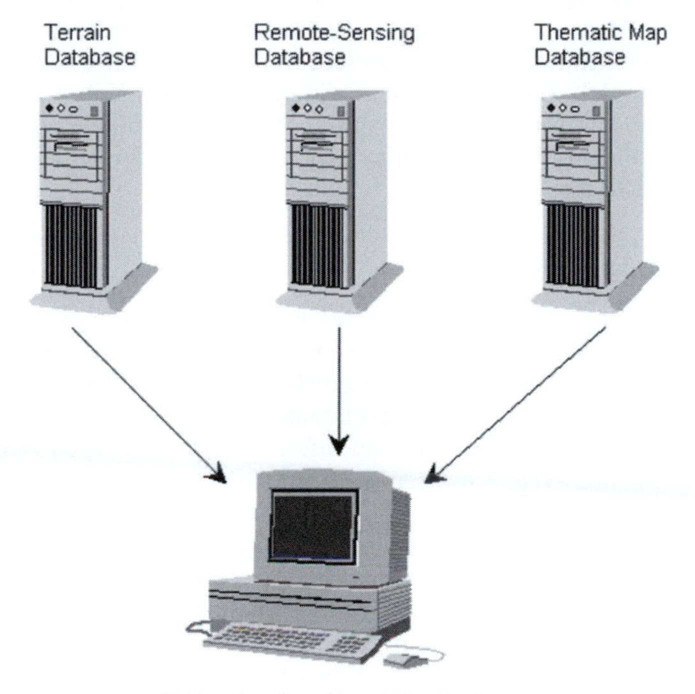

Figure 11.2 The principle of the VR user-interface is that it allows simultaneous 3D visualization of information from a variety of sources across a wide-area network.

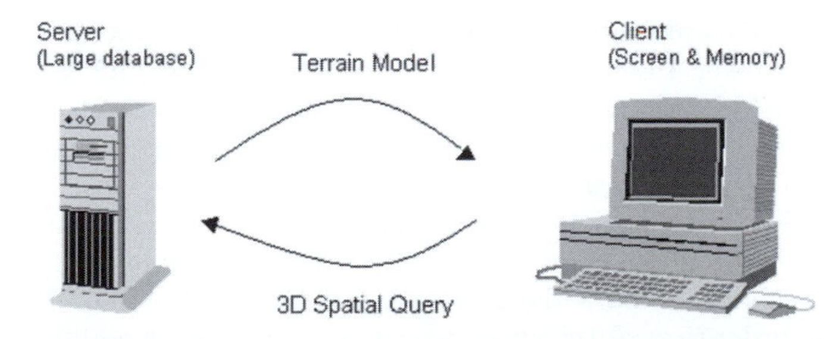

Figure 11.3 Extracting a 3D model from a large database involves selecting a subset of the original data points. Some of these are directly rendered on screen, because those parts of the model are currently visible. In addition, a variable amount of data can be stored in memory to speed up VR navigation as new regions of the model become visible during motion.

programs for directly exporting terrain models created in GIS packages to VR models on the Web; frequently these produce files which become too big for the limited bandwidth of the present-day Internet (although in many cases this is also due to inefficiency in the conversion routines). Hence, after explaining the functionality of our VR user-interface, we will go on to discuss how more effective data structures can be implemented for improved VR/GIS interaction.

A prototype VR user-interface

It is universally recognised that user-interfaces are a very important component of GIS for all levels of spatial data user (Frank, 1993). For the specialist, they provide a mechanism for quick and efficient interaction with datasets, sometimes through the use of complex queries. By contrast, for general-purpose users, they provide an intuitive metaphor to encourage further probing and greater awareness of the data and their information content. By developing a 3D user-interface, it is intended to facilitate both browsing of the multidimensional dataset and further specific exploration within the same visualization context (Brown, 1999).

Hence, a project was initiated to develop a VR user-interface to spatial datasets in South Wales with digital terrain models defining the shape of the surfaces. The principle was that the topographic surface provides an important reference to a wide variety of datasets, both directly (e.g. soils, geology, geomorphological hazards, etc.) and indirectly (highly-irregular terrain can act as a significant constraint on some cultural features). For example, in South Wales, geographical references to hills and valleys are very common in place names and administrative regions, and clearly imply a close semantic link with topography. Another objective was to allow a simultaneous index and visualization tool, because some datasets are restricted to certain small areas or are available at high resolution only for certain locations.

A working implementation has shown that the concept can be successfully applied (Brown, 1999). VRML was used to develop 3D terrain models which could be viewed through a VRML browser embedded in a standard Web browser (Figures 11.4, 11.5, 11.6, 11.7). The user-interface components were created by using the Java programming language which allows compiled code to be delivered through a Web browser as applets (Arnold and Gosling, 1998). Hence, an applet was developed in the familiar 'menu-type' layout to allow the user to interface with the 3D VRML environment in order to perform a wide variety of functions, in addition to the browsing enabled by virtual 'flying' and 'walking'. Options provided include the following capabilities:

- overlay datasets, such as satellite images and thematic maps,
- go directly to a chosen location or query 3D position,

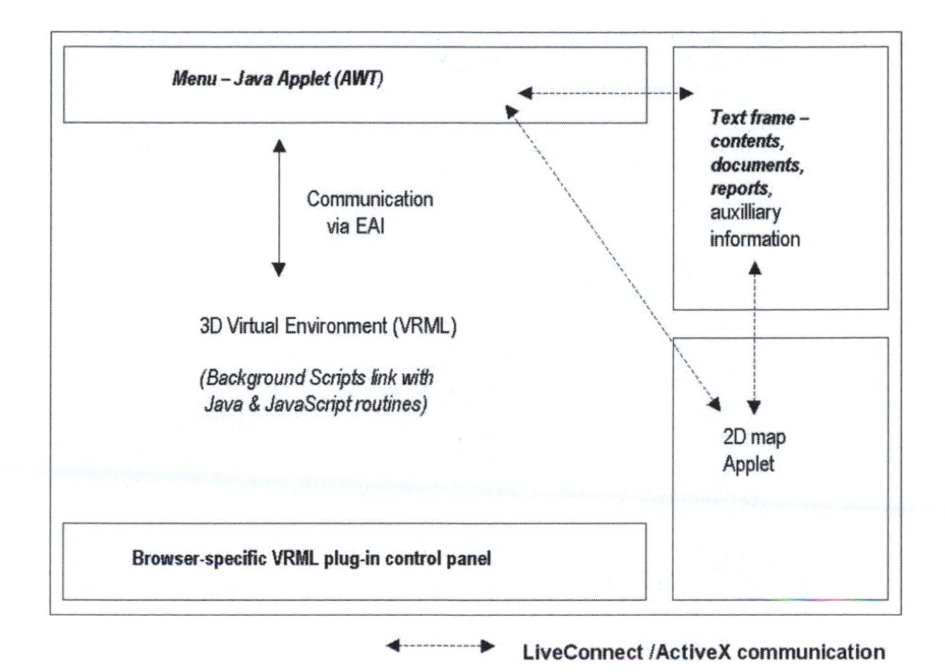

Figure 11.4 Web page design for the prototype VR user-interface. A VRML browser is embedded within a Web browser which has a menu applet (JAVA), a map applet (JAVA) and a key/contents listing, each linked to the 3D VR model for cross-reference.

- extract morphological attributes,
- change viewing parameters, such as lighting direction,
- query metadata information for chosen datasets,
- initiate an animated time-series via animation routines.

The Java applet and VRML browser are linked by using the External Authoring Interface (EAI), a series of library classes which allow communication through the medium of a Web browser (Brown, 1999). In addition, a 2D map was provided for additional user reference as it was found that, although features such as perspective distortion in 3D are useful in providing superior visualization, most end-users found that navigation was difficult, at least initially. Further auxiliary information is also provided in another frame of the Web browser, such as a key to a thematic map and educational material to increase awareness of the data content; this can include linking to other related sites on the Web.

The initial success of the project has encouraged further expansion, with several issues needing to be addressed so as to improve the quality of the interface for both small and large-scale versions. One of the most

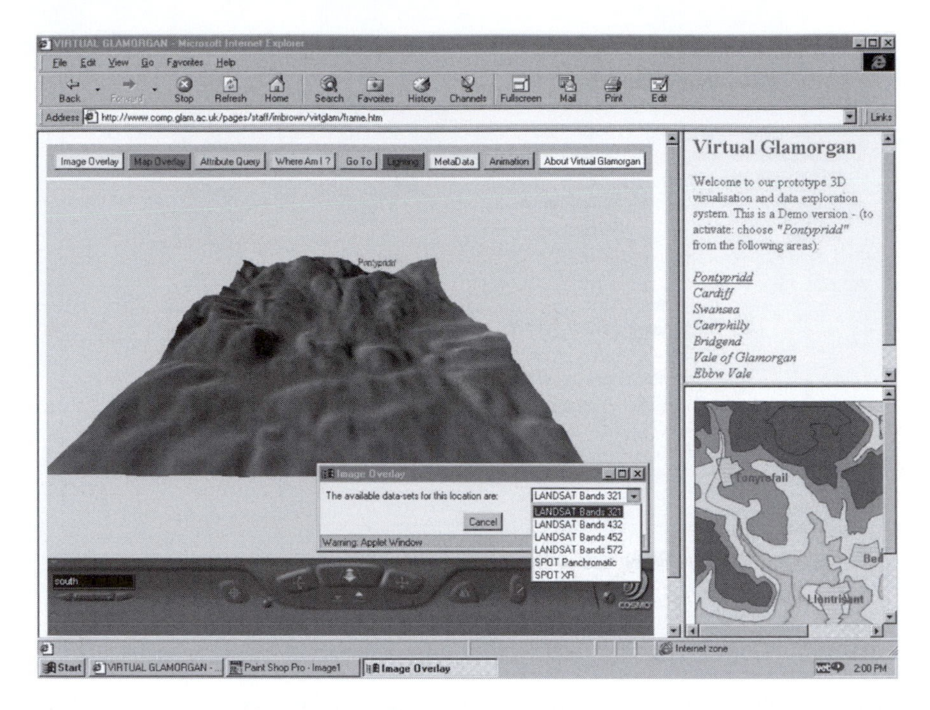

Figure 11.5 The initial view of the Virtual Reality User Interface showing the VRML plug-in browser (CosmoPlayer) embedded within a Web browser (Netscape Navigator). In this case, the user has selected 'Image Overlay' from the Java menu applet above the 3D scene and a sub-menu now shows the available datasets that can be draped over the terrain.

important of these is the underlying data structure in which the model is represented, since this has a critical role in efficiently visualizing the 3D data.

Data structures for VR models

Maintaining virtuality

Navigating in a VR system over a 3D landscape relies on obtaining a fast frame rate to maintain the impression of virtuality. Although a rate of over 20 frames/second is necessary to obtain real-time VR with a perfect illusion, a value over 10 frames/second is generally considered adequate for most purposes. However, at lower rates, a degree of frustration is common due to the lag time between moving the pointing device, such as the mouse, and the response appearing on the screen. This frustration becomes increasingly conscious to the user as the frame rate declines.

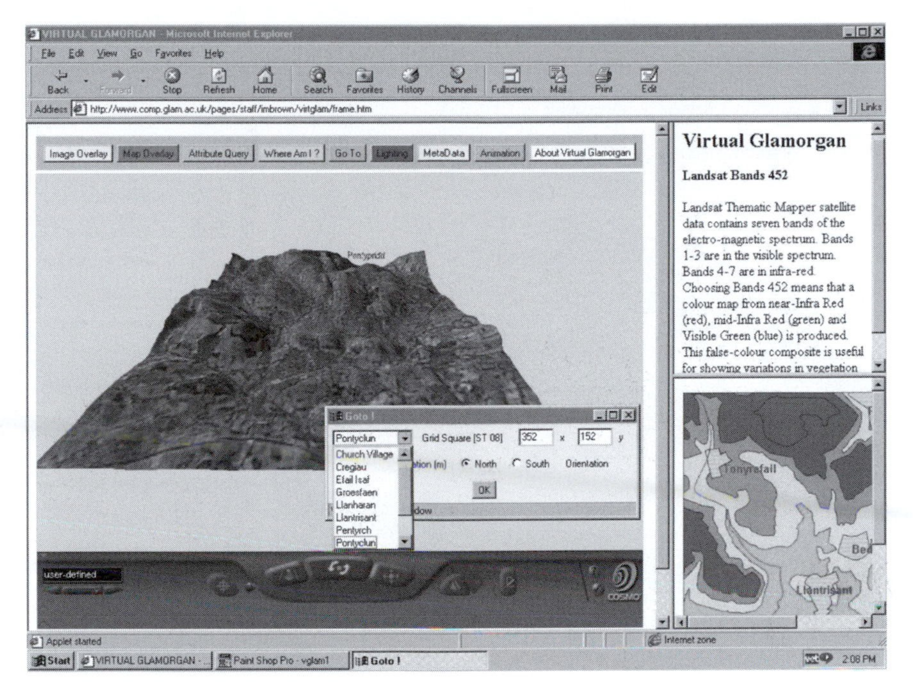

Figure 11.6 Following on from Figure 11.5, an image overlay has been applied to the virtual world with further information on the particular dataset displayed in another frame. The user has now selected the 'Go To' button from the top menu, which will move to a set location in the 3D scene allowing a more detailed visualization. These options are available in addition to the standard VRML browser navigational controls.

Another significant concern is the time taken to download the virtual environment before it becomes 'active'; it is important to avoid long download times which might deter users from exploring the environment. Loading from a CD ROM or hard disk means we can afford files of 1 MB or larger in size. However, when connected to the Internet, file size becomes much more critical. With an ISDN link or a fast modem (56 Kb/sec), the evidence suggests that the maximum file size should be 200 Kb. When we consider that many people on the Internet still rely on a 28.8 Kb/sec modem, then the maximum acceptable size decreases drastically to about 15 Kb. Using Java applets in conjunction with VRML files, using the EAI, can further erode the speed of interactivity.

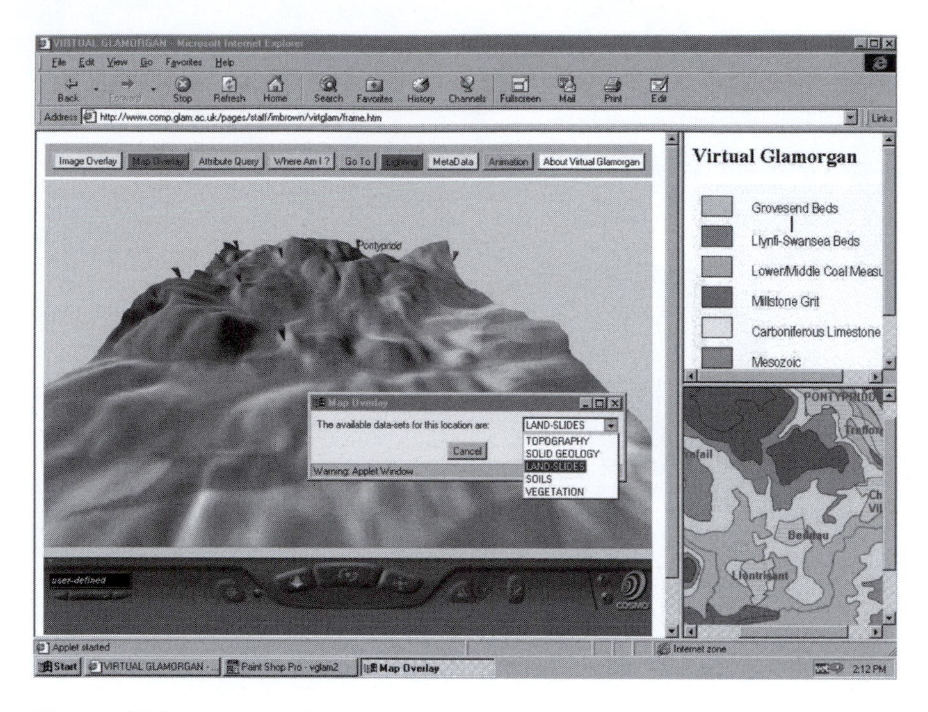

Figure 11.7 Two further data-sets overlayed on the terrain model to be explored. First, a geological map has been draped as an image. Second, landslide data, here portrayed as point symbols, have been 'toggled' on at their respective locations. This shows how vector datasets can also be visualized (together with attribute data, if required, in the 'key' frame).

GIS data structures

To maintain virtuality in a 3D user-interface based on terrain models, it is necessary to consider the data structures used to construct the models. The quality of interaction required brings a new perspective on the perennial GIS debate over whether a regular grid or triangulated irregular network (TIN) is better. With a regular grid, the height values are sampled at fixed intervals in a rectangular mesh. By contrast, a TIN consists of an irregular mesh of triangles, with the height values attached to each node. A TIN is more adaptive to the terrain, allowing more points in complex areas and less in flat areas; hence, by removing redundant data, fewer points are needed for a set degree of 'accuracy'.

However, the efficacy of VRML models for 3D interaction is strongly related to compact file sizes, both regarding download time and then subsequent interaction with the model (Brown, 1999). In this case, the implicit structure of a regular grid has a considerable advantage over a TIN because, once the dimensions and spacing of the grid are declared, only a sequence of height values need to be specified. Conversely, with a TIN, explicit declaration of all the (x,y) coordinates and height values need to be provided, as well as a list of the connections between vertices which produce the triangulated 'mesh'.

Typically, terrain data are also distributed in regular grid format (because of the compact size), therefore any potential greater accuracy of a TIN is limited because we do not have the original survey points. In our first illustrated case, a TIN is produced from the regular grid using a particular tolerance value. This value defines the degree to which the two surfaces can differ and dictates which grid points are included in the TIN. Under these circumstances, the TIN is unlikely to produce a more accurate representation of the terrain (Kumler, 1994).

Table 11.1 shows that to create a TIN with similar storage requirements to a regular grid requires a reduction in the number of points to about 10 per cent. This is often achieved by accepting a relatively large tolerance value. However, frame rates, which show the quality of interaction, are noticeably higher for a TIN structure. This is because the graphics primitives used in rendering are triangle-based. On a local area network, therefore, VR performance can be very good with a TIN. Unfortunately, the download time for a TIN increases considerably for long-distance connections across the Internet, making it less practical unless the number of points is reduced. Whether or not a reduction in the number of points produces models that are sufficiently accurate will depend on the specific application. In either event the model is often not as aesthetically smooth as with the regular grid.

Several studies have addressed the issue of decimating triangle meshes,

Table 11.1 Comparison of a regular grid data structure against a TIN for a test area of 100 km^2 in South Wales (field of view = 90°). Accepting a higher tolerance results in a more compact data structure similar in size to the regular grid. Tests were performed on a Pentium II (333 MHz/128 Kb RAM) with OpenGL graphics (ATI 3D RAGE PRO 8MB card) using Netscape 4.06/CosmoPlayer 2.1 at maximum screen size (1024 × 768)

	Regular grid	*TIN*	*TIN*
Size-compressed/uncompressed (kB)[a]	12 (37)	66 (193)	21 (68)
Tolerance (m)[b]	–	3.4	10
Points	10,201	3,775	1,214
Facets	10,000	7,382	2,618
Typical frame rate (centre-screen)	3.5–5.0	5.0–6.5	7.5–9.5
Internet transfer – local (sec)[c]	4	3	1.5
Internet transfer – international (sec)[c]	22	121	45

Notes

a Compression with gzip (supported by most browsers).

b TIN tolerance relative to the original grid used for data acquisition.

c Internet transfer times are typical results intended for relative comparison rather than literal values.

and the selection of critical points from a surface. For the simplified TIN in Table 11.1, a VIP algorithm (Fowler and Little, 1979) was used to retain the main features of the surface by identifying extremes, but other methods have been suggested to offer more efficient data storage or to minimise the loss of information (Jones, 1997).

Optimisations

The frame rates in Table 11.1 could be improved, at the expense of longer initial download times, by a variety of techniques, as used by specialist VRML 'optimisers', such as dividing up the mesh into smaller units or explicitly defining colours or texture values in the model, rather than in a supplementary graphics file. However, the general observation that the frame rates fall just below the level of true VR interaction appears to be very common amongst developers. This slight delay in response is probably not as critical for geographical data, however, as with pure entertainment VR, because the educational and scientific value rests with appreciating the subtleties of the 3D landscape as navigation occurs.

A more worthy objective would be to maintain the frame rate at an acceptably high, constant level. For instance, one strategy that can be used is to link the frame rate to the field-of-view of the observer, so that when the rate drops then the field-of-view is narrowed to compensate. Another strategy involves setting a visibility limit such that distant detail is not rendered, although this can lose some important contextual information.

Current situation

The terrain models used in our current VR user-interface are based on a regular grid data structure. The regular grid was chosen because it offers greater compactness and requires less pre-processing (as the data are supplied in this form). However, the potential advantages offered by a triangle-based data structure warrant further investigation These advantages include superior frame rate and the opportunity to integrate vector data accurately with the terrain model. The blocky structure of a regular grid means that frequently vector data do not precisely locate on the underlying terrain surface; this becomes apparent on close viewing when the object either 'floats' above or is embedded within the surface. By contrast, with a TIN, vector features can be integrated exactly with the surface facets to produce a smooth junction (Ware and Jones, 1997). Another advantage of TINs over regular grids when covering larger areas is that grids do not take account of the curvature of the Earth and spherical/elliptical projection systems, whereas TINs can (Reddy *et al.*, 1999) – although a semi-regular spherical grid would be a suitable compromise (Florinsky, 1998). Furthermore, as explained later, TINs may be a more suitable structure for developing multi-resolution models.

Variable scale and level of detail

The VR user-interface previously described has demonstrated that a fast interactive gateway to 3D data is viable on the Web. However, scaling up the concept to cover a much larger area introduces some difficult problems which imply that further refinement is required. For instance, if we increase the size of the area from 100 km^2 to 10,000 km^2 and retain the same spacing between sample points, the grid size and resultant file size will be significantly increased. As a consequence, the critical component of interaction and virtuality is reduced, especially across the Internet. Retaining virtuality for a larger area therefore necessitates a larger grid spacing, but this in turn means that, when 'zooming-in' to a chosen sub-area, the resolution of the scene becomes very poor. Clearly, the solution requires an adaptable mesh of points that adjusts to the scale of the model as it fills the screen.

For example, it is possible to extract a model of a set resolution (related to the current scale of view) from a large terrain database using a Common Gateway Interface (CGI) script and GIS software on the remote server holding the database (Figure 11.8). However, such continuous interchange of information between a client (local) computer and remote server on the congested network of the Internet generally proves too slow and inefficient to guarantee the level of interaction required for a VR user-interface (Brown, 1999).

A more practical solution is to have discrete distance thresholds within the virtual environment which, when they are crossed, move the viewer to a new level of detail (LOD). The LOD node in VRML allows such thresholds to be set, but because the language was not designed for large-scale virtual environments the results are presently far from optimal for terrain models. For instance, most VRML browsers are designed to pre-load all associated 'inline' data that may be required later, such as the different resolution levels in an LOD node, and this results in a bloated file size and a very slow initial download time. Significant improvements have been made by extending VRML using 'prototypes', notably the QuadLOD node of Reddy *et al.* (1999) which sub-divides the grid into a quadtree-like structure so that those cells that are outside the current view are excluded. The QuadLOD node certainly provides superior performance for large-terrain models but problems still remain. For example, when one 'zooms-in' to the centre of the model, then often more than one quad-element at the next level of the hierarchy becomes visible, and the large size of this combined grid accumulating in memory produces a poor frame rate.

It is also possible to surround the virtual environment with a series of hierarchical 3D proximity sensors, which act as invisible thresholds similar to the LOD control. These have the practical advantage that they can be used to trigger the loading of a different LOD when the bounding box defined by the sensor is crossed, using the *Switch* node, and not before. We have implemented such a 'load-on-demand' system using the JavaScript

The simple LOD representation presently existing in VRML has the major disadvantage that each level is stored independently, and hence the number of LODs must be small, otherwise the amount of available memory is exceeded. However, more sophisticated multi-resolution models have been developed that can produce a much higher number of LODs based on a single compact representation of the mesh. This allows smooth *morphing* between different LODs to avoid the disconcerting 'popping' effect that can occur when switching between LOD representations of differing scales. As well as allowing distinct parts of the same scene to be at different resolutions, these new modes also feature two important developments for VR interaction: *progressive transmission* and *view-dependent meshes*.

Progressive transmission

By encoding successive LODs within the multi-resolution algorithm, data can be incrementally transferred across the Internet, with a low-resolution model constructed first, then high-resolution data added for extra detail and accuracy (Hoppe, 1996). This has the advantage that, if the Internet connection is slow, at least a low-resolution model will be quickly available on screen; it is analogous to the method by which 2D images are now transmitted on the Web. Virtuality can be maintained because, during VR motion, a lower-resolution model may be used if necessary; then, when the viewer becomes stationary at a particular area of interest, processing power is freed to fill in the local detail with higher-resolution data. Furthermore, if a suitable user-interface control is provided, the end-user can also interactively control the LOD by changing the number of points used to build the model.

Generally, the sample points are ordered based on their significance in contributing to surface morphology, allowing extraction of lower-resolution models with a limited number of points that still give a good surface representation. Typically, for a terrain model we would wish to maintain ridges and channels as highly-significant points (Fowler and Little, 1979), and with a multi-resolution TIN, this can be specifically defined. Although a considerable amount of pre-processing may be required for the TIN structure, this occurs before the data are stored on the server and is not encountered by the end-user exploring the virtual environment.

The current version of VRML does not have this capability and although frameworks for introducing progressive transmission have been developed, they rely on external routines (in Java), which, although effective, prove cumbersome and impose a performance limit (Guziec *et al.*, 1999). Other 3D 'streaming' technologies, such as MetaStream, are superior to VRML in this respect in that they can ensure the frame rate remains at a set level (by limiting the number of triangles rendered), thus guaranteeing a certain quality of interaction.

View-dependent meshes

Selective refinement of the terrain model can be adapted so that the LOD is continuously variable through space, depending on the current position of the observer in the virtual world. Hence, the mesh is produced at a much finer resolution in the foreground compared to the background where extra detail is superfluous (Cignoni *et al.*, 1997; Hoppe, 1997). The algorithm needs to ensure that no 'cracks' appear between those areas of the surface at different resolution by blending the adjacent height values together. Furthermore, any portion of the surface outside the current view frustrum should not be rendered.

VR implementation

Navigation through the virtual world requires that input parameters to the algorithm producing the view-dependent progressive mesh be dynamically updated so as to reflect new viewing conditions. This has been achieved for large-terrain models on local workstations (Hoppe, 1998; Koffler *et al.*, 1998), but not, as yet, for querying and transmitting this 3D information across the Internet; the current version of VRML does not facilitate it.

Multi-resolution algorithms can be classified into two main classes:

- *tree-like models*: these are based on the recursive subdivision of a region, normally using triangles, into a nested series of smaller regions, which exactly fit inside it.
- *historical models*: by storing the evolution of a mesh by refinement or simplification routines, the process can be easily reversed when required.

Early results in comparing the efficiency of these two classes of algorithm suggest that, for a VR-GIS application, the historical models may be most effective because of their greater adaptability and reduced computational complexity, making extraction of a particular multi-resolution mesh from the original database more efficient. However, some forms of VR exploration require topological information also, which is usually directly encoded into the tree-like models. Further research is required to provide more conclusive results, particularly as in some GIS applications specific error thresholds are required to be related to the LODs. Strict performance criteria are not the only consideration.

Conclusions

Although VR shows great potential for visualizing large multidimensional GIS databases across the Internet, the degree of user-interaction may be considered quite basic compared to specialist 3D visualization systems (e.g. AVS/Express, Iris Explorer). By itself, VRML can only provide a limited user-interface, but integration with Java applets in a Web browser can considerably improve both the functionality and usability. There is a need for improved awareness of the strategies employed by GIS users in navigating and exploring virtual environments so that a 3D user-interface can be better linked to our knowledge of the principles of spatial cognition.

An even more critical aspect for visualizing large virtual environments is the need for more efficient data structures than those presently used in GIS. The need for a high frame rate to maintain quality VR interaction with the datasets is essential. This is also highly significant when developing an interface to visualize 4D spatio-temporal datasets. Although animation routines have been included in the current project, their speed across the Internet at present is often so slow as to make simultaneous VR interaction impractical.

A related issue is that there is a strong need for better conversion routines between GIS models and VRML which include these improved data structures. This is particularly important for creating a large system with many diverse data types. Ideally, there should be little need directly to use VRML to create 3D models, as occurs with some other standards (e.g. POSTSCRIPT or HTML). However, efficiency requirements currently necessitate considerable post-processing after conversion. Hopefully, the next generation of GIS and VRML (or future 3D equivalent) technology will enable this to occur more transparently.

Integration of multi-resolution models within the VR user-interface provides the key to developing the next generation of large-scale virtual environments. Progressive transmission of 3D models also introduces the possibility of integrating maps, images and other 3D structures (e.g. buildings) directly with the geometrical structure of the terrain model. Currently, for practical purposes related to Internet transfer, texture mapping of images on to terrain models is often limited to a poor size resolution (e.g. 256 × 256 pixels), or relies on individual mapping of colours to vertices (or faces) which creates very large files. By interleaving the 'streaming' of images and geometry across the network, the possibility of 'zooming-in' to a virtual environment and concurrently visualizing higher-resolution terrain data together with further sub-units of the thematic dataset being explored, can be fully realised.

There is, therefore, a strong link here between 3D multi-resolution modelling and the generalisation of datasets obtained from multi-scale databases. In this context, the VR user-interface can truly become the visualization gateway to a vast array of spatio-temporal datasets for the chosen region. Significantly, the biggest restriction on liberating this data into the public domain and developing new information systems and educational tools may be the copyright restrictions enforced on the distribution of the base data.

Acknowledgements

Acknowledgements are due to the various organisations who have supplied data used in this project and illustrated in the various figures. These include the Environment Agency (LiDAR data), the Ordnance Survey (1:50,000 scale PANORAMA and 1:10,000 scale PROFILE DTMs), GeoInformation International (Cities Revealed photography) and the National Remote Sensing Centre (LandSat and SPOT imagery).

References

Abel, D.J., Taylor, K., Ackford, R. and Hungerford, S. 1998. The exploration of GIS architectures for Internet environments. *Computers, Environment and Urban Systems*, 22, 1, 7–23.

Arnold, K. and Gosling, J. 1998. *The Java Programming Language*. 2nd Edition. Reading, MA: Addison Wesley Longman.

Barclay, T. *et al.* 1999. Microsoft TerraServer. http://terraserver.microsoft.com/.

Brodlie, K. 1994. A typology for scientific visualization. In Hearnshaw, H.M. and Unwin, D.J. (eds) *Visualization in Geographic Information Systems*. Chichester: John Wiley & Sons, pp. 34–41.

Brown, I.M. 1999. Developing a Virtual Reality User Interface (VRUI) for Geographic Information Retrieval on the Internet. *Transactions in GIS*, 3, 3, 207–20.

Carey, R. Bell, G. and Marrin, C. 1997. VRML97 Specification. http://www.vrml. org/home.html.

Cignoni, P., Puppo, E. and Scopigno, R. 1997. Representation and visualization of terrain surfaces at variable resolution. *The Visual Computer*, 13, 199–217.

De Floriani, L., Marzano, P. and Puppo, E. 1996. Multiresolution models of topographic surface description, *The Visual Computer*, 12, 7.

Dorey, M., Sparkes, A., Kidner, D., Jones, C. and Ware, M. 1999. Terrain modelling enhancement for intervisibility analysis. In Gittings, B. (ed.) *Integrating Information Infrastructures with GI Technology: Innovations in GIS 6*. London: Taylor & Francis, pp. 169–84.

Flood, M. and Gutelius, B. 1997. Commercial implications of topographic terrain mapping using scanning airborne laser radar. *Photogrammetric Engineering & Remote Sensing*, 63, 4, 327–66.

Florinsky, I.V. 1998. Derivation of topographic variables from a digital elevation model given by a spheroidal trapezoidal grid. *International Journal of Geographical Information Science*, 12, 8, 829–52.

Fowler, R.J. and Little, J.J. 1979. Automatic extraction of irregular network digital terrain models. *Computer Graphics*,. 13, 1, 199–207.

Frank, A.U. 1993. The use of geographical information systems: the user interface is the system. In Medyckyj, D. and Hearnshaw, H.M. (eds) *Human Factors In Geographical Information Systems*. London: Belhaven Press, pp. 3–14.

Goodchild, M.F. 1997. Towards a geography of geographic information in a digital world. *Computers, Environment and Urban Systems,* 21, 6, 377–91.

Guttman, A. 1984. R-Trees: A Dynamic Index Structure for Spatial Searching. Proc. of ACM-SIGMOD, June, pp. 47–57.

Guziec, A., Taubin, G., Horn, B. and Lazarus, F. 1999. A framework for streaming geometry in VRML. *IEEE Computer Graphics & Applications*, 19, 2, 68–78.

Hoppe, H. 1996. Progressive meshes. *SIGGRAPH '96 Computer Graphics Proceedings*. ACM Press, pp. 99–108.

Hoppe, H. 1997. View-Dependent Refinement of Progressive Meshes. *SIGGRAPH '97 Computer Graphics Proceedings*. ACM Press.

Hoppe, H. 1998. Smooth View-Dependent Level-of-Detail Control and its Application to Terrain Rendering, http://research.microsoft.com/~hoppe/svdlod.pdf.

Jones, C.B. 1997. *Geographical Information Systems and Computer Cartography*. Harlow, England: Longman.

Jones, C.B., Kidner, D.B. and Ware, J.M. 1994. The implicit triangulated irregular network and multiscale spatial databases. *The Computer Journal*, 37, 1, 43–57.

Koffler, M., Gervautz, M. and Gruber, M. 1998. The Styria Flyover – LOD Management For Huge Textured Terrain Models. *Proceedings of Computer Graphics International 98*. Hannover.

Kumler, M.P. 1994. An intensive comparison of triangulated irregular networks and Digital Elevation Models (DEMs). *Cartographica,* 31, 2, 1–99.

Li, R. 1998. Potential of high-resolution satellite imaging for national mapping products. *Photogrammetric Engineering & Remote Sensing*, 64, 1165–9.

Lindstrom, P. *et al.* 1996. Real-Time Continuous Level of Detail Rendering of Height Fields. *SIGGRAPH '96 Computer Graphics Proceedings*. ACM Press, pp. 109–18.

Reddy, M., Leclerc, Y., Iverson, L. and Bletter, N. 1999. TerraVision II: Visualizing massive terrain databases in VRML. *IEEE Computer Graphics & Applications*, 19, 2, 30–8.

USGS. 1999. GTOPO30: Global 30 Arc Second Elevation Data Set, http://edcwww.cr.usgs.gov/landdaac/gtopo30/gtopo30.htm.

Ware, J.M. and Jones, C.B. 1997. A Data Model For Representing Geological Surfaces. *Proc. of the 5th Int. Workshop on Advances in GIS*. ACM Press, pp. 20–3.

12 Visualizing the structure and scale dependency of landscapes

Jo Wood

Introduction

This chapter considers how we can use the metaphor of virtual reality and the process of geographic visualization to understand something of the structure and scale dependencies inherent in models of landscape. The ideas discussed here draw not only from long traditions of scientific visualization and cartography, but also from the more recent ideas of immersive navigation within virtual environments. Landscape models of one kind or another frequently form the backdrop to VR environments providing a spatial context to navigation. Here we will consider how exploring such an environment can aid our understanding of the landscape itself.

The visual depiction of landscape form has a long history, dating at least as far back as the images of mountains scratched onto earthenware in Mesopotamia over 4,000 years ago (Imhoff, 1982). The subsequent development of cartographic terrain representation has traditionally involved attempts to symbolise multiple perspectives of three-dimensional surface form on a static two-dimensional medium. Solutions have included early mimetic symbolisations of relief (e.g. Abraham Ortelius' 1585 map of Iceland); the use of hachuring to symbolise lines of steepest descent (e.g. Ordnance Survey of Great Britain's first series topographic maps); the widespread use of contour lines to represent slope normals, and the more recent use of automated shaded relief calculation from Digital Elevation Models (e.g. Yoeli, 1983). A problem faced when interpreting all of these representations is that surface form varies with both scale and perspective. The apparent shape of a patch of land will differ depending on the relative distance and direction of the observer.

Several solutions have been adopted to address the inherent subjectivity of relief representation. The historical transition from mimetic to abstract symbolisation (from pictures of hills, to hachures, to contour lines) can be seen as a move from the personally subjective to the more universally objective. Other solutions have included the incorporation of multiple perspectives within the same map through the use of insets or non-linear projections (e.g. Wainwright's guides to the Lakeland Fells

(Wainwright, 1992) show mountain paths from both a bird's-eye and oblique perspective view simultaneously). Alternatively, the scope and perspective of a map's relief representation may be made specific and explicit, such as the cartographic conventions adopted in geomorphological mapping (Evans, 1990). The approach described in this chapter uses the metaphor of virtual reality to allow viewers to combine subjective viewpoints of a landscape in order to understand its structural characteristics.

When is visualization virtual?

There is a danger that a book of twenty-five chapters on virtual reality will provide twenty-five definitions of the term, and it is not the intention of this Chapter to define another. However, the process of visual landscape understanding described here enables us to identify two aspects of the use of visual representation of landscape that help in clarifying what we mean by VR. To paraphrase McCormick *et al.* (1987), we can define the process of scientific visualization as 'understanding through graphical exploration of data'. What distinguishes the process of visualization from simple graphical presentation is the emphasis on exploration as an aid to understanding. Someone engaged in scientific visualization may not, at the outset, know what it is they may discover from the visual representation. The emphasis on 'graphic ideation' (McKim, 1972) and discovery are also key aspects to interaction with virtual environments, yet VR suggests a particular set of characteristics and metaphors to enable such discovery.

Boyd-Davies *et al.* (1996) provided an extensive characterisation of VR and suggests it forms 'a representation of a space in which users move their viewpoint'. Central to his and others' definitions (e.g. Heim, 1998; MacEachren *et al.*, 1999) is the notion of navigation within the same (three-dimensional) space that is being explored. The metaphor of 'immersion', whether or not it is enhanced through the use of immersive hardware (e.g. headsets, CAVEs, etc.), is a dominant one in VR. This appears especially so when exploring landscapes. The terrain provides a spatial backdrop for navigation, giving the immersed viewer a sense of position, scale and perspective.

The distinction between VR and other forms of scientific visualization is illustrated by the two modes of three-dimensional navigation available within the Cosmo VRML browser. The user-interface allows the user to change between 'examine' and 'movement' modes (Figure 12.1). The former allows the viewer to use the mouse to rotate, move and scale the object under investigation. In this mode the viewer is apparently objective, outside of the viewing space. The graphical representation is of an object that can be manipulated by the viewer and is consistent with many of the representations used in scientific visualization. The second viewing mode available in the Cosmo plug-in appears to hold the object under

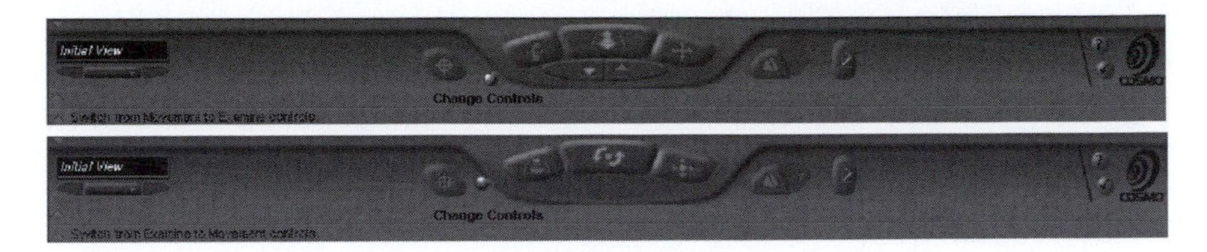

Figure 12.1 Two modes to the Cosmo, VR browser plug-in.

investigation still while the viewer moves in response to mouse control. Here the view is very much more subjective. We appear to be part of the same space that is being explored, not removed from it. This more immersive experience helps to characterise the VR approach to visualization (there are, of course, other characteristics such as MacEachren's *et al.* (1999) identification of the '4 Is' – interactivity, immersion, information intensity and intelligence).

The spatial transformations required to represent three-dimensional information on a two-dimensional computer screen are very similar for both 'examine' and 'movement' modes. The differences are in the way that the projection is changed in response to movements of the mouse or control from the computer keyboard. One of the consequences of engaging in immersive movement over landscape models is that navigation appears more responsive when the apparent viewer position is close to the surface. The surface model provides a fixed point of reference with which to compare one's position. A consequence of this type of navigation is that notions of spatial scale and distance become critical in interpreting the visual representation. The closer the viewer appears to be to the surface, the greater the difference between representations of near and far, between foreground detail and background generalisation. It is this very difference we can exploit when attempting to understand the structural scale dependencies that exist within landscapes.

Perspective rendering

The historical transition from mimetic to symbolic landscape representation in maps can be seen as a move towards a more universal objective representation. This has obvious advantages in terms of the applicability of the visualization in a variety of contexts. For example, a contour map effectively provides a 'reference source' of implied spot heights on a landscape. The information content is much higher than an equivalent hill-shaded representation. However, in losing the subjective viewpoint, we lose the sense of immersion within a landscape that can be vital in the generation of ideas. Importantly, we also lose much of the multi-scale representation inherent in subjective symbolisation.

Consider the problem of visually representing the slope at an arbitrary point on a landscape. The case is illustrated with a (random) fractal landscape in Figures 12.2 and 12.3. The slope may be inferred in only the most general terms from Figure 12.2(a). The colour of any point gives an indication of elevation, and the rate of change in colour over the image gives some idea of slope. We might infer from this image that the slope surrounding the peak towards the right of the image is somewhat steeper than that of the peninsula towards the centre. An alternative view of the same surface is given in Figure 12.2(b), which combines the elevation coded hue with shaded relief using a simple Lambertian lighting model (Foley *et al.*, 1993). Here shading is calculated based on the surface normal calculated by comparing each raster cell with the elevation of its eight neighbours. We now get a better indication of locally steep slope at the cost of the broader-scale picture. No longer is it clear that the slope surrounding the peak to the right is any steeper than that of the peninsula. Indeed, it would appear that the peninsula is the roughest part of the surface exhibiting the steepest slope. Finally, Figure 12.2(c) shows slope calculated analytically and shaded as indicated by the key. As with the shaded relief, slope is calculated locally on a cell-by-cell basis resulting in a picture of local surface variation at the expense of a more regional view.

Figure 12.3 shows the same surface as represented in Figure 12.2(c) but using a three-dimensional perspective projection. The images show snapshots from an interactive 'fly-through' over the surface where the viewer is immersed within the viewing space itself. Here it is possible to gain both a detailed 'large-scale' view of the surface in the foreground simultaneously with a generalised 'small-scale' view of the background. By allowing the user to control the imaginary camera position interactively, the relationship between cell-by-cell measures (as indicated by the shaded surface)

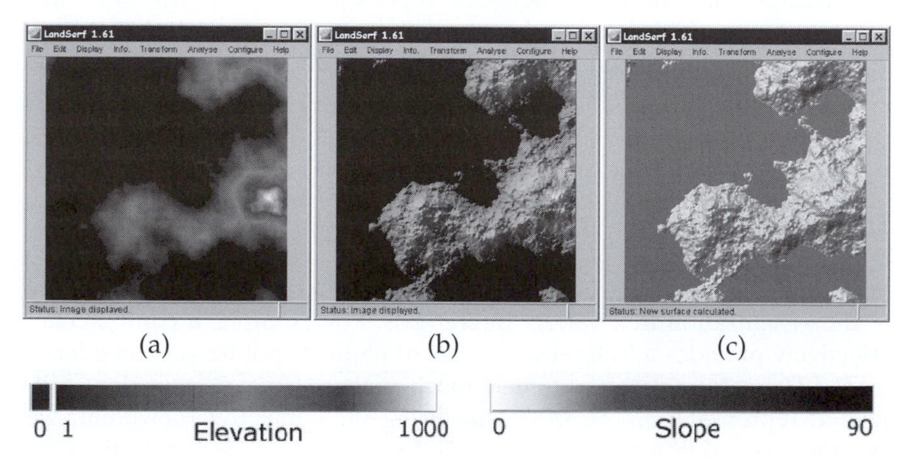

Figure 12.2 Three two-dimensional representations of a fractal surface (see text for discussion).

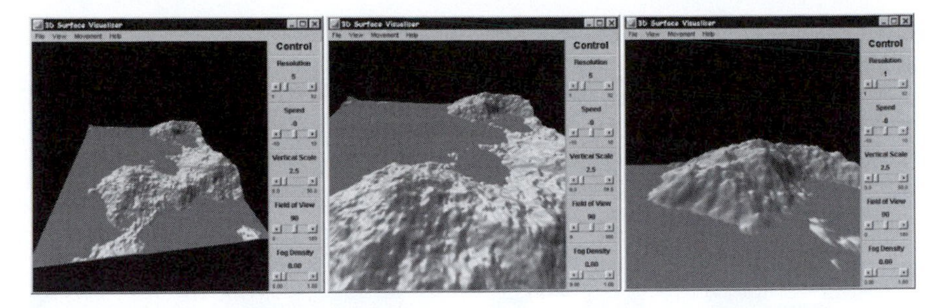

Figure 12.3 Three three-dimensional perspective representations of a fractal surface (see text for discussion).

and the more regional morphometry of the surface may be investigated. While by its very definition this style of interactive visualization is subjective, it offers advantages over static perspective views in that multiple viewpoints can be explored with ease. For example, horizon profiles can be particularly effective at representing scale tendency as shown in the final frame of Figure 12.3. By rotating viewing direction, different parts of the surface may be viewed in such a manner.

Controlling the various projection parameters required to produce a perspective view allows the viewer to explore the relationship between large- and small-scale features. After viewing position, the parameter that has most visual effect on this relationship is the image's field of view (FOV). Figure 12.4 shows the same fractal surface shown in Figures 12.2 and 12.3 with a chequered control surface draped over it in order to assess the scaling effects of the projection. Figures 12.4(a)–(c) show the effect of increasing the FOV combined with movement of the viewing position such that the hills in the background are rendered at approximately the same scale in all four images. Increasing the FOV tends to exaggerate the differences in scale between near and far features, the effect of which is to increase the sense of immersion within the landscape. Low FOVs tend to imply a more objective rendering that diminishes any scale dependencies within the surface. Finally, Figure 12.4(d) further increases the distinction between near and far views by adding a distance-dependent fog effect and a local shaded relief representation. The danger in adopting the style of visualization shown in Figure 12.4(d) is that the amplified subjectivity of the rendering results in very different views of the same spatial location as perspective changes. The morphometric character of the landscape appears to have a low 'map stability' (Muehrcke, 1990). Yet, when combined with interactive exploration, this variation becomes its very strength for DiBiase's private 'visual thinking' rather than public 'visual communication' (DiBiase, 1990).

Figure 12.4 The effect of viewing parameters on perspective projection. (a) 25 degree field of view; (b) 55 degree field of view; (c) 110 degree field of view; (d) 80 degree field of view with local shaded relief and fog depth cueing.

Scale tendency in landscapes

We can define scale dependency as the notion that the characteristics of a point on a surface vary when measured over different spatial extents or different levels of detail. We will also consider the more neutral term 'scale tendency' which describes the effects that may or may not be present when measuring a property at different scales (Goodchild and Quattrochi, 1997). Although related to the idea of map scale, scale-dependent properties are a function of two separate characteristics – measurement *resolution* and measurement *extent*. We define resolution as the smallest spatial extent on the ground that may be measured and modelled as part of a visualization, and extent as the area on the ground used to measure a surface property. The idea of scale dependency is an important one as it allows us to identify critical scales of analysis and representation (such as form-process relationships). It is also an important part of the surface generalisation processes such as resampling, aggregation and TIN vertex selection which are, in turn, necessary for efficient surface visualization.

Within geomorphology there is plenty of evidence that surface form exhibits scale dependency. Many of the empirically-derived quantitative relationships identified in the 1950s and 1960s were in the form of non-linear scalings of surface form (e.g. Drainage Area–Mainstream Length of Hack (1957), Alluvial Fan Area–Drainage Basin Area of Bull (1964)). Such relations have been termed *allometric relations* by Bull (1975), and have been placed in a wider quantitative framework by Church and Mark (1980). The scale dependencies observed by these authors are largely a response to spatial extent rather than spatial resolution. The development of fractal geometry theory (e.g. Goodchild, 1980; Mandelbrot, 1977) suggest that resolution has an influence over measured properties, even those generated by scale-independent processes. The more recent development of multifractals (e.g. Schertzer and Lovejoy, 1994) suggest that even fractal dimension, which predicts how measurement changes with resolution, itself varies with scale (Pecknold *et al.*, 1997). This is supported by geomorphological investigation that suggests that a unifractal model of most landscapes is inappropriate (Evans and McClean, 1995).

Further evidence of the importance of considering scale dependency is provided by those who have examined the effect of resolution on computer-generated measurements of surfaces. Hodgson (1995) shows that slope and aspect measured from a raster Digital Elevation Model (DEM) actually describe the (different) slope and aspect properties at a resolution 1.6–2.0 times the DEM resolution. Chang and Tsai (1991) demonstrated that where the underlying shape of a surface varied at a resolution significantly finer than a DEM, results measured from the DEM will show statistical bias in favour of lower slope values. The nature and quality of surface features measured from DEMs have also been shown to be dependent on the underlying DEM resolution (Martz and Garbrecht, 1998; Wood, 1998).

Together, the evidence presented suggests that scale tendencies are a fundamental part of measurement, and it is suggested it applies equally to the visualization and interpretation of surfaces. This is particularly the case when visualization is used to generate ideas for hypothesis testing (McKim, 1972) and exploration within a virtual environment. Without an appreciation of scale tendencies of representation, we may find it harder to synthesise the multiple subjective viewpoints we encounter when using the VR metaphor.

Do landscapes contain pits?

The process of landscape exploration using VR can be illustrated by considering the existence of *pits* or local depressions within landscapes. There are two traditions in the analysis and characterisation of surfaces. The most widely adopted is that of hydrology and geomorphology where landscape form is considered to be both a function of, and a control over, the movement of water over the surface. In this tradition, landscapes are rarely considered as containing local pits or depressions. If such pits were to exist, they would prevent the passage of water over the surface, or else would pool to form lakes (notable exceptions being limestone or Karst landscapes characterised by sink holes and networks of sub-surface caves and passages). Computer-based hydrological models, which involve the transfer of water over and ultimately off a surface, can be prevented from functioning by pits in the landscape. Pits that appear in a landscape model are therefore usually regarded as 'spurious' and tend to be removed from the model (Morris and Flavin, 1984).

A second tradition in landscape characterisation places a greater emphasis on the topology of the landscape rather than its surface geometry. This approach dates back at least as far as the mid-nineteenth century when Cayley (1859) and Maxwell (1870) characterised the topology of contour lines on a map. The approach taken in this tradition is to state that any continuous surface contains so-called *surface-specific points* that characterise local minima and maxima. Typically, these comprise local peaks, passes (or saddle points) and pits. The topology of a surface is identified by connecting these points via a network of channels and ridges. For mathematical convenience, the type of connections permitted are usually well defined, so for example, the Weighted Surface Network of Wolf (1984) states that each pass must connect to two peaks and two pits. In such a characterisation, local pits form an important and necessary part of the landscape model.

The apparent contradiction between the two approaches to landscape characterisation in the valid existence of pits can be explored through the three-dimensional visualization of landscape form. Figure 12.5 shows a typical two-dimensional view of a DEM of the English Peak District. Figure 12.6 shows some snapshots of an exploratory fly-through the same landscape.

Amongst other advantages, the three-dimensional view allows some comparison to be made between the classification of the surface features and the shape of the landscape (e.g. the misregistration of the channel and the valley floor can be seen in Figure 12.6(f)). Additionally, as any given location moves from the far to near view, we see its changing scale context. The fragmented channels interspersed with apparently spurious pits can be seen in Figures 12.6(a)–(c). As we move closer to the valley floor in Figure 12.6(c) we are provided with visual conformation that the dark pits in the valley floor would indeed prevent a hydrological model from functioning correctly (we would expect a channel to transport water of the bottom-right corner of Figure 12.6(c)). However, if we change the scale of feature classification (see Wood, 1998 for a description of how this

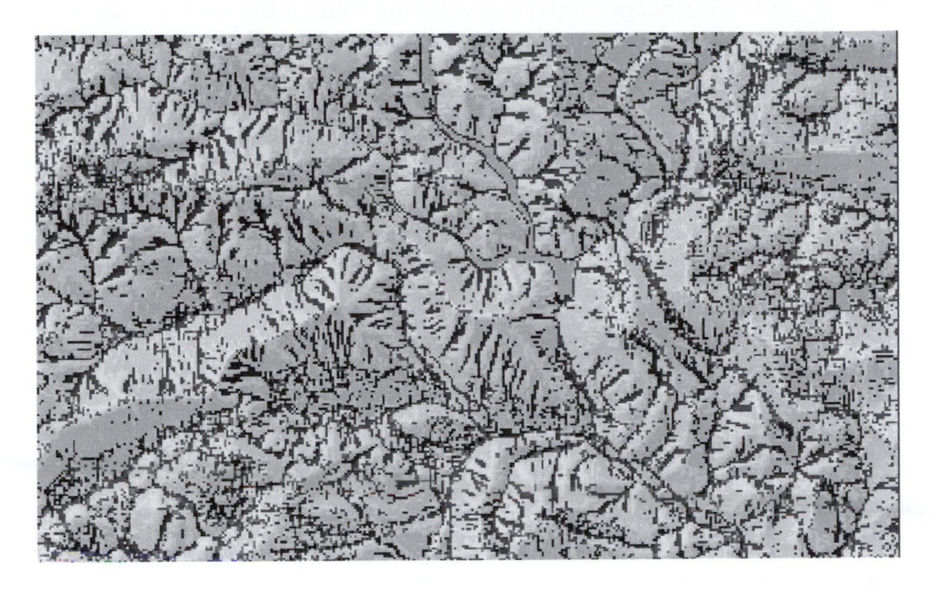

Figure 12.5 A conventional 3D representation of landscape form. The surface model shows a fragmented network of channels and ridges and surface-specific points over a 50m DEM of part of the Peak District, UK.

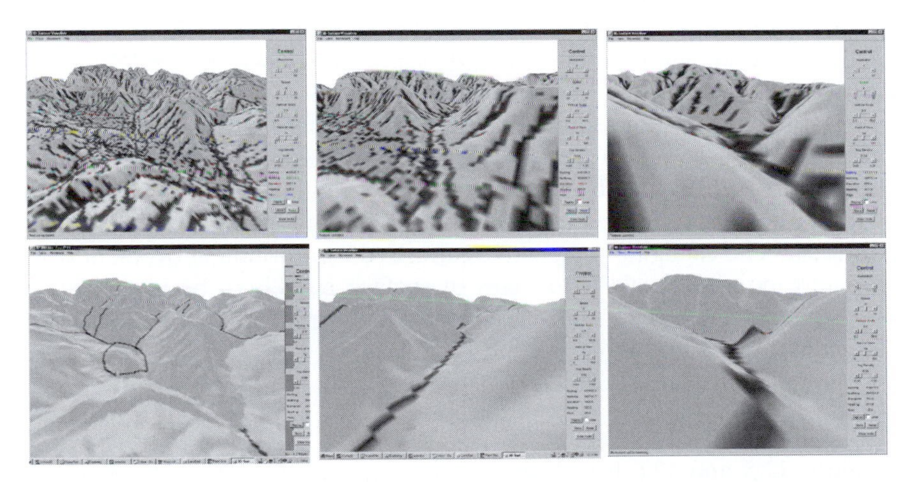

Figure 12.6 Snapshots of a three-dimensional exploration of the Peak District DEM shown in Figure 12.5. Figures (a)–(c), top row show feature classification, bottom row on a 3×3 kernel; (d)–(f) are based on a 25×25 kernel.

may be achieved), and visualize the relationship between the classified surface and terrain (Figures 12.6(d)–(f)), we see that pits tend to occur at channel bifurcations. This allows us to suggest an alternative cause of their appearance. The point of channel bifurcation frequently forms an inverted pyramid in the landscape, surrounded by the bifurcating spur on one side, and the valley sides on the other two. At some broad scale this inverted pyramid will appear as a local minimum, since the outflow channel is expressed at too fine a scale to be identified morphometrically. Pits appear more frequently at finer scales because we do not model sub-pixel hydrology where, presumably, there exist small channels able to transport water downstream. Such a hypothesis clearly needs to be validated by other means, but the important point in this context is that the visual explorations afforded by the virtual environment can generate ideas that perhaps we would otherwise not have been exposed to.

Conclusions

There is no doubt that advances in hardware and software technologies have presented us with new ways to visualize and interact with surface models. The metaphor of exploring a virtual environment based around some underlying surface model provides a continuous series of subjective views of a landscape. Unlike earlier cartographic representations, the multiplicity of these viewpoints allows us to build up an aggregate view of the landscape. Importantly, the VR metaphor allows the viewer rather than the 'cartographer' to control many aspects of the nature of each representation. While any one of these images may have a relatively low map stability, their combination provides us with insight into the scale dependencies inherent in surface models.

It has been argued here that central to the VR metaphor is immersive navigation within some spatial realm, and that the landscape model provides the viewer with a reference for that navigation. The style of navigation with a virtual environment tends to emphasise spatial notions of distance and scale. The representations of near and far are frequently very different. The changing perspectives afforded by real-time movement over a landscape allow scale tendencies of surface models to be investigated. Investigations which, in some cases, provide insight that would be difficult to gain via other means.

Acknowledgements

Figures 12.5 and 12.6 based on the Ordnance Survey 50m DEM of the Peak District, Crown Copyright.

References

Boyd-Davis, B.S., Lansdown, J. and Huxor, A. 1996. The Design of Virtual Environments, *SIMA Report No. 27*, London.

Bull, W.B. 1964. Geomorphology of segmented alluvial fans in western Fresno County, California. *United States Geological Survey, Professional Paper 352E*, 89–129.

Bull, W.B. 1975. Allometric change of landforms. *Geological Society of America Bulletin*, 86, 1489–98.

Cayley, A. 1859. On contour and slope lines. *The London, Edinburgh and Dublin Philosophical Magazine and Journal of Science*, XVIII, 264–8.

Chang, K. and Tsai, B. 1991. The effect of DEM resolution on slope and aspect mapping. *Cartography and Geographic Information Systems*, 18, 1, 69–77.

Church, M. and Mark, D. 1980. On size and scale in geomorphology. *Progress in Physical Geography*, 4, 342–90.

DiBiase, D. 1990. Visualization in the earth sciences. *Earth and Mineral Sciences, Bulletin of the College of Earth and Mineral Sciences, Pennsylvania State University*, 59, 2, 13–18.

Evans, I. 1980. An integrated system of terrain analysis for slope mapping. *Zeitschrift fur Geomorphologie Supplementband*, 36, 274–95.

Evans, I. 1990. Cartographic techniques in geomorphology. In Goudie, A. (ed.) *Geomorphological Techniques*. London: Routledge, Ch. 2.9, 97–108.

Evans, I. and McClean, C. 1995. The land surface is not unifractal: variograms, cirque scale and allometry. *Zeitschrift fur Geomorphologie Supplementband*, 101, 127–47.

Foley, J.D., van Dam, A., Feiner, S. and Hughes, J.F. 1993. *Computer Graphics: Principles and Practice*. Wokingham: Addison-Wesley.

Goodchild, M.F. 1980. Fractals and the accuracy of geographical measures. *Mathematical Geology*, 12, 2, 85–98.

Goodchild, M.F. and Quattrochi, D.A. 1997. Scale, multiscaling, remote sensing and GIS. In Goodchild, M.F. and Quattrochi, D.A. (eds) *Scale in Remote Sensing and GIS*. London: CRC Press, 1–11.

Hack, J.T. 1957. Studies of longitudinal stream profiles in Virginia and Maryland. United States Geological Survey, Professional Paper 294B, 45–94.

Heim, M. 1998. *Virtual Realism*. Oxford: OUP.

Hodgson, M.E. 1995. What cell size does the computed slope/aspect angle represent? *Photogrammetric Engineering and Remote Sensing*, 61, 5, 513–17.

Imhoff, E. 1982. *Cartographic Relief Presentation*. New York: De Gruyter.

McCormick, B., DeFanti, T. and Brown, M. 1987. Visualization in scientific computing. *ACM SIGGRAPH Computer Graphics 21:6.*

MacEachren, A.M. 1995. *How Maps Work: Representation, Visualization and Design*. London: Guilford Press.

MacEachren, A., Edsall, R., Haug, F., Baxter, R., Otto, G., Masters, R., Fuhrmann, S. and Qian, L. 1999. Exploring the potential of Virtual Environments for Geographic Visualization. In *Abstracts of the Annual Meeting of the Association of American Geographers, Honolulu*, p. 371.

MacEachren, A., Kraak, M.-J. and Verbree, E. 1999. Cartographic issues in the design and application of geospatial virtual environments. In *Proceedings of the International Cartographic Association Conference, Ottawa*, pp. 108–16.

McKim, R.H. 1972. *Experiences in Visual Thinking*. California: Brooks/Cole.

Mandelbrot, B.B. 1977. *Fractals: Form, Chance and Dimension*. San Francisco: W.H. Freeman.

Martz, L.W. and Garbrecht, J. 1998. The treatment of flat areas and depressions in automated drainage analysis of raster digital elevation models. *Hydrological Processes*, 12, 843–55.

Maxwell, J.C. 1870. On contour lines and measurements of heights. *The London,*

Edinburgh and Dublin Philosophical Magazine and Journal of Science, 40, pp. 421–7.

Morris, D. and Flavin, R. 1984. A digital terrain model for hydrology. In *Proceedings of the 4th International Symposium on Spatial Data Handling* (International Geographical Union, Zurich), 250–62.

Muehrcke, P. 1990. Cartography and geographic information systems. *Cartography and Geographic Information Systems*, 17, 1, 7–15.

Pecknold, S., Lovejoy, S., Schertzer, D. and Hooge, C. 1997. Multifractals and Resolution Dependence of Remotely Sensed Data: GSI to GIS. In Goodchild, M.F. and Quattrochi, D.A. (eds) *Scale in Remote Sensing and GIS*. London: CRC Press, Ch. 16, 361–94.

Schertzer, D. and Lovejoy, S. 1994. Standard and advanced multifractal techniques in remote sensing. In Wilkinson, G., Kanellopoulos, I. and Megier, J. (eds) *Fractals in Geosciences and Remote Sensing.* Joint Research Centre, Report EUR 16092EN, 11–40.

Travis, M.R., Elsner, G.H., Iverson, W.D. and Johnson, C.G. 1975. VIEWIT: Computation of seen areas, slope and aspect for land-use planning. U.S. Department of Agriculture Forest Service General Technical Report PSW 11/1975. Pacific Southwest Forest and Range Experimental Station, Berkeley, California.

Wainwright, A. 1992. *A Pictorial Guide to the Lakeland Fells* (7 volumes). London: Michael Joseph.

Wolf, G.W. 1984. A mathematical model of cartographic generalization. *Geo-Processing*, 2, 271–86.

Wood, J. 1998. Modelling the continuity of surface form using digital elevation models. In Poiker, T. and Chrisman, N. (eds) *Proceedings of the 8th International Symposium on Spatial Data Handling*. Vancouver: International Geographical Union, 725–36.

Yoeli, P. 1983. Shadowed contours with computer and plotter. *The American Cartographer*, 10, 101–10.

13 Providing context in virtual reality

The example of a CAL package for mountain navigation

Ross Purves, Steve Dowers and William Mackaness

Introduction

The rapid advances in cheaply-available technology over recent years mean that the provision of interactive VR environments is no longer solely the province of specialised software houses. In particular the advance of environments such as VRML and Java3D have allowed those whose primary interests are in applying 3D technology, rather than developing it, to produce customised software for a variety of applications including visualization of change for planning proposals (Lovett *et al.*, Chapter 9; Miller *et al.*, Chapter 10), educational software (Moore *et al.*, 1999; Moore and Gerrard, Chapter 14), the visualization and interpretation of archaeological data (Gillings, Chapter 3) and medical training (El-Khalili and Brodlie, 1997; Brodlie and El-Khalili, Chapter 4).

The changing profile of those developing VR software has resulted in a revisiting of many issues related to user interaction with VR. In this chapter we discuss a Computer Aided Learning (CAL) package developed for teaching mountain navigation and use it as an example to explore issues pertaining to navigation through a 3D environment. In particular we discuss how context, rather than photo-reality, is an important quality in facilitating navigation through a 3D landscape.

This chapter is centred around the concept of providing levels of context to allow users to navigate through a CAL package for teaching mountain navigation. Context, in this sense, is taken to be information which aids in the interpretation of, and navigation through, a space. That space could be the metaphorical one of CAL, where the user interacts with a number of different information sources including text, pictures, a 3D landscape, a navigational toolbox, and panoramic photographs. Equally it could be the virtual space of the landscape used in delivering the CAL curriculum. In this chapter we focus on providing context within the virtual space of the 3D landscape.

The chapter examines, in brief, how one navigates in the real world, with reference to the wealth of geographical literature on this topic. Then we describe the rationale for, and the implementation and evaluation of, a CAL package for teaching mountain navigation. With reference to the CAL package we then explore issues enabling users to navigate within the 3D space used in our package, and attempt to relate these issues to navigation through real spaces.

Navigating through spaces in the real world

It is only by having some understanding of the mechanisms that are used to navigate through real spaces that we can properly address how we facilitate navigation through a virtual landscape. Darken and Sibert (1996) asserted that knowledge about wayfinding in the physical world could be directly applied to virtual worlds. Even the most cursory examination of navigation through real spaces indicates that the navigational metaphor (Golledge, 1992) used varies according to a variety of factors, including:

- has the location been visited in the past?
- is the journey being made with a specific purpose?
- what navigational tools are available to aid the journey through the space?
- by what means is the journey being carried out?

In this research, we are interested in the way an individual moves through or along a mountain path on foot, but we will first examine a number of other navigational metaphors in order to illustrate some of the differences between them. If, for example, one is travelling by train, then the primary tool used in navigation is likely to be a schematic map. Such maps generally show the location of one point in relation to a network around it, rather than positioning points within some scaled space. To navigate to a destination by train, one need only know which point on the map was last visited and the direction of travel. We need to know where we are in order to work out how to get where we are going. Navigating through a town on foot is a quite different process. On our first visit to a new location, we might use a street map and actual street names to orient ourselves and navigate through the town. Very quickly, the position of certain 'landmarks' is learnt and the navigator starts to use visual cues which may or may not be on the map. The type of landmark used is again dependent on the navigator's knowledge of the area. For example, a tourist in Edinburgh would be likely to use the castle as a reference point, whereas a local would be much more likely to give directions via a local pub or café!

In developing a CAL package for mountain navigation it was important that we understood as much as possible about how people navigated through mountain environments. It quickly became apparent that, just as those who know a town well navigate in a different way to the newcomer, so experienced mountaineers navigate in a different way to the majority of 'recreational hill users'. The experienced mountaineer relies primarily on ground shape and a constant check of position in order to move through space, and uses other landmarks as secondary features for orientation themselves. However, many people use a combination of point, line and area features such as cairns, buildings, streams, roads, forests and lakes as landmarks (or navigational handrails) for orientation, and pay relatively little attention to shape. Some 'landmarks' are seasonally ephemeral (e.g. a dried river bed) and others are artefacts of prior mapping (old triangulation pillars, harvested forests, or afforestation programmes). To some degree these landmarks 'compete' with one another, as the walker tries to resolve differences between the map and the landscape. In the former case, the navigator visualizes a landscape, with a number of landmarks upon it. In the latter case the navigator sees simply a flatland (Figure 13.1) (Tufte, 1990) with a series of landmarks which are used as handrails to navigate between. In the design of our CAL package we wished the user to view the landscape in a way more similar to that of the experienced mountaineer: the shape of the landscape, rather than the features lying upon the flatland is highlighted. This emphasis, in turn, has an impact on the way in which we provide contextual information within the package. In the next section, we discuss in more detail the rationale for developing the CAL package, and explain how it was implemented.

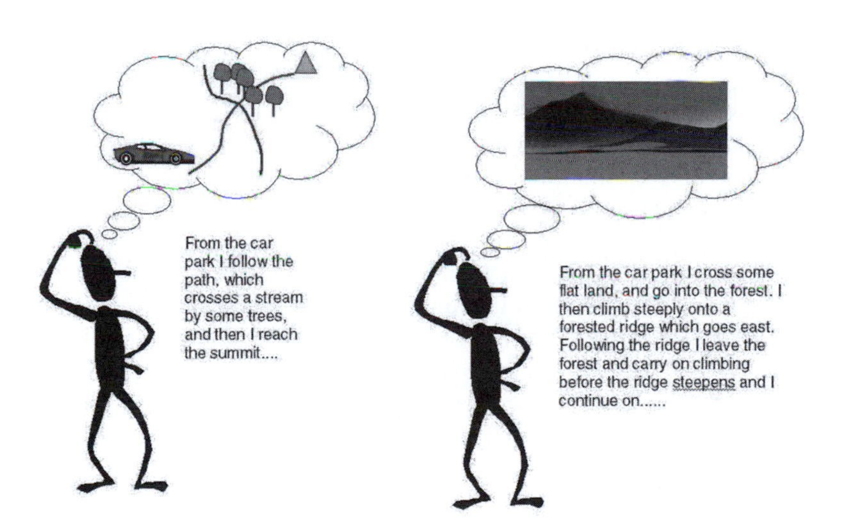

Figure 13.1 Alternative world views: the *flatlander* and the mountaineer who visualizes shape: see CD ROM for a colour version of this figure.

A CAL package for mountain navigation

Background

A recent report on Scottish mountain accidents (Anderson, 1994) reported that 23 per cent of incidents between 1989–93 were directly attributable to poor navigation. The CAL package described in this chapter was developed in response to the Scottish Mountain Safety Group's (SMSG) concern that inability to visualize terrain correctly from the map was a major factor in navigational errors. Given the developments in 3D technology described elsewhere in this volume (see pp. 35–46) it was felt that an affordable CAL package could be developed to attempt to address this problem. As a result the authors were commissioned by the Scottish Sports Council (SSC) on behalf of the SMSG to investigate the use of 3D visualization in teaching mountain navigation. The resulting package is described in detail in Purves *et al.* (1998) and is summarised here. The development of the CAL package was stakeholder-led: that is to say that experts, representative bodies and potential users were heavily involved in development at all stages, and not simply used in the evaluation of the final package. As a result the curriculum was arrived at by a broad consensus amongst the groups consulted and further developed through an iterative process.

During the course of this project it was widely stated by mountaineering instructors that the biggest area of discrepancy between the perceived and actual skills of those that they instructed in navigational techniques lay in the inability of many people to visualize 3D shapes from a simple contour map. Many people were relatively competent in more 'technical' skills, for example taking bearings between two points. In this section we discuss the implementation and evaluation of the CAL package focusing on our desire to teach users about shape.

The CAL package was designed to take the form of a journey up the mountain of Driesh, starting from Glen Doll in the eastern Highlands of Scotland. The curriculum focused throughout on the importance of ground shape as well as discussing other important skills in mountain navigation. Figure 13.2 illustrates the form of the journey and the introduction of different curriculum elements as the user progresses on their journey up the mountain.

Implementation

The requirements for this project included a desire that any resulting software be multi-platform and have negligible costs associated with it. Thus the CAL package was implemented using Web technology and is, in principle, multi-platform, though in practice all development and testing was carried out on PCs using Netscape Navigator and equipped with the

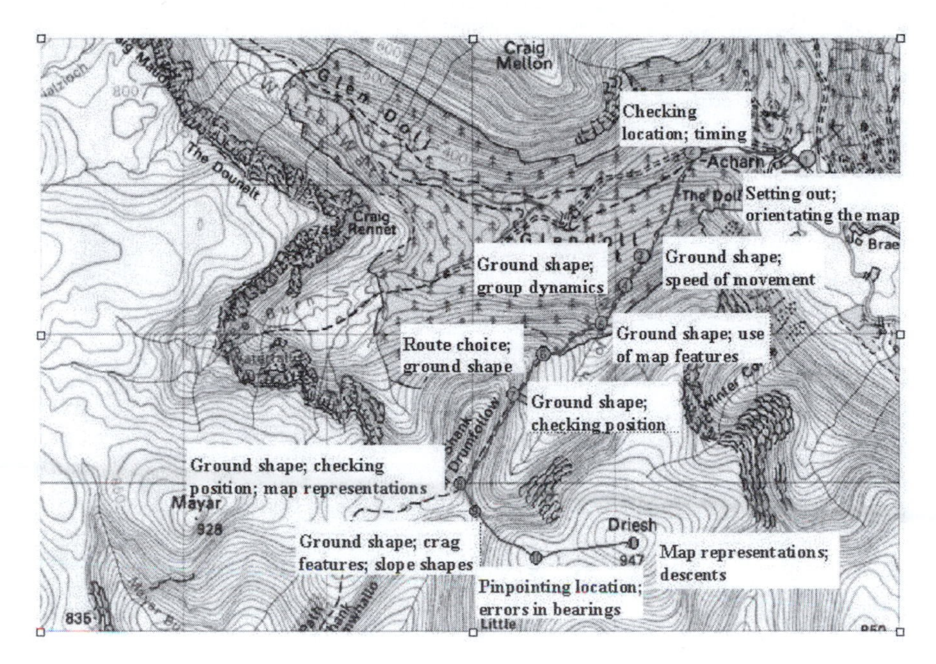

Figure 13.2 The curriculum contained within the CAL package: see CD ROM for colour version of this figure. Reproduced from Ordnance Survey maps by permission of Ordnance Survey on behalf of the Controller of Her Majesty's Stationery Office, © Crown Copyright. ED 243949.

appropriate plug-ins. The package was designed with a simple interface as shown in Figure 13.3. The user can move about within the 3D landscape shown at the bottom right, where clicking on a waymarker will result in them being teleported to the next waymarker on the route, at the same time textual information for that location is also displayed. The user can also click on hypertext hotlinks within the document to read subsidiary information about a variety of related topics.

The basic CAL interface is shown in Figure 13.3 and uses HTML (Hypertext Mark-up Language) to provide textual content, and 3D scenes are displayed using VRML (Virtual Reality Modelling Language) (Ames *et al.*, 1996; VRML, 1997) with the associated navigational toolbox provided through a Java applet. VRML allowed us to produce geo-referenced 3D landscapes, produced using Ordnance Survey 50m DEMs, through which the user navigates and which contains further 3D objects and hotlinks. VRML also allows us easily to incorporate geo-referenced drapes over the landscape as discussed in the next section. The navigational toolbox and the 3D VRML scene are linked by the External Authoring Interface (EAI, 1999) which allows Java to trap events from the 3D world. This linkage means that changes in the 3D world, as the user rotates or

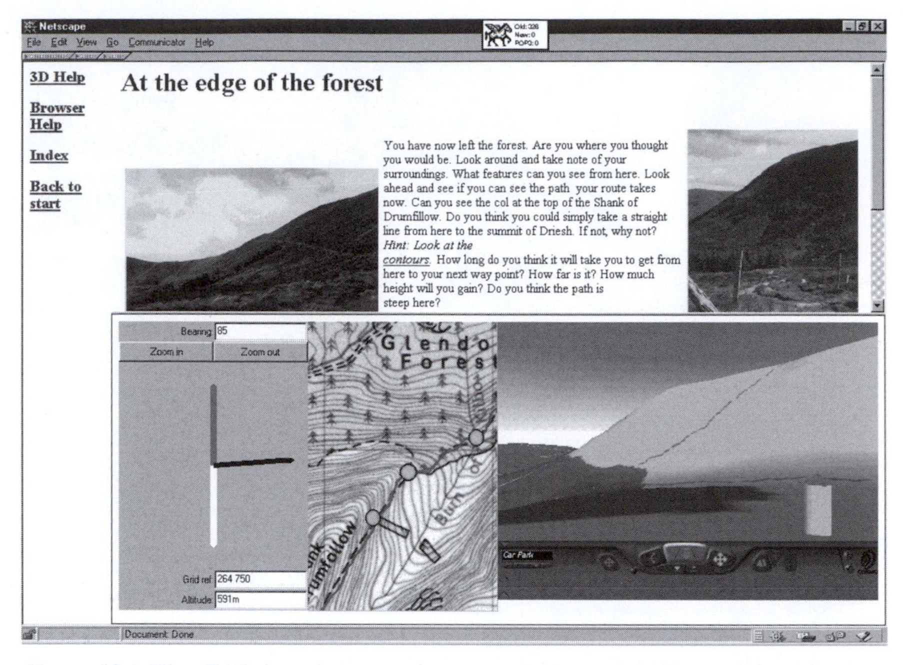

Figure 13.3 The CAL interface showing textual elements, photographs, 3D scene (with Landline-based drape and waymarker) and the navigational toolkit of scrolling map, compass and associated information: see CD ROM for colour version of this figure. Reproduced from Ordnance Survey maps by permission of Ordnance Survey on behalf of the Controller of Her Majesty's Stationery Office, © Crown Copyright. ED 243949.

moves through space, are reflected in the toolbox, with the compass arrow pointing in the same direction as the user looks and the map scrolling to maintain the user's position in its centre. This map is only to help users fix their position within the landscape; we would intend that users should employ a paper copy of the Ordnance Survey 1:50,000 map for the area during the exercises. A QuickTime VR panorama is also used, allowing the user to scroll around a 180° panorama of photographs taken from the col (Waymarker 9 in Figure 13.2). Several still pictures are used, allowing users to compare and contrast several representations of the real world, namely photographs, the VRML landscape and the map.

Evaluation of the CAL package

As was explained in the background to this section, this project was stakeholder-led which involved much feedback throughout the development. However, when a working prototype was available we carried out a more formal evaluation in two main ways:

- by demonstration and discussion with mountaineering instructors and representative groups,
- by potential users being asked first to fill in a questionnaire detailing their mountaineering skills, then to use the package in a hands-on session and finally to fill in a further questionnaire assessing the usability, performance and educational content of the package.

The sessions with the instructors were extremely successful, with very positive feedback. Several specific concerns were expressed and where possible rectified or made open for further discussion. A general consensus was reached that such a package would be a useful way to aid in teaching the understanding of ground shape. Furthermore, those present indicated that the level of detail in the 3D scenes was sufficient, and that a representation of the hills where shape was the primary feature was more useful than a more 'photo-real' image.

The feedback from potential users was also very positive, but with some interesting differences. Most of these users already considered themselves to be experienced. However, more than 70 per cent reported that they better understood the importance of ground shape in mountain navigation after using the package, which was our primary aim. Furthermore, most users were happy with the basic usability and design of the package. However, many users felt that insufficient detail was present in the landscape to allow them to move through it easily. Since the navigational window gave them information about exactly where they were, it was clear that we needed to provide more context to the landscape in order to allow them to move around in a more realistic fashion. In the next section we discuss how we went about adding the necessary levels of context, whilst still attempting to keep a representation of the landscape in which the primary aspects of ground shape remained the most important design goal.

Context within the CAL package

Several levels of context were provided within the package. For example, if the user navigated from point to point using the waymarkers shown in Figure 13.3 then they had several levels of information available to them. The text window would display information appropriate to their location and might describe what features could be seen at a particular location in space. The title of the text window, for example, 'At the edge of the forest', might give them further information about their location. The 3D view in the VRML window would give them another level of detail, allowing them to move around and rotate, looking at the shape of the ground around them. Finally, the navigation panel gave them an exact location in several different ways: as a cross on a map; as a grid reference; as a height; and as an indication of the direction in which they were looking.

From our evaluation, it was clear that the level of information given was sufficient to locate users within the 3D world, and indeed to allow them to 'walk' from one identifiable point, such as a waymarker, to the next. It did not, however, allow them to feel confident about where they were on the basis of the 3D view alone. Thus we decided to examine means to add a suitable amount of extra contextual information to the 3D scene in order to address this problem.

Adding context in 3D

In this section we explore different techniques for adding context to our 3D world in order to facilitate navigation through the world. We explore a number of different levels of enhancement including the addition of virtual waymarkers, through the use of a range of different drapes, to the extraction and insertion of 3D objects into the virtual landscape. We discuss the advantages and disadvantages of each of the different levels of context in a number of different ways:

- how visually effective were they within the landscape?
- did they improve the navigational metaphor within the landscape?
- was the resulting landscape more realistic?
- do the additional data cause a significant change in performance?

Adding waymarkers

Waymarkers were introduced to the landscape at an early stage. These were positioned at strategic points and were used to introduce the user to different elements of the curriculum as illustrated in Figure 13.2. They consisted of tall green marker posts with red tops which were easily visible from a distance. By clicking on a waymarker the user could advance to the next stage in the walk, and the related text was loaded in the appropriate HTML window. The waymarkers allowed the users to synchronise their position in the text with the landscape in a visible manner. However, they are virtual waymarkers – such features do not exist in the real landscape. The ability effectively to 'teleport' from one location to another using the waymarkers by clicking on them does not sit well with our need to emphasise the important elements of navigation, but does provide inexperienced users with a simple navigation mechanism. Navigating through the landscape in this manner is much closer to our description of a train journey, where one needs only a schematic map and a knowledge of the direction of travel, than to the process of navigating in a mountain environment. Pleasingly, through observation of users, it was noted that many quickly progressed to walking between the waymarkers as their level of skill at using the interface improved.

Height-based texture drape

The simplest attempt to improve the visual quality of the landscape used a height-based texture drape. The elevation data in the DEM were loaded into ARC/INFO's GRID module and each cell in the grid was classified with a colour according to its height. The resulting image could then be used as a texture to drape over a 3D landscape. Figure 13.4 shows the results of draping such an image over the landscape. The resulting image demonstrates that the landscape looks more realistic – however, the information does not add any context in navigational terms to the user, except at the most basic level. In fact, this drape was used during the evaluation and clearly does not provide sufficient information to enhance navigation through the landscape. The use of raster drapes has some effect on performance, but the package still performed at an acceptable level.

Map-based texture drape

A very simple, yet effective drape can be produced by simply using the raster-scanned version of the 1:50,000 colour Ordnance Survey map, which is also used in the navigation panel as a drape for the landscape. Although

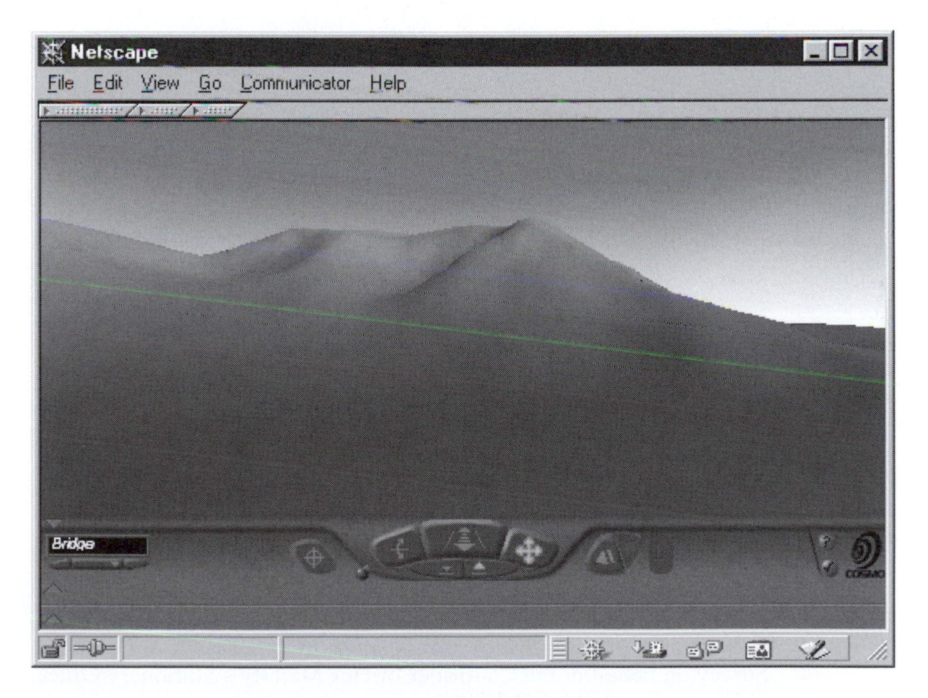

Figure 13.4 A height-based drape: see CD ROM for colour version of this figure. Reproduced from Ordnance Survey maps by permission of Ordnance Survey on behalf of the Controller of Her Majesty's Stationery Office, © Crown Copyright. ED 243949.

at first glance this appears very effective, this drape (Figure 13.5) is unsuitable for a number of reasons. First, we are trying to teach people about the link between the map and a landscape. By draping the map over the landscape we remove any need for the user to attempt to interpret contours – we simply make the landscape look like the map. Users would probably have little difficulty in navigating through this landscape, since they can simply cross-reference the local features on the landscape with their relative position on the more global map. This provides a powerful tool for illustrating what contours are. Our aim, however, was to move the user beyond this, in attempting to visualize position in the landscape using interpretation of shape as the primary tool. When presented with a visualization of contours in 3D most people have little trouble understanding the concept – it is the more abstract process of taking the 'flatlands' of a contour map and visualizing them in 3D where the inherent difficulty appears to lie. Thus, we considered it important to make the link implicit and provide an opportunity to compare a map to a representation of reality, rather than merely produce a 3D surface with a map-based texture.

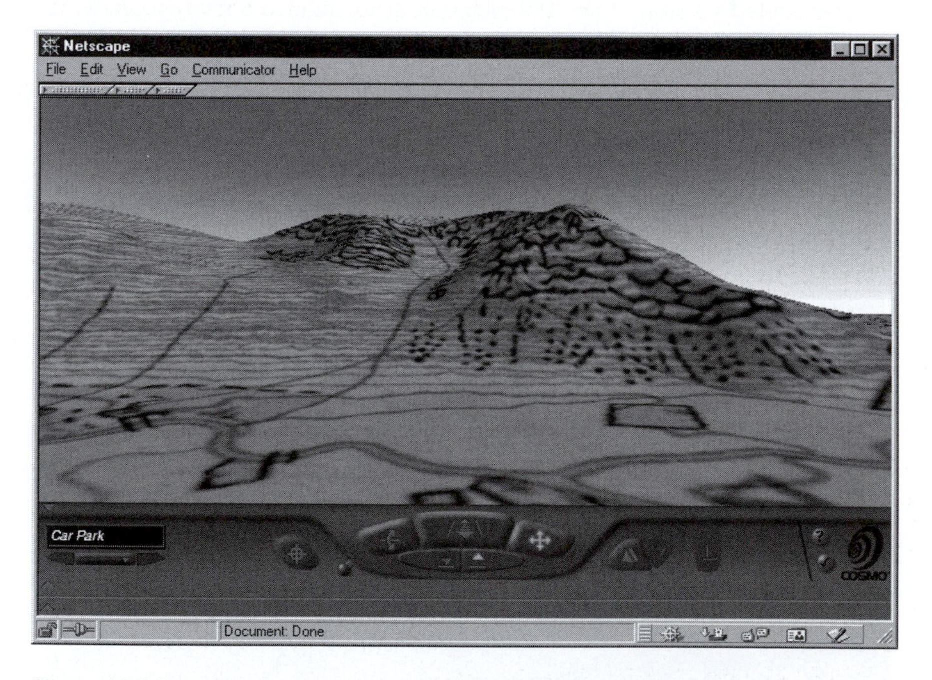

Figure 13.5 Draping the raster map: see CD ROM for colour version of this figure. Reproduced from Ordnance Survey maps by permission of Ordnance Survey on behalf of the Controller of Her Majesty's Stationery Office, © Crown Copyright. ED 243949.

Raster drape extracted from Landline

In order to produce a drape which introduced real features, such as forests and streams, and to provide more contextual detail, we next produced a drape based on vector data from the OS's Landline vector data set. In mountainous areas the source of these data is the 1:10,000 map series. We processed the data to produce a series of labelled polygons and lines representing the forested areas and streams within the landscape. Using these data we then generated a raster drape, as shown in Figures 13.3 and 13.6, illustrating forested areas and streams. This drape was of good quality and provided the users with an extra level of information beyond the basic ground shape with which to orient themselves, but it did not make the landscape look like the map. Although this information does successfully add an appropriate level of context to our landscapes it does not address these equally in all areas. There are unlikely to be forests or streams on summit plateaux, and thus we still have a region which is simply monochromatic. Furthermore, in these regions there is the least variation in shape so orientation using ground shape is at its most difficult. In fact, many people 'navigate' in higher regions on hills by simply following paths. At this point, we must recognise the limitations of a virtual

Figure 13.6 A drape based on Landline vector data: see CD ROM for colour version of this figure. Reproduced from Ordnance Survey maps by permission of Ordnance Survey on behalf of the Controller of Her Majesty's Stationery Office, © Crown Copyright. ED 243949.

landscape as a representation of space, and instead attempt to fully inform the user of these very real difficulties.

Forest extracted as vector data from Landline

The most realistic landscape which one could produce would contain real 3D objects. It is possible in VRML to build 3D objects and place them within a landscape, performing any transformations as appropriate. We decided to explore the possibility of adding forested areas to the virtual landscape in 3D. Two datasets were used for this work, the 50m DEM and Landline vector data. Forestry polygons were manually identified and extracted as sets of coordinates in a GIS. A Java program was written which took such a polygon and the vertices of the DEM used to create the landscape and found the intersection between these two objects. Intersections inside the polygon with the landscape were found at specified intervals, thus producing a set of 3D coordinates at which trees could be placed. The resulting technique was used to 'forest' large areas of the landscape. An example of this afforestation program can be seen in Figure 13.7(a).

On examination of Figure 13.2 it is clear that a large percentage of the

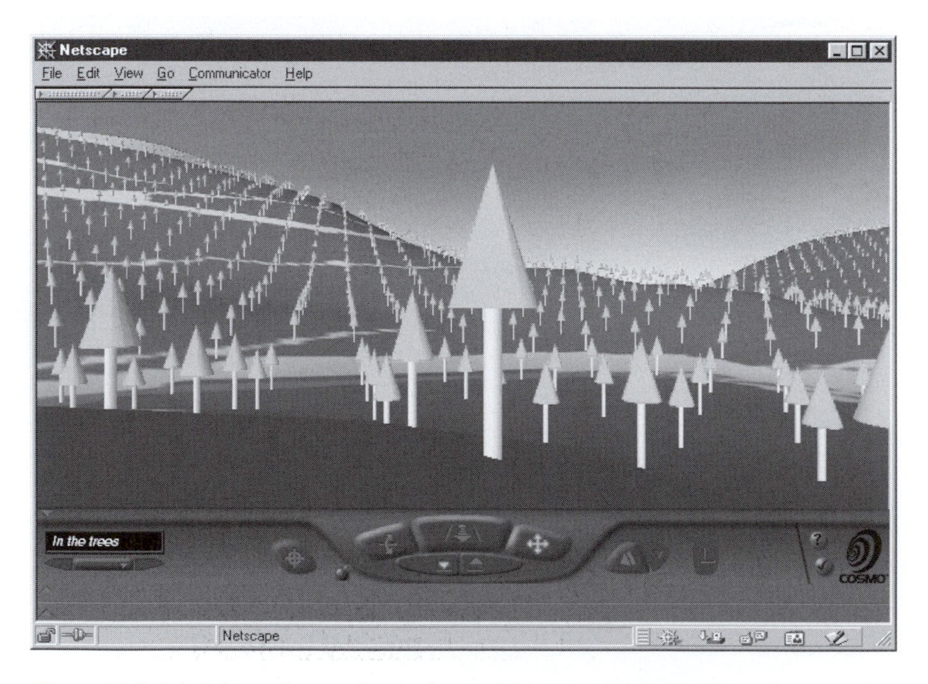

Figure 13.7 (a) A large forested area from within: see CD ROM for colour version of this figure. Reproduced from Ordnance Survey maps by permission of Ordnance Survey on behalf of the Controller of Her Majesty's Stationery Office, © Crown Copyright. ED 243949.

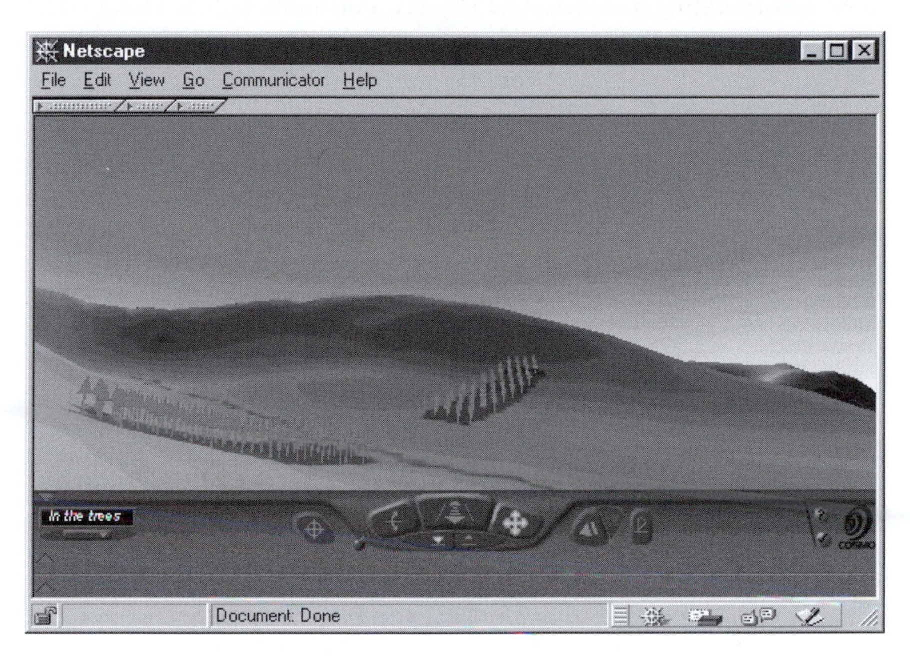

Figure 13.7 (b) A small forested area from without: see CD ROM for colour version of this figure. Reproduced from Ordnance Survey maps by permission of Ordnance Survey on behalf of the Controller of Her Majesty's Stationery Office, © Crown Copyright. ED 243949.

lower land in Glen Doll is forested. For example, if a 1 km square area is afforested with trees at a 20 m interval, 2,500 new objects will be placed in the landscape. Not only does this result in very severe performance implications for the VR software, but the resulting forested area has trees which are much too widely spaced and does not significantly enhance the 'reality' of the landscape (Figure 13.7(a)). If, however, we forest small plots, where we can place the trees very close together, the resulting zones do look relatively realistic (Figure 13.7(b)). The main problem in enhancing the third dimension of the landscape, rather than simply using drapes, is that sufficient detail to make a landscape realistic is very difficult to provide. For example, if we look at a forested area from a distant vantage point, it may look relatively realistic (Figure 13.7(b)). If, on the other hand, we walk through a forest where the trees are some 20 m apart (Figure 13.7(a)), the view is nothing like what one would expect in a real forested area.

This is perhaps a good illustration of the limitations of adding context to a 3D scene. In terms of information content, rather than photo reality, exactly the same data are used to derive Figures 13.6 and 13.7. However, the performance of the raster-based landscapes is considerably better than those incorporating numerous 3D objects. Furthermore, the visual effect in Figures 13.3 and 13.6 is arguably more pleasing than that in Figure 13.7.

Conclusions

In this chapter we have discussed issues pertaining to the development of a CAL package for teaching mountain navigation. Through the use of a number of different elements including interactive 3D scenes, the CAL package attempts to teach mountaineers the importance of understanding shape above all else. In the course of evaluation it became clear that a dichotomy existed between those teaching navigation who were essentially satisfied by the representation of a landscape denuded of clutter, and those learning to navigate who desired a wealth of secondary information to facilitate their movement through the scene.

In adding context to the scene we attempted to minimise the addition of unnecessary clutter, whilst facilitating movement through the 3D landscape. The use of height-based drapes added little information, while simply draping a map over the landscape turned it into another representation of the map, albeit a visually appealing one. However, by adding a drape extracted from Landline vector data we were able to select what we considered to be appropriate extra contextual information, namely forests and streams. Creating a 3D version of these data added no new information, whilst significantly degrading performance. We used waymarkers to indicate significant points in the journey up the mountain, and noted that the 'teleporting' navigational mechanism was easy to use, but perhaps inappropriate to our application.

The use of aerial photographs as drapes would provide another means of adding context to our scene; however, colour images were not available for the study area. We should also perhaps question why we are trying to add information to our scene. In the case of this CAL package, we are not trying to produce a landscape which most closely mimics the map, or even reality, but rather a virtual landscape where the shape of the ground is the primary element in the scene.

Acknowledgements

This research was funded by the Scottish Sports Council. Anne Salisbury of Moray House, the members of the Scottish Mountain Safety Group and the staff of Glenmore Lodge, together with all those 'guinea-pigs' who helped in the evaluation are gratefully acknowledged for their contribution to this research.

References

Ames, A.L., Nadeau, D.R. and Moreland, J.L. 1996. *VRML 2.0 Sourcebook*. Chichester: Wiley.

Anderson, C.M. 1994. *The Scottish Mountain Rescue Study 1964–1993*. Edinburgh: HMSO.

Darken, R.P. and Sibert, J.L. 1996. Wayfinding strategies and behaviours in large virtual worlds. In *Human Factors in Computing Systems, Proceedings of ACM CHI 96*, Vancouver, BC, Canada, 13–18 April, pp. 142–9.

EAI. 1997. *The Virtual Reality Modeling Language (VRML) – Part 2: External authoring interface.* Committee Draft ISOIEC 14772–2: xxxx, http://www.web3d.org/WorkingGroups/vrml-eai/Specification/

El-Khalili, N.H. and Brodlie, K.W. 1997. Distributed VR training system for endovascular procedures. In *Proceedings of the Fourth UK VRSIG Conference,* 1 November 1997, Brunel University, Uxbridge, UK, pp. 110–19.

Golledge, R.G. 1992. Place recognition and wayfinding – Making sense of space. *Geoforum*, 23, 2, 199–214.

Moore, K., Dykes, J. and Wood, J. 1999. Using Java to interact with geo-referenced VRML within a Virtual Field Course. *Computers & Geosciences*, 25, 1125–36.

Purves, R.S., Mackaness, W.A. and Dowers, S. 1998. Teaching mountain navigation skills using interactive visualization techniques. In *Visual Reality. Proceedings of the International Conference on Multimedia in Geoinformation*, Bonn, Germany, 16–18 March 1998, pp. 173–83.

Tufte, E.R. 1990. *Envisioning Information*. Cheshire, Connecticut: Graphics Press.

VRML. 1997. Specification: www.vrml.org/Specifications/VRML97/index.html.

14 A Tour of the Tors

Kate E. Moore and John W. Gerrard

Introduction – virtual environments in education and geography

The employment of virtual environments, in a variety of forms and complexities, is escalating in the spheres of geography and education. Multimedia technologies are now commonplace on the desktops of spatial scientists and students. Geographical Information Systems, visualization software, multimedia Computer Aided Learning packages and, most recently, virtual reality browsers are available and used in teaching and research. These threads have largely developed independently but the technologies are now converging and the challenge is to integrate them successfully and purposefully.

Grounded partially in a background of map making, geographic visualization has transformed cartography from a communication device to an exploratory tool. It is part of the 'cognitive processes involved in transforming data into information' (DiBiase, 1999: 90). However, since the recognition of 'visualization' as a knowledge formulating process visualization tools have been made for the novice as well as the expert and most recently its application in geographic education has also been acknowledged (DiBiase, 1999). One of the primary features of all VFC component software is this use of visualization methods within the pedagogic context (Dykes *et al.*, 1999). Here the role of visualization in understanding or providing insights into spatial data has been applied to fieldwork. One of the main functions of fieldwork was seen to be development of observational skills, which is akin to the visualization process. So it is envisaged that visualization methods may be used to extend this skill. Within the sphere of visualization is a realisation that the conjunction of multiple views of the world and/or data (multiple representations) stimulates a greater understanding or insight into that data. It is this synthesis as it is exploited in the Virtual Field Course (VFC, 1999) that enhances the learning environment for the student. The availability of multiple representations: media, map data and virtual reality, of the same geographical region, each of which offers a different perspective of the place, improves the scope of student learning.

For education in general, Computer Aided Learning environments have proven multimedia to be a valuable resource that helps to synthesise ideas and stimulate understanding. Capturing and analysing an instance of reality in the form of images or of time intervals in the form of video is an accepted learning aid. Similarly narration and commentary by an expert in a discipline on the region or topic being studied promotes understanding. Using multiple and varied forms of representation in a learning environment harnesses different forms of intelligence in such a way as to attain high student achievement (Freundshuh and Helleviks, 1999) as well as offering greater enjoyment and interest. Interestingly, computer-assisted learning has been found to be of greater value for short-term studies (Clark, 1985) that could, by implication, include the concentrated learning experience of residential field courses.

Virtual reality is being used increasingly within many spheres of education and it has been acknowledged to be a useful resource. 'VR ... has the potential ... to provide a very rich learning environment ... VR is an emerging technology that has demonstrated potential as a dynamic and effective teaching tool ... it allows us to experience a body of knowledge interactively' (Kalawski, 1995). At present this is partially because it still provides novelty of experience but it also provides a set of unique characteristics that may provide a foundation to improve student learning. Zeltzer (1992) identifies these as autonomy, when the environment has some behaviour of its own; presence, or a sense that the user is in the virtual space; and interaction, between the user and the virtual world.

Laurillard (1996: 174) recognises that multimedia offers 'the affordances of academic ideas, or precepts ... through both interactive experience of the phenomenon, and a correlated discussion of the formal description of that phenomenon.' These affordances, or the way we are subconsciously led to use or perceive an object, are germane to geographical virtual reality because of our natural experience of being in and moving around geographical spaces. Particularly in geographic education VR can overcome many of the restrictions imposed by using 2D maps as VR uses natural semantics (Mikropoulos, 1996), so reducing the shift in thinking when using traditional maps. Students are quickly able to walk or fly through natural-looking virtual landscapes. It is then a small step to navigate and understand alternative depictions of reality (Figure 14.1).

Virtual reality when combined with other forms of media may offer even greater potential for a cognitive approach to education. Models of reality often fall short in detailing minutiae that may be of importance (one person's clutter may be another's data). Virtual reality has the potential for delivering data at a variety of scales but the delivery of this in a seamless manner presents technical difficulties. There is a trade-off between level of detail attainable and speed of navigation and many methodologies are being developed to provide efficient download and navigation of this type of virtual space. As an alternative solution,

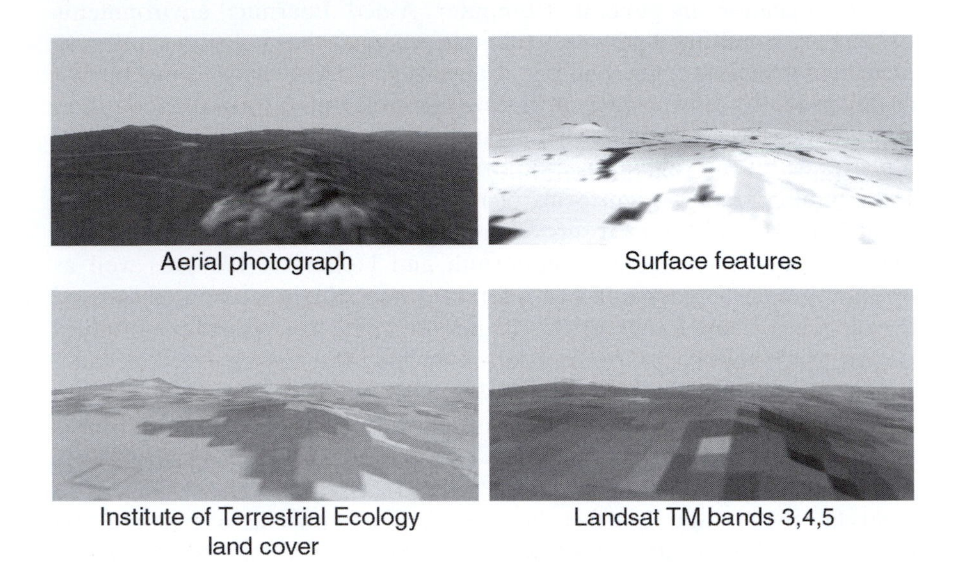

Figure 14.1 Near-realistic and alternative views of reality.

Dykes (2000) uses a geographical virtual environment based on panoramic imagery that can deliver instances of data at a variety of scales. However, by combining VR with geo-referenced multimedia the benefits of both can be harnessed. Virtual reality provides user-controlled navigation and a variety of reality models and linked media can supplement this with specific detail or information. Multimedia field courses have been developing, particularly as Web-based resources or CAL material (CLUES, 1999). With the addition of interactive visualization and virtual reality, a powerful and extremely useful holistic environment may be developed.

The development of traVelleR software described here has been concurrent with several other projects using VRML and Java that have investigated Web-based solutions for delivering terrain applications. This can be referenced particularly in areas of visualization such as fieldwork teaching (Moore, 1999), training for mountain navigation skills (Purves *et al.*, 1998; Chapter 13), forestry assessment (Miller *et al.*, Chapter 10) and visualizing massive terrain databases (Reddy *et al.*, 1999). The latter is an example of the significant work undertaken by the GeoVRML Working group of the W3D consortium (1999) focusing on aspects of modelling real-world, high-resolution terrains. An integrated terrain navigation system can be seen in the work of Purves *et al.* (1998; Chapter 13) and, like traVelleR, this work uses the External Authoring Interface to link VRML and Java to produce VR dynamically linked to a 2D map and an HTML system. It also forms a structured Computer Aided Learning package for the specific purpose of teaching mountain navigation skills.

To summarise, multimedia, virtual reality and pedagogic input provide data of which spatial location is the prime key and visualization the method for converting this data into information. This relationship has been exploited in the development of the Virtual Field Course materials. The following sections describe how the data is used through a variety of teaching modes in one software environment and an illustrated example is described in the form of a Tour of the Tors.

Why a Tour of the Tors?

The South Devon region, including Dartmoor, is the most popular fieldwork destination in the UK by Earth Science disciplines from HE institutions. From a survey of all Geography, Geology and Earth Science Departments in the UK, taken in 1998, 30 per cent of the sixty-five returnees visited the South Devon area. This area has been used as one of the two main prototype regions for the Virtual Field Course (VFC) project. The granite tors that punctuate the horizon and slopes of the moors are one of the key geomorphological features of Dartmoor. These structures are a focus for many fieldwork activities including assessment of landscape evolution at a regional level, investigation into tor structural factors at a localised level and assessment of human impact on the tors and related areas. This area is used as an example to illustrate the work of the Virtual Field Course project in general with specific reference to one of the virtual reality software components used by students in conjunction with fieldwork.

The VFC is a three-year research and development initiative to investigate the potential for using virtual environments to aid in the teaching of fieldwork. Four elements can be shown as being the quintessence of this work: fieldwork, pedagogy, data and technology (Figure 14.2).

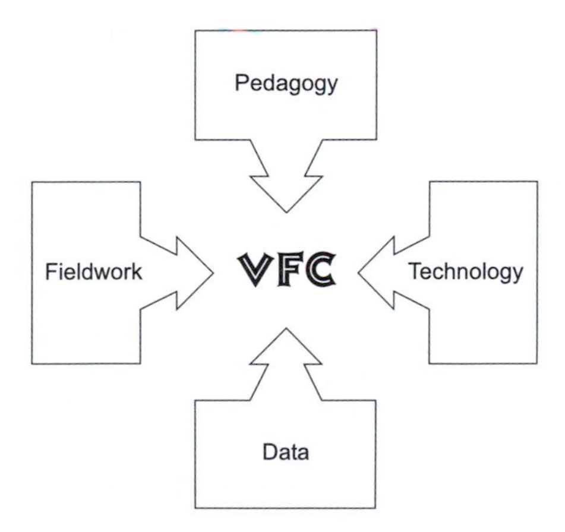

Figure 14.2 Elements of a Virtual Field Course.

First, fieldwork itself provides the motivation as it is a key part of many academic disciplines in the Higher Education sector and it is an effective, enjoyable form of both learning and teaching (Kent *et al.*, 1997). Fieldwork provides the opportunity to try new ways of course delivery and the VFC targeted using IT to support, rather than replace, fieldwork as the key aim of the project. Computers are increasingly being integrated into the general curriculum including field courses, but the fieldwork itself still remains of prime importance. Working in real-world locations provides learning experiences not achievable by classroom teaching alone.

Second, the pedagogic rational for fieldwork and the inclusion of IT in fieldwork has been carefully examined (Williams *et al.*, 1997). Working in the field is a stimulating experience but to be of true value it needs a sound educational framework. This is critical when introducing computer technology. The pedagogic aspects have been addressed by the provision of appropriate teaching and learning materials to enhance all fieldwork activity. Within the VFC project this has been realised primarily with the development of worked examples called Virtually Interesting Projects (VIPs). The VIP illustrates how VFC software may be used to enhance a specific educational theme in fieldwork. It is usually developed in conjunction with a teacher or expert, and the Tour of the Tors is one such project.

Third, data and information is used to enrich the fieldwork, whether it be the primary data collected by students as one of the prime motives for working in the field or secondary data provided to students by which they learn and evaluate their own observations. Data can take many forms: samples from the field, interviews, photographs, videos, text, census information, maps and virtual reality models. In disciplines that study real-world phenomena, human and physical geography, geology and other spatial sciences, a key element in understanding is developing a grasp of the spatial relationships of the subjects under study. Spatial location therefore becomes a central issue and the traditional use of maps in fieldwork underlines this importance. In response all VFC data carries spatial references and a generic metadata structure (Dykes *et al.*, 1999; VFC, 1999) has been developed whereby data may be accessed principally by its geographical location.

The final component of the VFC is the technological structure that is the vehicle for delivering the information to improve the academic potential of fieldwork. The VFC has developed an object-oriented approach to the technological structure and software design. This includes a series of user components that can access data from a distributed database via a central hub mechanism (Dykes *et al.*, 1999). Multimedia and virtual reality have been incorporated into this overall strategy. The VFC component traVelleR was developed to assess ways of integrating geo-referenced media and virtual reality, and designed as a teaching tool within the fieldwork context.

Using 'A Tour of the Tors', a VIP aimed to explore the location and development of tor features, this chapter illustrates how virtual reality, multimedia and geographic information have been linked to produce a flexible learning environment and how different teaching methodologies have been applied in a single software environment. From this it illustrates how students in a specific fieldwork situation can build on expert knowledge and their own empirical studies in the real and virtual environment and become tour leaders themselves.

Applying virtual environments to fieldwork projects

traVelleR

traVelleR is a virtual environment integrating virtual reality models, maps and media referenced to geographical locations (Figure 14.3). It has been designed to provide a spatial interface to information about fieldwork locations. It is a generic system around which a variety of field-based projects can be designed or supported. To meet the objective of using affordable and accessible software, equipment and data resources, the Virtual Reality Modelling Language (VRML) standard was considered as appropriate for development and also has the potential of delivering easily available software over the Web. traVelleR was developed using this

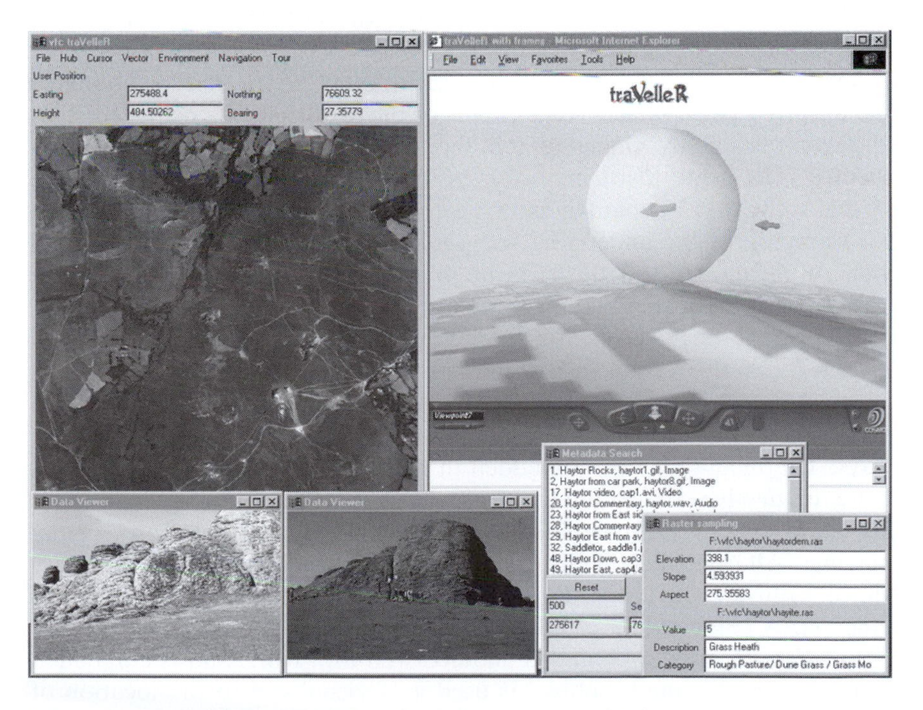

Figure 14.3 traVelleR, a virtual reality interface to spatial data.

technology in conjunction with Java (Javasoft, 1999) and the External Authoring Interface (EAI).

The software is designed to interface to the VFC geo-referenced database and provides a wide range of functionality to interrogate both spatial thematic data and spatially-referenced media. The generic nature of the software creates a shell in which to centre specific fieldwork projects tailored to individual requirements. The functionality of traVelleR has been developed with the aim of supporting many of the objectives of traditional fieldwork. It is also intended to enhance the interest and knowledge-building of students in aspects of fieldwork preparation and debriefing.

The first task in designing this software was to reference VRML models to real-world coordinates. Research into the optimisation of modelling geo-referenced terrains is described in Moore *et al.* (1999). When referenced to a common coordinate system the model can be dynamically linked to other objects such as a map to provide navigation cues, geographical thematic data or geo-referenced media data. However, the transformations when using virtual reality are complex and three-way transformations are required in this system. Real-world coordinates provide the key to the data but both screen and model coordinates and directions have to be interpolated. Once geo-referenced the linking of map and 3D views provides a degree of navigation not available by navigating the virtual scene alone. Movement through the scene may be by conventional mouse movement through the virtual scene or by jumping to locations indicated on the linked map.

traVelleR provides views of different realities so an understanding of the different geographical components of the scene may be achieved. This is accomplished by changing the thematic map drape on the terrain surface. The same maps may also be viewed in the map window. The virtual scene may be spatially interrogated at the click of the mouse button for 'samples' of the underlying geographical data. This last function provides a means to drill down through several data layers and may be applied as a surrogate to taking samples in the field.

Mapped scenes may also be viewed in conjunction with geo-referenced media. Again the terrain surface may be 'sampled' for media data within a given search radius consequently, the terrain surface is the interface to media data. Virtual reality and media may be combined in two different ways. The media may be embedded in the scene, providing a more complete, immersive context but currently this is inefficient as download times are slow due to the combined overhead of VR rendering and media playing. Alternatively media can be played in custom or proprietary viewers outside the scene. For video clips particularly, this proves more efficient. A combination of the two approaches has been found to be the most effective answer. Sound is handled well in VRML and is embedded in the scene. A sound 'bubble' is used as a visual cue to the location of commentary and is triggered either by user proximity or by touching the bubble. Video and HTML documents use proprietary viewers. Images are displayed in custom windows but may also be used within HTML pages.

One of the main features of traVelleR, used for the Tour of the Tors, is the facility to define annotated routes through a scene and play them back at a later date. The user first opens a dialogue to input information for the 'Tour'. The route is digitised on the 2D map and for each point along the route media objects may be added that illustrate some key features of that location. This tour may then be animated in the VR window with the appropriate media object shown or played at the appropriate point. At the simplest level a route may be digitised without supporting media. This is analogous to a GPS waypoint file and may be played back as such. The route may be followed at a constant height flying over the terrain or at surface level as though walking (although at much accelerated speed) through the landscape. This file may be saved as a GPS file. GPS waypoint files may be loaded to retrace actual routes taken in the field. VRML has the capacity to interpolate between each point using a PositionInterpolator node and a TimeSensor node that defines the total time taken for the path and key time intervals for each point along the route. Time is defined in the tour dialogue as travel time from the previous node and pause time to view the associated media. A PROTO node has been developed which contains the combined OrientationInterpolator, PositionInterpolator, TimeSensor and Viewpoint parameters (Figure 14.4).

From this initial work a VFC journal file structure has been produced (Figure 14.5) initially based on a simple text file. It has been further developed by Wood (Chapter 12) into a format compliant with the new eXtensible Markup Language, XML (World Wide Web Consortium, 1999) so foreseeing future development of meta-information of this type.

The Cook's Tours

The Cook's Tour is a term derived from the first public tour organised by Thomas Cook of Leicester. It refers to the time taken on most field courses to introduce students to an area they may otherwise be unfamiliar with when students are driven around a region with short stops at different localities accompanied by teachers pointing out features of the area. This form of fieldwork is often extremely passive and such lectures could equally well be done in a classroom environment.

Some of the physical limitations of the normal Cook's Tour may be overcome using a virtual tour. For instance different routes can be devised following a non-linear order other than that available by following local roads. Site visits may be ordered by temporal or subject context. A student-led exercise on field courses is often found to be of much greater academic value and it follows that if students devise and construct their own tour they get a greater involvement in the fieldwork activity (Kent *et al.*, 1997). The development of the tour journal is analogous to the

```
PROTO FlyBy [
        exposedField SFTime cycleInterval 1.0
        exposedField MFFloat positionKey []
        exposedField MFVec3f positionKeyValue []
        exposedField MFRotation orientationKeyValue []
        exposedField SFBool enabled FALSE
        exposedField SFTime startTime 0
        exposedField SFTime stopTime 0
        exposedField SFVec3f startPosition 0 0 0
        exposedField SFRotation startOrientation 0 0 1 0
        exposedField MFString media []
        exposedField SFInt32 mediaObjects 0
        exposedField SFString mediaFile ""

] {
        #the viewpoint node to be animated
        DEF ViewTransform Viewpoint {
                position IS startPosition
                orientation IS startOrientation
                description "Tour"
        }

        #The time sensor to control the speed of movement
        DEF Clock TimeSensor {
                enabled IS enabled
                startTime IS startTime
                stopTime IS stopTime
                cycleInterval IS cycleInterval
                loop TRUE
        }

        #The PositionInterpolator that stores key positions
        #or tour stops along the route
        DEF Path PositionInterpolator {
                key IS positionKey
                keyValue IS positionKeyValue
        }

        #The OrientationInterpolator that stores key view orientation
        # angles along the route
        DEF Direction OrientationInterpolator {

                key IS positionKey
                keyValue IS orientationKeyValue
        }

        ROUTE Clock.fraction_changed TO Path.set_fraction
        ROUTE Clock.fraction_changed TO Direction.set_fraction
        ROUTE Path.value_changed TO ViewTransform.set_position
        ROUTE Direction.value_changed TO ViewTransform.set_orientation
}
```

Figure 14.4 The Fly-by PROTO node.

fieldwork trail that can form a basis or aid to fieldwork and an extremely valuable transferable resource. 'Well-constructed fieldwork trails can capture something of the experience of seeing an area through the eyes of an expert' (Jenkins, 1997: 27). This concept of the guided tour and tour construction parallels the metaphors of the Sage (teacher) and the Storyteller (student) discussed by Cartwright and Hunter (1999) for enhancing

Journal file:
Tour name
Series of Tour Stops:
  Stop reference
  Location: northing easting height
  View orientation: degrees east of north
  Travel time to stop
  Pause time at stop
  Media:
    Filename
    ObjectType

TourName Tour of the Tors
  StopNum 1
  276214.44 78932.22 550.0 270.0
  TravelTime 0.0 *StopTime* 5.0
  Haytor1.jpg Image
  Haytor.wav Audio
  StopNum 2
  274830.0 78838.89 550.0 196.33604
  TravelTime 20.0 *StopTime* 5.0
  StopNum 3
  274651.12 77073.336 550.0 90.0
  TravelTime 20.0 *StopTime* 5.0
  Houndtor.wav Audio
  Houndtor.avi Animation
  HoundtorJoints.gif Image

Figure 14.5 Preliminary structure of a journal file.

geographical resources using multimedia. It also applies a constructivist approach to education that 'provides open-ended tasks and simulation exercises, which require students to play a more active role in hypothesis generation and problem-solving' (DiBiase, 1999: 92).

The 'Tour of the Tors' provides just one example of how a virtual field course can play several roles in the fieldwork experience. In this VIP a teacher constructs a Tour complete with spatial and multimedia data to lead the student through a surrogate 'lecture in the field' but with added value of combined real-world views and secondary data such as soils, geology or land cover maps. Students can therefore make comparisons between seen and unseen features in the landscape. This function is useful as a priming and familiarisation mechanism for work in the real field. Experts in specific disciplines, regions or topics may construct other pre-constructed tours.

The same software may be used independently by students to explore a region. This mode acts as a familiarisation process. When combined with visits and data collection at the real location students may be asked to make a critical comparison between mapped and surveyed data. They use the software in an exploratory visualization mode to discover and create information from the data. As an example students are asked to explore the virtual reality model of Dartmoor and gather information about the general regional geography: the geological features, soil, drainage, climate and vegetation cover. On the real Cook's Tour they are then able to spend more productive time taking localised samples and observations by which to verify or criticise the virtual fieldwork. In certain circumstances the virtual field trip may provide a better overview of the real landscape, particularly in British fieldwork locations such as Dartmoor where the actual landscape is often obliterated through extreme weather conditions such as fog.

A third mode of operation of a virtual field course is to enable students to enter their own data and construct their own tours of the region. This may function as a learning mode, a method of assessment or as data for use by other students in the future. The Virtual Field Course has developed a centralised database structure and comprehensive mechanism for indexing and retrieving spatially-referenced data from a shared database. This enables students to share data resources, not just in a passive interrogation mode, but by adding their own data to the database. The journal file extends this facility so each student can share their view, experience and interpretation of the fieldwork area with others. In essence s/he becomes a Storyteller by creating a structured thread or tour from a set of data.

A Tour of the Tors

The Dartmoor Tors

The features that best characterise the landscape of Dartmoor are the granite outcrops called tors. They hold a particular significance in the fieldwork associated with the area. As one of the literally outstanding features in the area they are often used as a focus for field studies. This is primarily in the study and analysis of landform and geomorphological processes. However, they also generate potential for studies of human impacts on the region as they act as focii for visitors to the area. The associated recreational implications of these and surrounding areas is of concern to Dartmoor National Park.

Gerrard (1978: 204) says, 'Granite landforms in general and those of south-west England in particular have attracted a great deal of attention.' Their distinctiveness in defining the landscape also makes them a key feature in virtual models of the region. They are easily recognisable and their spatial location and distribution is of importance. Therefore they make an excellent basis from which to explore the ways virtual environments can be used to aid studies of such features.

Several related properties of tors and the surrounding granite-formed landscape are available for study and debate. Various 'models' of tor formation have been proposed, the most comprehensive being those of Linton (1955) and Palmer and Nielson (1962). These ideas have formed the basis for most subsequent analysis and discussion. However, the suggestion that retreat of scarps produces tors with pediments at their bases (King, 1958) is also worth considering. The significance of Linton's paper has been assessed recently by Gerrard (1994). Linton proposed a two-stage model. First, deep chemical weathering under warm humid conditions produced a thick regolith with core stones occurring where joint planes were most widely spaced. Second, the products of weathering (the regolith) were removed by mass wasting processes to leave the sound granite and core stones as upstanding tors. The tors were probably exhumed under periglacial conditions during the Pleistocene. Periglacial action may also have modified the tors themselves. Thus, Linton defined a tor as 'a residual mass of bedrock produced below the surface level by a phase of profound rock rotting effected by groundwater and guided by joint systems, followed by a phase of mechanical stripping of the incoherent products of chemical action' (Linton, 1955: 476).

Following work on the 'gritstone' tors of the Pennines (Palmer and Radley, 1961), Palmer and Nielson (1962) developed a single-stage periglacial mechanism to explain the nature and evolution of Dartmoor tors. They suggested that the tors, plus the mass of boulders or clitter that mantle many Dartmoor slopes, were formed by the frost action and solifluction which occurred during the Pleistocene. The tors were 'upward projections of solid granite left behind when the surrounding bedrock was broken up by frost-action and removed by solifluction (Palmer and Nielson, 1962: 337). Therefore they might be termed 'palaeo-arctic' tors. These contrasting theories have been debated extensively and have provided much fruitful discussion and exercises for student field courses. Field projects can incorporate elements of measurement, observation and analysis and provide opportunity for related work in the field, in the laboratory and in the classroom. A more detailed review of the Dartmoor geomorphology and tor formation can be found in Campbell *et al.* (1998).

A Virtually Interesting Project

In the Tour of the Tors VIP (Figure 14.6) the aim of the project is for the students to construct their own hypotheses related to the structure and

VIRTUAL FIELD COURSE

vip

A Tour of the Tors

Author(s): John Gerrard, University of Birmingham
Kate Moore, University of Leicester

Date: January, 1999

Location: Dartmoor, UK

The South Devon and Dartmoor region is the most popular fieldwork destination in the UK by Earth Science disciplines from HE institutions. One of the key geomorphological features of Dartmoor are the granite tors that punctuate the horizon and slopes of the moors. These structures provide a focus for many fieldwork activities including assessment of landscape evolution at a regional level, investigation into tor structural factors at a localised level and assessment of human impact on the tors and related areas.

The 'Tour of the Tors' is an example of how characteristics of the landscape may be examined concurrently with multimedia representations such as video and spatially referenced sound. Students may either follow a structured animated 'Tour' under the guidance of an expert or pursue empirical investigations through freeform interrogation of the database. Students are asked to generate and test their own hypotheses on tor formation or tor characteristics. Use of VFC software can enhance student appreciation and understanding of the area, through inclusion and comparison of their own field observations, or as a fully integrated assessment exercise whereby students follow the predefined 'Tour' then identify and describe further similar sites within the region.

educational objectives

- To introduce students to the geomorphology, particularly tor formation, of the Dartmoor landscape
- To formulate hypotheses related to structure and evolution of the tors
- To develop the ability to measure and analyse tor characteristics through empirical study

technical requirements

Hardware: Pentium II processor 233MHz or above, 64MHz RAM, 3D graphics card preferable, Multimedia capabilities: sound, and video preferable

Software: VFC traVelleR, Media Player, Spreadsheet or Statistical analysis software

Figure 14.6 Details for the Tour of the Tors Virtually Interesting Project.

Documentation: VFC traVelleR user guide,

Data:
Coverages relating to a basic geography: height (DEM), topographic map, surface slope and aspect (derived from DEM), soil type , geology/rock type, land surface cover or land use.
Georeferenced multimedia data on the region and tors in particular.
Statistical data on tor joint spacings and orientations.

Field equipment: Tape measure, compass,

task definitions

In the Lab:
Students use the preconstructed tour by John Gerrard to gain a general familiarity with the landscape features: the granite tors.
Students formulate hypotheses that they wish to test and decide what data they need to collect in the field.

In the field:
Measurements are taken of joint spacings and orientation, grain size and structure at one or several sites

In the lab:
Distribution statistics are calculated from field data and histograms produced and added to the VFC database.
Through independent interrogation the shared database is used to verify the hypotheses.
Students finally use traVelleR to develop their own tour to illustrate their findings and present their conclusions. This may be used as a fieldwork assessment exercise.

time required

1 x 1hour sessions in lab (plus introductory 2 hour session if necessary)
1 x 2hour session post-fieldwork

1 day in the field

teacher preparation

Prepare DEM, and image maps for regions
Construct pre-defined introductory tour(s) of suitable sections of Dartmoor
Supervise data collection in field

potential problems

Weather on Dartmoor

assessment

A final tour conducted by students is used to present findings and conclusions for assessment.

other resources

None required

evolution of the tors and to use both the Virtual Field Course software and empirical study to evaluate the hypothesis. They start by studying background information to help formulate a hypothesis. Students use the tour preconstructed by the authors, with commentary by John Gerrard, to gain a general familiarity with the landscape feature themselves. The tour provides a fly-through of the Haytor Down area composed of travelling between a series of 'Tour Stops'. At each stop a series of media files are registered. As the user approaches, a click on the pointer in the VR scene will allow the set of media files to be shown or played. The tour has been devised to provide a general overview of the characteristic elements of the Dartmoor landscape as well as to provide information of the specific nature and location of tors. Photographs and maps cannot explore this interrelationship of features adequately. Many of the tors rise from gently convex summits, which have been variously interpreted as remnants of subaerial planation surfaces or possibly etchplains. The general nature of the landscape can be viewed admirably in the fly-through tour. The 'Tour Stops' have been devised to illustrate the variety of tor types and location. Gerrard (1974, 1978) has classified tors as (a) summit tors, (b) valley-side and spur tors; and (c) small emergent tors cropping out on the flanks of low convex hills. Each of these types is represented in the Haytor Down area. Detailed measurements of the characteristics of these groupings were made which included relationships between tors and terrain features such as slope and the spacing and orientation of the joints that are a feature of tors. Both summit and valley-side tors possess relatively closely-spaced vertical joints whereas those of the emergent tors are much more widely spaced. A very similar classification of tors was produced by Ehlen (1994). She demonstrated significant differences between the groups. Summit tors possessed high relative relief; the rock was closely crystalline and possessed the widest joint spacing. Summit tors appear to be controlled by vertical joints or by vertical joints and horizontal joints combined. Spur tors possessed narrower vertical joint spacing and horizontal joint spacing was intermediate. Rocks on spur tors were more finely grained. Valley-side tors possessed narrow joint spacing and the rocks were also finely grained. The Tour of the Tors will enable students to assess these ideas as it provides a brief introduction to the area and features being studied. Students may then explore other data in the database. At this point students formulate hypotheses that they wish to test and decide what data they need to collect in the field.

The hypotheses may be based on previous work or on ideas generated by the flyover. Hypotheses can be based on data in the system that need testing on other tors or in other areas; examples could be 'Is Haytor a typical tor?' or 'Are relationships established from the database applicable to other parts of Dartmoor?' It is often difficult to formulate specific hypotheses without preliminary field data. The feasibility of hypotheses can be assessed from the flyover and the associated database. A number of operational issues may be resolved with the help of traVelleR. Prior knowledge is usually required to produce a rigorous and efficient sampling and measurement methodology. The data available plus the views of the tors can be used to assess issues such as how many joints need to be measured and how the measurements are to be obtained. Operational problems, such as what is a joint, can be resolved in virtual reality.

After this introduction in the laboratory, groups go out into the field and undertake practical work. Measurements are taken of joint spacings and orientation, as in an assessment of grain size and structure for different tors at one or a variety of sites. On return to the field base or lab these values are entered into a statistical analysis package or spreadsheet. Distribution statistics are calculated and supporting graphics such as histograms generated. This part of the exercise follows the same structure as any standard field exercise but the findings may be entered into the VFC database. Students are asked to develop their own hypotheses on the formation and development of the tors. Then, through independent interrogation of the database using the traVelleR software together with taking their own observations and measurements in the field, they are able to verify their hypotheses. Secondary data such as slope, aspect, geology and climatic data can also be used. The shared database can enrich the available experience of students by showing examples of other sites not accessible or restricted.

Finally, they use the software to develop their own tour with accompanying commentary or HTML documents that illustrate their findings and present their conclusions. This use of the VFC as a fieldwork assessment exercise empowers the student to develop their own 'lecture in the field' that may be used in subsequent years by future fieldwork students.

Conclusions

The ability to navigate and travel freely around virtual scenes provides the means to visualize alternative realities and gain knowledge not always available from fieldwork alone. The integration of different media types, virtual reality and geographical information provides multiple representations that are a useful basis for visualization. Using virtual environments in conjunction with real fieldwork extends students' engagement in project work and provides a more active, dynamic mechanism for the follow-up to data collection – the analytical and presentation stages. However, it is also possible for staff and students to share their personal visualization experience and knowledge gained by constructing simulated tours of a region. In this way the Tours of the Tor is one example of how the construction of illustrative projects such as this provides a resource for future users of the Virtual Field Course.

References

Campbell, S., Gerrard, A.J. and Ghreen, C.P. 1998. Granite landforms and weathering products. In Campbell, S., Hunt, C.O., Scourse, J.D. and Keen, D.H. (eds) *Quaternary of South-West England*, Geological Conservation Review Series. London: Chapman and Hall, pp. 73–90.

Cartwright, W. and Hunter, G. 1999. Enhancing geographical information resources with multimedia. In Cartwright, W., Peterson, M.P. and Gartner, G., (eds) *Multimedia Cartography*. Berlin: Springer-Verlag, pp. 257–70.

Clark, R.E. 1985. Evidence for confounding in computer-based instruction studies: analysing the meta-analyses. *Educational Communication and Technology Journal*, 33, 249–62.

CLUES. 1999. Centre for Computer Based Learning in Land Use and Environmental Sciences. http://www.clues.abdn.ac.uk:8080/headpage.html.

DiBiase, D. 1999. Evoking the visualization experience in computer-assisted geographic education. In Camara, A.S. and Raper, J. (eds) *Spatial Multimedia and Virtual Reality*. London: Taylor and Francis, pp. 89–101.

Dykes, J.A. 2000. An Approach to Virtual Environments for Visualization Using Linked Geo-Referenced Panoramic Imagery, *Computers, Environment and Urban Systems*, 24 pp. 127–52.

Dykes, J., Moore, K. and Wood, J. 1999. Virtual environments for student fieldwork using networked components. *International Journal of Geographical Information Science*, 13, 4, 397–416.

Ehlen, J. 1994. Classification of Dartmoor Tors. In Robinson, D.H. and Williams, R.B.G. (eds) *Rock Weathering and Landform Evolution*. Chichester: J. Wiley and Sons, pp. 393–412.

Freundshuh, S.M. and Helleviks, W. 1999. Multimedia technology in cartography and geographic education. In Cartwright, W., Peterson, M.P. and Gartner, G. (eds) *Multimedia Cartography*. Berlin: Springer-Verlag, pp. 271–80.

Gerrard, A.J. 1974. The geomorphological importance of jointing in the Dartmoor granite. In Brown, E.H. and Waters, R.S. (eds) *Progress in Geomorphology, Institute of British Geographers Special Publication*, 7, 39–51.

Gerrard, A.J. 1978. Tors and granite landforms of Dartmoor and eastern Bodmin Moor. *Proceedings of the Ussher Society*, 4, 204–10.

Gerrard, A.J. 1994. Classics in physical geography revisited: Linton, D.L., 1955, The problem of Tors. *Progress in Physical Geography*, 18, 559–63.

JavaSoft. 1999. Sun Microsystems, http://java.sun.com/.

Jenkins, A. 1997. *Teaching More Students 9, Fieldwork With More Students*, The Oxford Centre for Staff Development, p. 27.

Kalawski, R. 1995. Keynote speech to The Potential of VR for UK Higher Education – AGOCG Workshop Report http://www.man.ac.uk/MVC/SIMA/vr_wshop/forward.html.

Kent, M., Gilbertson, D.D. and Hunt, C.O. 1997. Fieldwork in geography teaching: a critical review of the literature and approaches. *Journal of Geography in Higher Education*, 21, 3, 313–30.

King, L. 1958. Correspondence: The problem of tors. *Geographical Journal*, 124, 289–91.

Laurillard, D. 1996. Learning formal representations through multimedia. In Entwistle, N., Hounsell, D. and Marton, F. (eds) *The Experience of Learning*, 2nd edition, Scottish Academic Press, pp. 172–83.

Linton, D.I. 1955. The problem of tors. *Geographical Journal*, 121, 470–87.

Mikropoulos, T.A. 1996. Virtual geography. *VR in the Schools*, 2, 3, 3.

Moore, K. 1999. VRML and Java for interactive 3D cartography. In Cartwright, W., Peterson, M.P. and Gartner, G. (eds) *Multimedia Cartography*. Berlin: Springer-Verlag, pp. 205–16.

Moore, K., Dykes, J. and Wood, J. 1999. Using Java to interact with geo-referenced VRML within a Virtual Field Course. *Computers and Geosciences.*

Palmer, J. and Nielson, R.A. 1962. The origin of granite tors on Dartmoor, Devonshire. *Proceedings of the Yorkshire Geological Society*, 33, 315–40.

Palmer, J. and Radley, J. 1961. Gritstone tors of the English Pennines. *Zeitschrift für Geomorphologie*, 5, 37–52.

Purves, R.S., Mackaness, W.A. and Dowers, S.D. 1998. Teaching mountain navigation skills using interactive visualization techniques. In Hauska, H. (ed.) *Visual Reality. Proceedings of the International Conference on Multimedia in Geoinformation*, Bonn, Germany, pp. 173–83.

Reddy, M., Leclerc, Y., Iverson, L. and Bletter, N. 1999. TerraVision II: visualizing massive terrain databases in VRML. *IEEE Computer Graphics and Applications*, 19, 2, 30–8.

VFC. 1999. The Virtual Field Course. Department of Geography, University of Leicester, http://www.geog.le.ac.uk/vfc/.

The Web3D Consortium. 1999. http://www.web3d.org/.

Williams, N., Jenkins, A., Unwin, D., Raper, J., McCarthy, T., Fisher, P., Wood, J., Moore, K. and Dykes, J. 1997. What should be the educational functions of a virtual field course. In *Proceedings of Cal '97, Exeter, UK.* Exeter: University of Exeter, School of Education, pp. 351–4.

World Wide Web Consortium. 1999. Extensible Markup Language. http://www.w3.org/XML/.

Zeltzer, D. 1992. Autonomy, interaction and presence. *Presence*, 1, 127–32.

Part III

Virtual cities

15 Introduction

Michael Batty, David Fairbairn, Cliff Ogleby, Kate E. Moore and George Taylor

Defining virtual cities and built environments II

The landscape metaphor central to the Part is less appropriate to urban environments that are largely composed of built forms. In digital terms, built environments lie at the intersection between geography and geometry. Their definition is largely one involving the three-dimensional (3D) representation of buildings and streets where the underlying two-dimensional (2D) map structure is usually implicit rather than explicit. What geography brings to these virtual realities is an explicit link between the 2D map and the 3D form. Currently this relationship is best seen in the way GIS software is beginning to incorporate the third dimension, notwithstanding the problems posed by different data structures that dominate the 2D and 3D worlds discussed in earlier chapters in this book. The types of environment that form the material of this section of the book are admirably illustrated by the numerous contemporary digital realisations of complex urban scenes, one of which – population density in Westminster in central London – is illustrated in Figure 15.1.

The historical antecedents which have led to such virtual urban environments are rich and varied. Four themes can be distinguished:

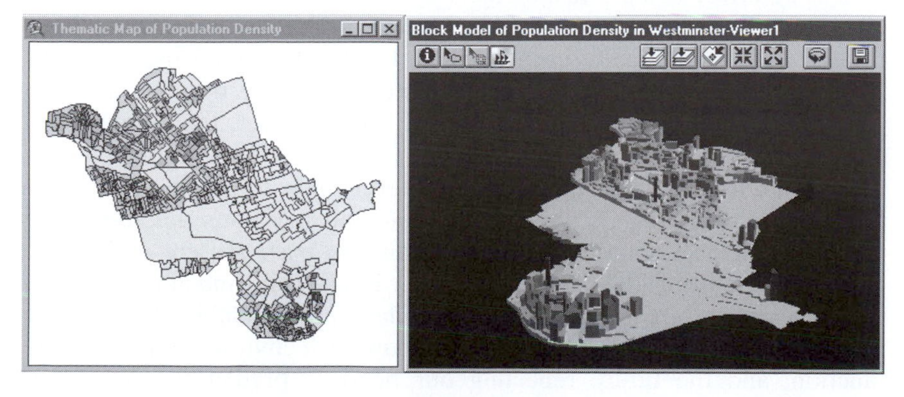

Figure 15.1 Virtual Westminster: a 3D block model from a 2D thematic map developed using ArcView 3.1.

computer-aided design (CAD), cartography, computer games and simulation. CAD models represent the heartland of 3D built environments with computer-aided architectural design and drafting systems providing much of the momentum for their development. Although such systems still dominate the production of 3D models of cities through which users can walk and fly, there are still very few links back to the 2D world. This is in contrast to the current effort in GIS to move forward towards the 3D world as embodied in the various extensions to GIS packages such as the *3-D Analyst* extension designed for the desktop system *ArcView* which was used in Figure 15.1 representing Virtual Westminster.

Cartography has provided many of the concepts that give designers the ability to scale, abstract, generalise and colour 2D maps that can be transposed to 3D scenes (Moore, 1999). Nevertheless, it has been the role of computer games that has been the impetus for the development of rapid viewing of 3D graphics, user interactivity and level-of-detail rendering in the development of virtual environments that has made 3D mapping feasible. Lastly, although the scientific simulation of urban processes such as traffic flow, location modelling and energy consumption is something that is only now being considered as essential to 3D built environments, simulation in general terms has provided considerable momentum to such developments. This is seen particularly in games, for example *SimCity*, but also through the general development of flight simulators and industrial process engineering. More recently, all these developments which lie at the heart of virtual urban modelling have been accelerated by the easy availability of high-speed computer graphics, the ability to share data over the Internet, and the interaction with data via the World Wide Web and other multimedia interfaces. To provide some perspective on these antecedents, this heritage is illustrated using the time line in Figure 15.2.

Virtual cities portray urban areas usually composed of 3D scenes through which users can move and interact. In addition, there is now increasing effort being put into the way we can represent how such 3D environments change through time. The environments which have been created have traditionally been fashioned for professional use by architects, urban designers and planners. However, as they become easier to generate on the desktop and disseminate over the net, the emphasis on exclusively expert use is beginning to change in favour of more popular use for purposes of learning about and interacting within such urban environments. In contrast to this array of actual and potential users, the particular application of such tools pertains to past, present and future urban environments: the past reflecting our interests in, and abilities to, re-create historical scenes from whatever data are available, the present reflecting our interests in learning about how such environments currently function, and the future reflecting our need to predict the impact of planned changes on the built environment before such change is actually implemented. The techniques and technologies involved in such virtual

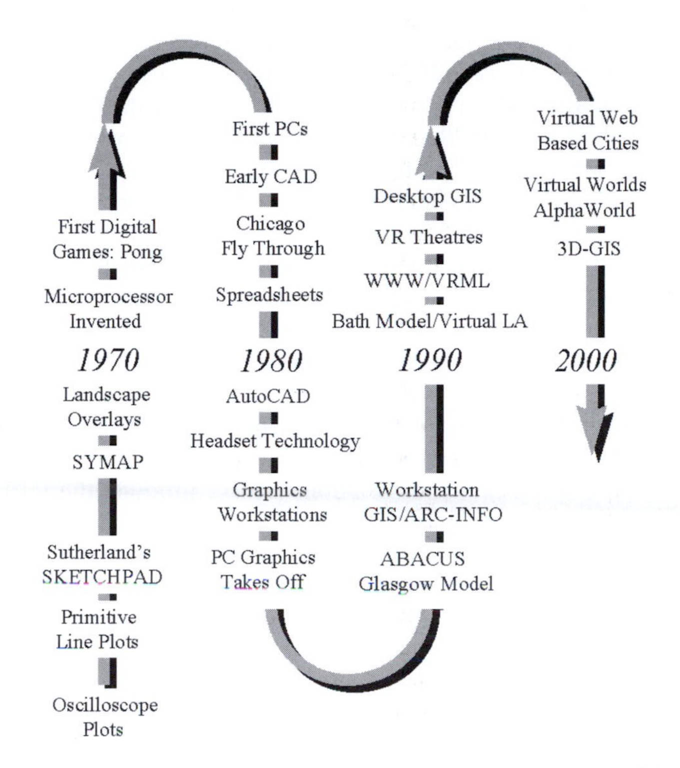

Figure 15.2 A time line for the development of virtual built environments.

realities facilitate user interaction in many ways, and provide rapid feedback on choices and decisions that users can make as they passively navigate or actively design such environments within the context of their applications.

Contemporary applications

This part of the book attempts to provide some examples of the current range of applications to the built environment. Urban design, or architecture-writ-large, is a major use although the transition from conceptual 3D design to subsequent 2D planning is significant, particularly in terms of what geographical science is bringing to the construction of these virtual cities, notably from GIS. Communication rather than analysis and design is an important driver of such models especially for public participation and community involvement in the planning process. In contrast, there is increasing use of such models for entertainment (as in the traditional computer games market), and also for more serious use in providing information about urban environments for tourism. The emphasis on the way users interact with such environments – from aspects such as their navigation through such scenes to ways in which they employ the user-interface to such products – provides an important research focus explicit in the chapters in this section, as well as throughout the book. The use of virtual realities in simulating transport (and movement generally) in cities, in the delivery of public services through interfaces which contain such models or variants thereof, and for managing routine processes which make cities work are all activities that are finding an expression in these applications. Some examples will make this clear.

Virtual cities probably represent the most glamorous of such applications. CAD models, often constructed on the desktop, can easily be translated into environments which users can access immersively through head-mounted displays, for example in CAVES and VR Theatres, as well as across the net using VRML browsers. The models constructed for Glasgow (Grant, 1993) and Bath (Day, 1994) show how realistic the rendering of such scenes can become. The construction of these models has invariably been motivated by practical and topical issues of public policy, although the form of the environments developed is rather superficial in that the urban processes, which drive the development of the city are largely absent. The emphasis here has been largely on representation rather than explanation, analysis or prediction. In contrast, the model built for Los Angeles by Ligget, Jepson, and Friedman (1996) has been constructed from the 2D map, using ARC-INFO, adopting and engaging much of the functionality that now exists within this package. A veritable cornucopia of such models now exist, particularly on the Web, and many fine examples can be accessed from the Planet 9 site, for example (www.planet9.com). An illustration of one of these – Virtual London – is shown in Figure 15.3.

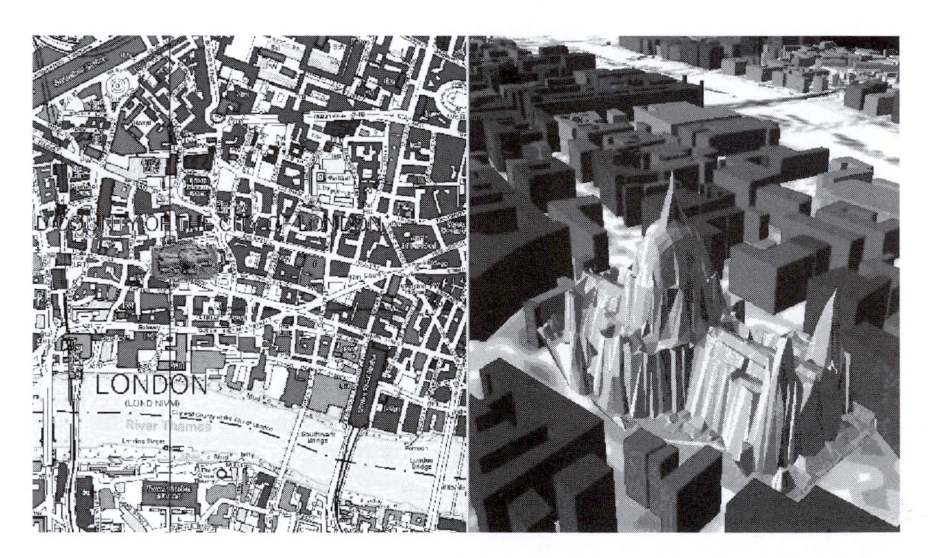

Figure 15.3 Flying over St Paul's Cathedral in the 'City' section of Virtual London.

Websites which display information about cities for purposes ranging from advertising to the delivery of services are also significant. Some of these scarcely fall into our implicit definition of virtual cities although many of them have elements of visualization which draw on developments in CAD and VR. Good examples of these are the websites for Bristol and Bologna, listed as part of a lengthy catalogue (Aurigi and Graham, 1998). Further, there are now some significant reconstructions of past urban environments in both the ancient and modern worlds. The classic examples, such as Lascaux and Stonehenge, are hardly urban although rather good examples exist for cities such as Rome (Davis, 1997). Virtual environments for tourism are appearing daily on the Web. Virtual campus maps and navigation aids are widespread, while city websites often incorporate simple interactive scenes linked to maps which make use of a variety of multimedia and digital photography to produce integrated walk-throughs: this enables visitors to get some real sense of how such urban environments function and what is contained within (Shiffer, 1992). Wired Whitehall, for example, enables users to walk around the most significant sites in central London from Buckingham Palace through Trafalgar Square to Covent Garden (Dodge *et al.*, 1997; Smith, 1997). The London Environmental Information System allows users to view a multiplicity of information sources on London's environment (Figure 15.4). Increasingly these kinds of VR are being used on the Web to enable electronic commerce to take place. CompuServe's site, Shopping City, is built around the VR software Superscape which enables users to enter particular streets which contain particular shops and to browse the wares and other information relevant to a virtual shopping trip. It is only a matter of time before these kinds of interface become directly connected to online purchasing.

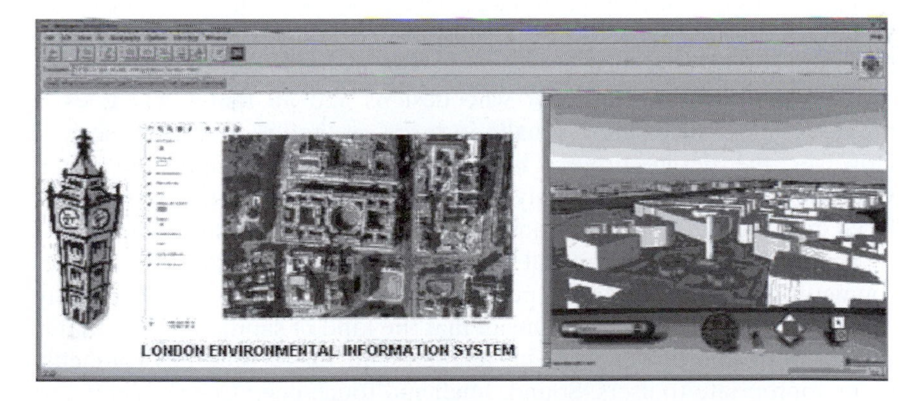

Figure 15.4 Delivering environmental data, multimedia and policy advice to users structured around a multimedia Internet GIS through the Web.

The key challenges

Although there are now some powerful visualizations of urban environments which are accessible as virtual realities, there are many problems with their development and there is an enormous research agenda. Such visualizations are still dominated by what the technology offers. They are still driven by an often exclusive emphasis on representation, with little thought being given to how such environments can be linked to the more abstract data and analysis so important to their use in planning and design. Very often, all these environments offer is a superficial set of objects through which users can navigate. Rarely can users access relevant information about the attributes of these objects. The sort of link between geometrical and geographic forms and their attributes that is central to GIS, for example, rarely exists in CAD and one of the biggest challenges of VR in geography is to impress and develop such functionality.

An extension of this need to populate digital urban environments with attributes and objects of serious importance is the requirement to develop appropriate processes of urban change and urban functioning which lie behind these representations. If such visualizations are to be useful analytically and also predictively, the methods and models needed to make sense of urban change must be embedded within them. The same problem has faced GIS. Building the requisite functionality, not only in terms of urban applications but more generically, into systems that are largely motivated by good digital representation is an issue that has faced this field over the last decade and now faces the development of VR. Part of the need to develop good models of processes also relates to the requirement to develop much more effective representations of the way users can interact with such models. The processes relevant to the way such visualization might be used must be researched and incorporated more effectively into such applications. For example, most CAD simply assumes that it is the designer who designs and all that CAD does is provide some visualization, usually in terms of 'before and after' scenes. However, the design process is considerably more complex in that a variety of tools are used in analysis and design which all need to be interfaced either directly, or more loosely coupled, with such visualizations.

There is also an important set of problems relating to making such environments more relevant to the purposes at hand. This does not necessarily mean more realism but it does mean that the kind of sanitised scene that is the norm in this field should be tempered to make such environments more appropriate to users. Sound, smell and touch need to be engendered within such environments, where appropriate, while better and more effective rendering is always on the agenda. However, one of the biggest challenges is to populate such environments with objects that increase the sense of realism. For example, although the problems of an external user or users interacting with such environments have been attacked quite effectively, passive users – people – existing in such scenes remain problematic. A criticism often made of architects' sketch drawings are that 'beautiful people' are positioned in unrealistic positions to show off the designs to the best effect. In CAD and VR, such environments are often bereft of people and users interacting with such scenes have the environment to themselves as they navigate through. The 'ghost town' image of virtual cities must be tackled.

This emphasis on better representation always leads to the need for deeper structures within such models. Although there is some debate about whether virtual environments should be built around data structures geared more to the 3D objects than the 2D map, there is the more basic problem of researching how processes of motion and of change can be represented and simulated within such scenes. What is needed is a shift from the emphasis on presentation and communication per se to one which gives much more weight to questions of simulation and the incorporation of process. This, in turn, would provide much greater realism, would engender greater acceptance and would ensure greater applicability of such techniques. It is the challenge of the next decade and it is a focus to which geographical science is uniquely able to respond.

The chapters

The four chapters presented here provide an overview of some of the current developments and applications of virtual reality, GIS, and simulation of the built environment. By no means do they address all the issues that have been raised in this introduction but they identify a number of important themes which are summarised in this section.

In their overview of the techniques of measurement science which can be appropriately used to obtain accurate data for virtual urban environment construction, Fairbairn and Taylor (Chapter 16) discuss the impact of traditional surveying and photogrammetric methods and note the important contemporary developments in such procedures. Increasing ease of use, speed of data capture and integration within a virtual-worlds flowline are characteristics of such methods. In addition, data capture using techniques such as the Global Positioning System (GPS) and non-conventional imaging are discussed. Further possibilities of accurate geometric and attribute data collection rely on methods outside the realms of experience of geomatics scientists. However, such methods are briefly addressed, along with views on the future of automated and semi-automated spatial data capture. The overview of methods is supported by example applications and projects which have (or may) lead to the successful creation of accurate and detailed virtual urban worlds. Ogleby (Chapter 17) in his chapter focuses on virtual cities of the past. He uses a variety of techniques to re-create the ancient capital of Thailand, including geomatic data collection techniques, historical data, narrative, live actors and architectural analogy.

Moore (Chapter 18) discusses mapping the least tangible components of the urban environment as a counterpoint to creating models of the 'realistic' urban scene in fine detail. Just as 2D cartography encompasses both topographic and thematic elements, visualization in virtual reality should include symbolic representations to enhance understanding of hidden geographies in the multidimensional context. The example software demonstrated here illustrates the dynamic creation of a 3D urban database. As Gillings *et al.* (Chapter 2) identify, the user and map creator roles can be united and the map user becomes an integral part of the map or urban scene that s/he is creating and navigating. This chapter therefore focuses on two themes: issues of cartographic design in a 3D urban context; and the development of virtual reality as a dynamic, interactive and engaging tool for visualizing spatial data.

The last chapter, by Batty and Smith (Chapter 19), brings these ideas full circle, back to basic ideas about virtual cities. They show how different conceptions of virtuality have been developed for cities ranging from Web pages through to virtual worlds. This chapter also charts a different but related development, reviewing how information technology is being used to wire cities, but as is now ingrained on our contemporary psyche, hardware defers to software in IT, and the hardware which constitutes wired cities – intelligent buildings, highways, concentrations of high band width optical fibre and so on – is gradually being dominated by software applications which embody the real applications of virtual cities (Batty, 1997a). There is a connection here to other geographies too. The new economic geography of cities is now focused around explaining how the local economy is becoming digital and how locational advantage and disadvantage are changing in a global world. Wired cities have much to say in such explanations. There are also links here to cultural issues for the development of cyberspace and the emergence of virtual communities embody new experiences of space and virtual space, and have important implications for social interaction. What Batty and Smith seek to show is how visual media and its current incarnation in terms of virtual realities are so essential to the portrayal and appreciation of the digital city.

References

Aurigi, A. and Graham, S. 1998. Virtual cities and the 'crisis' in the urban public realm. In Loader, B. (ed.) *Cyberspace Divide.* London: Routledge.

Batty, M. 1997. The computable city. *International Planning Studies*, 2, 154–73.

Davis, B. 1997. The future of the past. *Scientific American*, 278, 8, available at http://www.sciam.com/0897issue/0897review1.html.

Day, A. 1994. New tools for urban design. *Urban Design Quarterly*, July, 20–3.

Dodge, M., Smith, A. and Doyle, S. 1997. Urban science. *GIS Europe*, 6, 26–9.

Grant, M. 1993. Urban GIS: the application of information technologies to urban management. In Powell, J.A. and Day, R. (eds) *Informing Technologies for Construction, Civil Engineering and Transport*. Uxbridge, UK: Brunel SERC, pp. 195–9.

Ligget, R.S., Jepson, W.H. and Friedman, S. 1996. Virtual modeling of urban environments. *Presence*, 5, 72–86.

Moore, K. 1999. VRML and Java for interactive 3D cartography. In Cartwright, W., Peterson, M. and Gartner, G. (eds) *Multimedia Cartography*, New York: Springer-Verlag, forthcoming.

Shiffer, M.J. 1992. Towards a collaborative planning system. *Environment and Planning B: Planning and Design*, 19, 709–22.

Smith, A. 1997. Realism in modelling the built environment using the World Wide Web. *Habitat*, 4, 17–18.

16 Data collection issues in virtual reality for urban geographical representation and modelling

David Fairbairn and George Taylor

Introduction and context

Batty *et al.* (Chapter 15) have indicated that cartography and mapping have played, and still play, a major role in the development of urban models for virtual reality. The contribution of these activities extends from initial data capture by measurement of the real world through to efficient rendering of VR images by presenting spatial data using guidelines from cartographic practice.

The particular aim of this chapter is to address issues which pertain to data capture for the purposes of virtual world creation. The emphasis is on large-scale and high-resolution data of the type needed to create virtual city environments. An assessment is made of the data collection options available and the utility of each method, with illustrative examples showing the wide range of possible solutions to the problems of integrating urban data for VR applications.

As Part II of this book has indicated, much VR usage in geography is directed at the handling of small-scale data of natural environments, typically obtained from remote sensing imagery or digital topographic datasets. In the modelling of urban environments, however, different spatial data collection techniques can be employed, using, for example, ground survey, the Global Positioning System (GPS) and photogrammetry. Employment of such methods and subsequent use of the data obtained requires knowledge of the technical limitations of certain technologies (e.g. GPS), the reference framework within which the data is collected (including aspects such as map grids and projection), the desirable levels of resolution and accuracy involved, the three-dimensional nature of the city-scape (including underground services), the flowline from field to virtual world and the level of abstraction required. In all cases it is important, due to the wide range of potential methods which can be employed, to have a firm knowledge of the manner in which the data will be stored and the modelling and manipulation to which it will be subject.

Urban environments are invariably complex, and they can be represented by a complex range of data. Tempfli (1998) discusses the separation of the data required into positional information, information on extent (including shape and size) and topological information. Each of these categories has application to the fundamental components of the virtual city, which are the buildings and terrain faces which make up the model. The spatial primitives which will be captured represent inflexion and corner points, boundary lines and partitioning faces. Together, these boundary representations can be used to build the objects in the urban model. The assumption that discrete features will be collected, in terms of position, attribute and topological relationship, makes the tasks of data capture described in this chapter straightforward. In most cases positional information related to specific points is the crucial element: such points are connected to create arcs, which can be further manipulated to obtain planar faces. Topological information can be attached to these geometric primitives, as can semantic and attribute information (Gruen and Wang, 1998). Even continuous surfaces, such as the ground surface, can be captured efficiently and manipulated using contemporary terrain data-handling techniques.

Beyond these seemingly uncomplicated data capture tasks, we must appreciate the requirements imposed by the variety of applications for such VR representations (Tempfli, 1998). There is a need to address the level of resolution at which the data is collected, and the possible need to have 'multi-scale' representations. The city may be a complex model at one scale, but merely a point location at a smaller scale. Certain spatial analysis techniques to which the VR model may be subject will require complex topological coding: for example, enquiring whether a pipeline passes underneath a building may be a difficult query. Further, the maintenance and updating of the urban VR model must not be overlooked: the data model must be flexible enough to cope with future change.

For each data capture exercise, it is important to determine the data model which is being applied, the data structures which will be used and the integration of measured, coded, interpolated and relational data which is necessary. Only then can a realistic assessment be made of the data to be measured and the subsequent manipulation to be applied.

Data capture technologies

Figure 16.1 (Edwards, 1991) illustrates a conjectured scenario for a future urban utilities engineer. This prescient image was created a decade ago, yet indicates some of the contemporary aspects to be taken into account by those designing and implementing urban VR applications: the availability of hidden sub-surface data, the use of a positioning device, the integration of standard map data, and the development and linking of attribute databases.

In order to achieve this, and similar, types of large-scale, urban VR applications, the collection of raw positional and measurement data 'in the field' must be undertaken. An increasingly wide range of potential

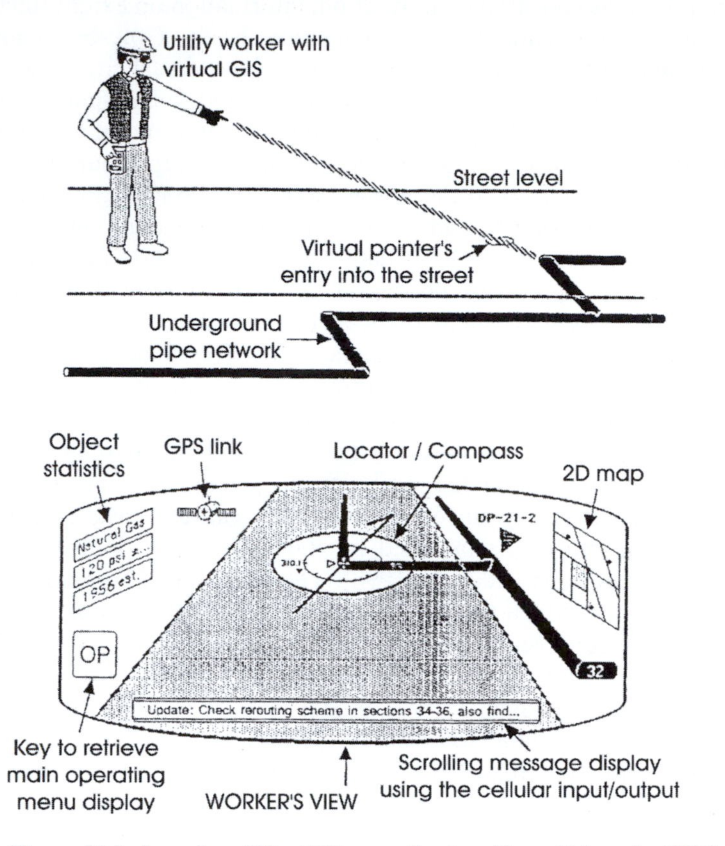

Figure 16.1 An urban VR utilities application (from Edwards, 1991)

technologies can be applied to such data collection tasks. Traditionally, such data observation and measurement has been the realm of geomatic scientists and engineers who have experience in the methodologies and can best appreciate the nature, quantity and accuracy of the data needing to be captured. In recent years, however, the possible scope of data collection procedures has widened to include sensors which are not familiar to geomaticians, such as ground penetrating radar, atmospheric monitors and land surface radiometers, and the ease of use of all technologies has improved such that non-specialists are capable of obtaining adequate data in a straightforward manner.

This section details the standard techniques available to a virtual-world creator for supplying spatial data to a world-building project. Such techniques can involve both fieldwork and office work, can require further enhancement to varying degrees, and can contrast considerably in cost, in level of accuracy and resolution, and in time taken for data capture.

GPS (and its Russian equivalent, GLONASS) is a simple and increasingly widely-used data capture method. Relying on a network of radio signal-emitting, earth-orbiting satellites whose exact position is known or predictable, an earth-bound observer can use a range of signal-receiving devices to record distances from the orbiting vehicles. Such receivers vary in cost (from approximately $100 to over $100,000), in reliability, and most importantly for urban data capture, in accuracy. Positional accuracy will range from 100 m up to 5 m for low-cost systems and from 0.05 to 0.01 m for high-cost systems. Sub-centimetre positioning is possible with sophisticated post-observation processing techniques. The method of GPS relies on an accurate knowledge of the time at which a signal was emitted by a satellite and an equivalent accuracy in determining the time the signal is received. Expensive receivers have more accurate clocks, and are able to receive signals from more satellites (which improves accuracy). In addition, they have access to radio communications technology which can be used for improving accuracy.

GPS is inherently a three-dimensional system, although it should be noted that the precision of positioning in the height/depth (z) dimension is considerably lower than that in eastings (x) and northings (y). It is possible to capture ground detail by walking along urban streets, and to establish building height by observing GPS from rooftops. In addition, terrain modelling, indicating the relative heights of the urban ground landscape, can be undertaken using GPS-sourced data.

As an increasingly 'black-box' technology, the operation of GPS observation is simple and the digital flowline from receiver to digital map data file (and graphical representation of this on screen) is straightforward. Data capture in the field can be achieved by 'field digitising' – walking or motoring around the urban landscape tracing a route, outlining a building 'footprint', determining locations of street furniture and assessing the vertical nature of the city environment by accessing rooftops. In many cases extremely 'high volume' data capture can be achieved with up to ten discrete points per second capable of being recorded. Such techniques can be as easily applied to extensive areas, for example in digitising the road network of a complete urban area, as to small land parcels, for example a building development site; and the surveying can be undertaken in all weather conditions and at any time of the day.

However, this description belies the fact that in many circumstances, particularly in densely-detailed urban areas, observation of GPS signals is fraught with difficulty and, further, that caution must be exercised in the use of unadjusted GPS data, which may exhibit considerable inaccuracy and inability to be merged effectively with other spatial data. First, it is important to realise that the city environment has a considerable impact on the signals passing from satellite to receiver. GPS signals cannot be sensed when tall buildings, or indeed significant tree cover, interrupt the signal path. In the 'urban canyons' of many contemporary cities, signals are completely blocked: this is clearly also the case in tunnels, in underground facilities and inside buildings. Occupying the corner of a building to ascertain its position may prove impossible, as up to one half of the open sky may be obscured and many satellites made invisible. When only a few satellites are visible, the accuracy of positioning deteriorates, and if the number of visible satellites falls below four, then no position can be determined. Complications arise when signals can be received, but only after being reflected by hard or metallic surfaces in the urban environment. The reception of such indirect signals is called 'multi-pathing', and there is no technique for determining whether a signal does come directly from the satellite or whether it has bounced off an adjacent wire fence, concrete building or road surface.

Inaccuracies exist in GPS as a result of deliberate degradation of the signal from satellites for civilian users. Selective availability (SA) effectively introduces an error in the clock of the satellite, although this signal degeneration has recently been discontinued. Other shortcomings in accurate positioning occur as a result of the effects of the atmosphere and ionosphere on the satellite signal. Such effects can be minimised by simultaneously observing the satellites from two separate locations – the point to be determined, and a reference base station at a known position. The technique of differential GPS observation relies on such a configuration of fixed reference station and roving receiver. Occasionally, the data from the fixed station is broadcast by radio such that real-time corrections can be applied to the receiver as it collects positional data. Increasingly, base stations are operated by commercial organisations which can broadcast corrections on standard radio frequencies or provide world-wide coverage using satellite broadcasts.

In all cases, the GPS receiver (and its human operator) must visit the points to be positioned: GPS is a direct measurement technique. A further disadvantage is that once the points have been determined, they may not be compatible with data from other sources, for example from national topographic map series or from remote sensing imagery. The conversion of GPS derived positions to a fixed national coordinate system is not a trivial task: in Great Britain (GB), the national mapping agency, Ordnance Survey, has only recently (1998) released transformations which can be used to accurately match, to a precision of 20 cm, GPS coordinates with National Grid (GB) coordinates.

The traditional method for determining three-dimensional positions in the urban environment has been land surveying. This is a high accuracy, standard technique for measuring locations, which again relies on considerable human input to occupy or observe locations whose position is required. Future developments in this technology will inevitably automate the procedures, but at high capital cost.

A major advantage of land surveying is its flexibility – it is as easy to undertake an internal building survey as it is to pick out positions in a street or in a park. The method is particularly appropriate for two-dimensional data capture of the urban fabric – street lines, building outlines, or lamp-post positions – although it can also be used to estimate building heights and pick out detail on façades. A major requirement and common task is to capture details of plane surfaces in the urban landscape – land parcels, vertical building sides, interior room floors, walls and ceilings – and to further calculate intersections of these faces to automate three-dimensional solid modelling for VR. Such operations may not require the physical occupation of positions to be determined, but they certainly need to be visible from known points over which survey equipment is set up. In most cases it is, in fact, necessary to place some type of target (often a reflecting prism which can return to the instrument an infra-red signal whose delay from emission to reception is used to calculate distance) at the point to be measured. Clearly, considerable amounts of practical fieldwork need to be undertaken. It may be that the major application of land surveying techniques for VR is in visualization of simple development sites (into which planned and potential structures can be inserted) rather than the re-creation of complete urban environments.

The fieldwork and measurement efforts are considerably multiplied in a complex urban area where a potentially enormous amount of data may need surveying. True three-dimensional capture may require the multiple observation of points from a variety of different known locations. These requirements, along with the practical difficulties of setting up instruments, obtaining clear and extensive lines of sight, and operating equipment within an environment where there is considerable human interference, often put considerable strain on the surveying procedures. The creation of three-dimensional objects for inclusion in a VR world-creating process may require further and considerable *post hoc* modification and data integration to successfully obtain correct and definitive features from a variety of disparate field measurements.

There are benefits evident in recent technological advances in land surveying technology. 'Reflectorless' technology relying on pulsed laser, rather than infra-red, distance measurement can yield accurate results without the requirement to occupy the position observed. Indeed the urban environment provides suitable surfaces for reflecting such pulsed laser signals. The automation of field data capture may be further enhanced by instrumentation which can sense (using projective geometry applied to automatically obtained measurements) faces and corners, and can then pick up (both inside and outside) walls, windows, doors, beams and columns (Stilwell *et al.*, 1997). The requirement that land surveying observations be undertaken using instruments sited at known locations may be pre-empted by devices with in-built GPS receivers which can determine their initial position with reference to the type of satellite signal already described.

A radically different approach to data capture for urban VR is to concentrate on visual, rather than geometric, data, in the form of images of a continuous nature. Here the prime data source is photography taken from an aerial platform or from the ground. Such images can be used to extract the geometrical information required for constructing boundary primitives. In addition, the imagery can be used to render textures onto planar surfaces and solid objects. The technique of extracting measured locations from such imagery is called photogrammetry, whilst the use of such images for rendering is texture mapping. Both techniques are inherent in digital photogrammetry – contemporary image handling and measurement using scanned photographic images (monochrome or colour) or digital images obtained directly using a charge coupled device (CCD) camera (Gruen, 1996). The rapid development of close-range digital photogrammetry has led to new applications in a broad range of industrial, scientific and architectural areas, additional to traditional three-dimensional data capture for mapping, CAD, GIS and virtual reality. Such techniques often use artificial intelligence, machine vision and multimedia graphics to extract three-dimensional object information, increasingly from non-conventional images (i.e. those which are not sampled using accurately calibrated, fixed format cameras capable of stereo image acquisition).

New methods to allow for the semi- (and potentially fully-) automated methods of constructing three-dimensional objects are described in Streilein (1994), whilst the creation of such objects from single (as distinct from stereo) images is becoming straightforward and common. The latter procedure requires an explicit model of the projective transform in the image, assuming either a parallel or, most commonly, central projection. In addition, single image methods make several assumptions about the environment which has been imaged and which is to be measured. For most man-made objects, particularly those in urban areas, there exist components which can be presumed to be parallel, co-planar, perpendicular or symmetrical. Identification of such properties can lead to the inclusion of constraints in the measurement process, or to the inclusion of 'fictitious observations' which can be used in the establishment of geometrical rigour (Dorffner and Forkert, 1998; van den Heuvel, 1998). Further algorithmic approaches have been developed which rely on edge detection, line segmentation (simplification) and grouping by co-linearity, proximity and parallelism. Cohen and Cayula (1984) give an algorithm that deconstructs complex three-dimensional objects into simple surfaces, such as planes and cylinders, producing contours of image intensities and using them for surface classification and shape estimation. The Phoenix algorithm described by Parsley and Taylor (1998) is a combined image- and geometry-based approach to the reconstruction of architectural models (Figure 16.2). Here, a central projection model is assumed and the semi-automatic identification of object vanishing points on the image allows for the establishment of the original camera focal length, position and orientation, relative to the object. A real-world distance is normally needed to scale the model and experimental results have shown that Phoenix gives maximum

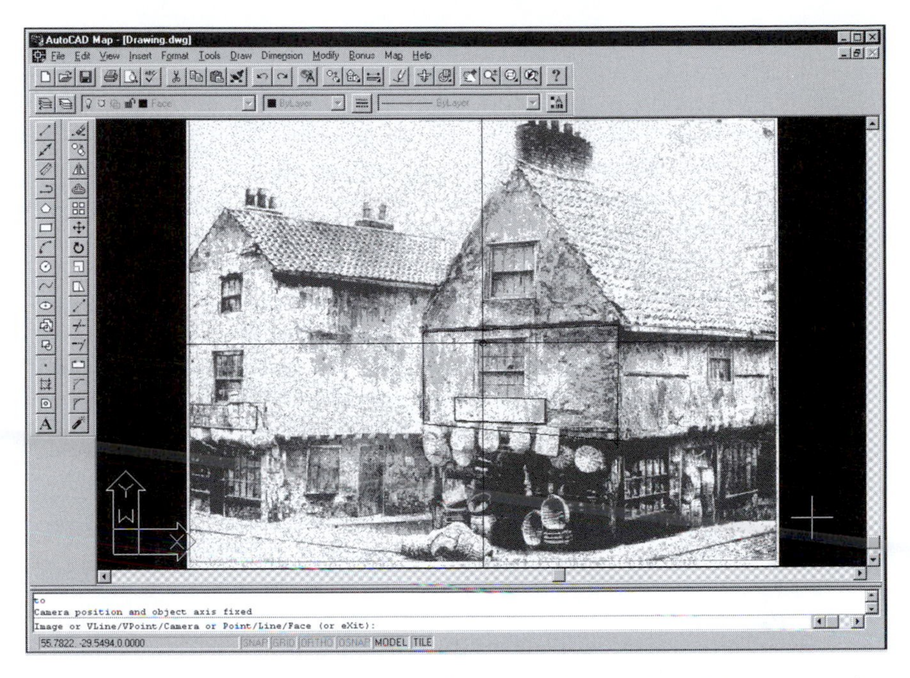

Figure 16.2 The Phoenix algorithm applied to architectural model reconstruction (a) monocular image with construction lines.

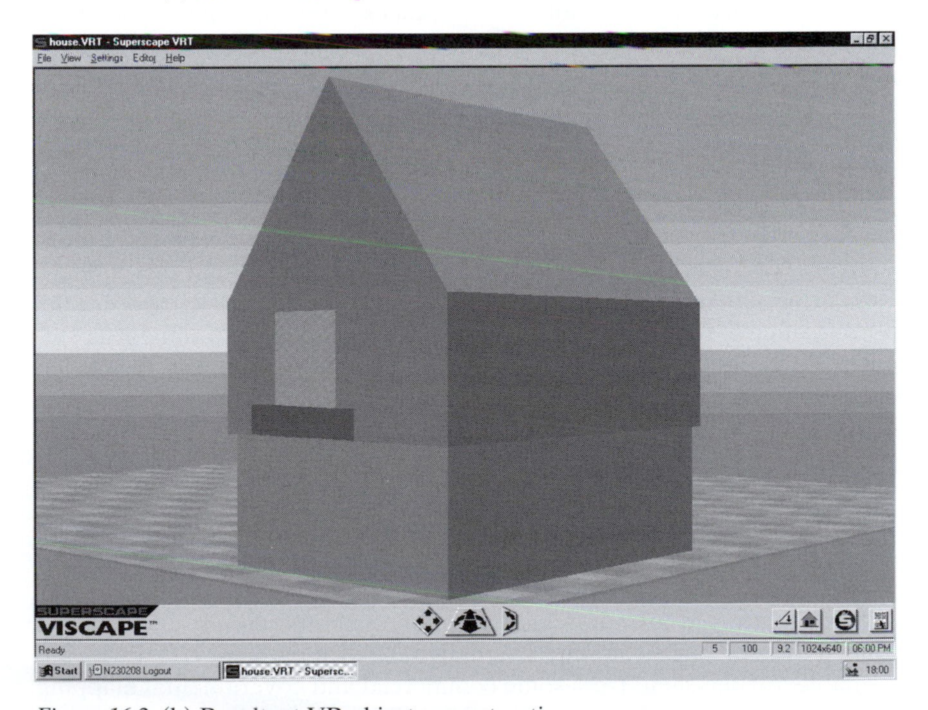

Figure 16.2 (b) Resultant VR object reconstruction.

discrepancies of <0.5 m when compared to traditional survey and photogrammetry, which is adequate for many urban applications.

The interpretation of real-world features, which in all data capture methods mentioned so far relies on the human visual sense, can be enhanced when digital imagery is being used, through the use of pattern recognition techniques. Such computer vision based methods, applied both to terrestrial photography and airborne imagery, have a role to play in three-dimensional feature construction. As was the case with the direct measurement techniques already mentioned, the physical creation of objects often relies on Constructive Solid Geometry (CSG), which combines simple primitives by means of Boolean set operators (union, intersection, subtraction) in order to obtain regular buildings, complex roof shapes and street furniture. Assumptions can be made that the ground plan of a building defines the borders of its roof and an 'extrusion' procedure can raise the two-dimensional plan outline to an appropriate height. Further assumptions, and least squares and other adjustment procedures, can ensure the construction of graphics primitives even when certain parameters (e.g. parallelism) are unknown or when feature parts are not visible. These methods of automated reconstruction lead to three-dimensional city-scapes that are useful virtual worlds for representation and modelling.

All the image-handling techniques so far mentioned exhibit reliance on efficient and user-friendly software packages, some specifically designed for precise photogrammetric restitution, some taken from more general graphics-handling disciplines. For example, the ability to 'stitch together' panoramic or 360 degree images into a seamless continuum is a fundamental function of software such as QuickTime VR. The handling of coordinates and three-dimensional 'wire-frame' objects can be undertaken in projective geometry-type software (such as PhotoModeller) which uses photogrammetric principles to extract accurate information from non-standard configurations of images (e.g. non-stereo, multi-directional). Rendering software, often relying on a base of such accurate 'wire frame' objects suitable for draping of imagery, has developed from standard computer aided design (CAD) roots. Draping of imagery can also be carried out over surface models which represent both terrain and the urban fabric: high-resolution imagery (at present from airborne sensors) can be used to prepare realistic visualizations Figure 16.3 illustrates the contemporary 'state-of-the-art' in such image draping software. From such a perspective, current demands are clearly for realism and speed of model creation, rather than for synthetic content which may take considerable manipulation to create.

The surface and terrain models mentioned above can also be obtained using photogrammetric methods, and digital elevation models (DEMs) are available 'off-the-shelf' from some commercial and governmental mapping suppliers (Intermap Technologies, 1999). Terrain modelling (represented

Figure 16.3 Image draping onto an urban surface model (Courtesy Super Soft Inc., Beijing, China).

by DEMs) is often neglected in urban modelling, where an assumption that the earth is flat tends to prevail. In addition, it is the third dimension of the city fabric (rather than the ground) – a digital surface model (DSM) – which is of prime importance,. Ground height, however, can contribute to enhanced realism and, in addition to photogrammetry, radar, airborne laser scanning and other methods of airborne altimetry can be used to obtain accurate DEMs and DSMs. Further image-based technologies applicable to urban data capture include video-based methods, either terrestrial or airborne, LIDAR, and methods of extracting accurate data using such techniques are already widely used in industry.

The major advantage of photogrammetry as a tool for urban data capture is that it is a non-contact technique which allows for the highly accurate capture of positional, attribute and image information. In addition, costs can be low, particularly when the same imagery can be used both geometrically to construct the urban model and render a surface image onto the objects created. There are problems, notably of invisible data in overhangs, dead ground, shadows and underpasses. The highest precision still requires considerable field control in the form of accurately located points using traditional survey techniques.

In addition to aerial surveys (which may also sense in non-visible wavelengths such as infra-red for heat surveys in urban areas), considerable amounts of image data can be obtained from satellite imagery. Due to the low resolution and small viewing scale of current satellite remotely sensed data, methods of extracting useful information for urban applications are limited. However, standard techniques of data classification can categorise surface characteristics with considerable accuracy, and there are also benefits in the repetitive coverage of the earth's surface which can ensure an added temporal dimension to the VR world. The techniques of draping image data over natural terrain and urban structures, mentioned above, may increasingly use higher-resolution satellite imagery newly available from commercial remote sensing missions. The wide area coverage of satellite scenes may soon be supplemented by more focused sub-metre resolution space imagery suitable for the extraction of accurate three-dimensional measurements and subsequent urban VR world visualization. Such data is likely to be extremely expensive, and it is clear that aerial surveys will not be superseded by space-sourced data in the short or medium term.

There is a considerable range of existing spatial data which can be efficiently used to initiate and enhance the creation of urban VR worlds. For the majority of such projects, and all geographical applications in VR, existing topographic mapping is the most obvious data source. Urban zones are invariably mapped at the highest level of detail, and their maintenance schedule ensures that revision is undertaken in urban areas with greater frequency than in other parts of the country. In general, the 'footprint' (outline) of buildings as portrayed on large-scale mapping is an invaluable source of information for initiating urban world creation. In many cases, the topographic mapping has been digitised and already exists in a form suitable for direct import into VR world-authoring software: possibly in a standard national format or in a commercial interchange format such as .DXF (from AutoCAD), .DGN (from Intergraph) and .SHP (from ESRI). In addition, the reference system for most topographic mapping is based on the national grid and projection which will form the basis for all further data integration.

The creation of urban features and the matching of this mode of data use to the structured model required are both enhanced by the object-oriented nature of vector topographic mapping which portrays discrete urban features in a 'featureless' background. It must be stressed, however, that the digital data sourced from such mapping may not exhibit such 'object-orientation', although such an ideal is being pursued by some topographic data suppliers (Cooper and Pepper, 1997). Furthermore, the standard topographic map has very limited information on the third dimension, or on temporal aspects, and, as these aspects form a major component in VR world-creation, particularly in the urban area, ancillary data is also required.

The supply of data by national mapping agencies, which are the major – and in many cases the sole – producers of such large-scale topographic mapping, is subject to considerable world-wide variation in pricing policy. In some countries, data costs for extensive vector coverage of urban areas can be extremely high. Such considerations may also apply to other spatial datasets capable of being used in database creation. Existing thematic datasets may, if available, offer environmental and socio-economic information of considerable importance to urban VR. Agencies connected with the handling of hydrological, geophysical, terrain, meteorological, cadastral, public utility and local authority information are worthy of detailed investigation to determine the possible contribution of their data. The availability of such data at low cost from co-operating researchers may preclude the need for otherwise expensive fieldwork. Once again, however, it is important to stress the need for accurate positional information to allow for data integration.

The minor contemporary role of actual physical models of urban centres should be mentioned. A 1:500 solid wood or plastic miniature representation of the cityscape, for example, gives advantages of multi-user viewing, easy navigation and viewpoint positioning and a wide field of view to give context. Photographic technology allows for images to be taken at 'street-level' within the model. It also introduces elements of familiarity which may be appealing to non-specialist viewers (e.g. city councillors) engaged in discussion with planning officials. Unfortunately, the common viewpoint is usually inappropriate (in effect, in mid-air); the model has a fixed scale and resolution; there are significant resource implications in updating the model; the only use is visualization, as simulation is difficult; and finally, the model cannot be distributed as it physically exists only in one place. Given contemporary technology, this is unlikely to be a viable technique in urban modelling in the future.

New, integrated and non-conventional technologies for spatial data observation

The methods so far described are primarily designed to capture objects in the urban landscape and the fabric of the urban environment. They are thus primarily directed towards the creation of physical entities which can be viewed in VR worlds. A considerable amount of further data is, however, available for capture and integration, and the whole area of attribute and semantic data collection is one that must be addressed.

Attribute data often needs to be obtained by direct measurement, just as positional data does. Portable spectrometers and radiometers, for example, which can determine heat emission and reflectance from urban surfaces such as walls and roofs, are commonplace. Such data may also be obtained from imagery, for example that sensed in the infra-red part of the spectrum. Further direct measurements which are regularly used, but which may be unfamiliar to a geomaticians or GIS scientist, include the results of seismic surveys, which can determine the nature of drift and solid geology; the use of ground penetrating radar (GPR) which is considered useful by archaeological investigators and those interested in man-made sub-surface features; techniques of pipe- and cable-following using inertial within-pipe devices ('pigs') or low power electrical charging and following of cables ('scientific dowsing'); and the widespread collection of sample data for soils, groundwater, vegetation, noise and air quality surveys, all of which offer considerable ancillary information capable of being integrated into the urban model. Civil engineering techniques used in site surveys of all types are a potentially fruitful source of semantic and attribute data (Clayton *et al.*, 1995), as are *in situ* meteorological observations. The former may involve instrumentation such as resistivity meters which emit and sense variable electrical signals and which can determine subterranean features, soil moisture content and geological outcrops; magnetometers, which can also locate distinct sub-surface features; piezometers, capable of detailed groundwater pressure measurement; and vibrating trucks with 'geo-phones', commonly used to undertake the localised equivalent of seismic survey.

In all cases, such data needs to be linked to a geometric object – at its simplest, a point in space – in order for it to be precisely positioned and capable of representation in the VR scene. Usually some form of potentially complex data integration is needed to ensure that, for example, the archaeological GPR output can be located in a reference system such as a national grid; or that the coordinates of weather station data are accurately located to ensure adequate portrayal of three-dimensional temperature information; or that the precise extent of different textures (say glass and concrete) on the face of a tower block building are recorded for adequate visualization.

Data characteristics

A hypothetical example application for urban virtual reality is outlined here to illustrate the need for high-quality preliminary data modelling, structuring and capture prior to efficient rendering, visualization and use of the created world. The aim proposed is the development of an urban energy usage information system which can be used to calculate, model and visualize the heat budget of a substantial city area for the purposes of proficient management of resources, prediction of demand, and design and creation of an economical built environment. More specific tasks might be to model illumination by sunshine and shadow zones; to predict and monitor three-dimensional airflow patterns; to assess exposure to prevailing wind and sunshine; to gauge energy consumption by land or building use; to determine heat loss from buildings; to calculate heat budget figures by considering reflectance and absorption characteristics; to record actual energy usage, property by property, over a significant time period; to simulate future energy needs with reference to development plans; and to check the effect of building adjacency on energy consumption.

Clearly an accurate three-dimensional solid model of the townscape is required, both for visualization purposes and for accurate calculations, for example of the total surface of glass-walled buildings. Such a model will rely on positional information potentially available from the range of sources already discussed. Standard horizontal and vertical reference systems must be determined to establish a common data framework. The inherent level of resolution of input data – reflecting pixel size (e.g. from scanned aerial photography or thermal aerial survey), survey and plotting precision and map generalisation – must also be addressed to ensure compatibility. The representation of the urban form – as combinations of points, lines, areas and volumes – needs to be established to match the appropriate level of resolution, and any 'level of detail' (successively scaled versions of the VR world with varying grades of representation) must be set up in the initial model. Further issues, such as the variable topological structure of different constituent data sets, may also need addressing to ensure geometrical fidelity.

A wide range of metadata characteristics, which describe the nature and properties of the data set, will need to be considered to ensure that added value (beyond visualization) emanates from the urban VR world. Much of this metadata is concerned with quality and subordinate factors such as accuracy, completeness, level of processing and currency.

A range of reflectance data is needed to create a surface albedo model and to develop accurate models of radiation flow (Lo *et al.* (1997) discuss the application of such high-resolution airborne remotely-sensed reflectance data for the study of urban heat islands). Actual consumption figures for energy supply, in the form of electricity, gas and oil, to all premises in the urban area might be required. Clearly, actual examples of energy use are important factors is determining a heat budget for the urban area and modelling the potential impact of different usage patterns and varying urban forms. In addition, the finer details of vegetation coverage (for example, grass and trees) within the city need to be captured as both evaporation from these surfaces and the impact of shadows from tree cover (as well as from tall buildings) are important components in heat loss modelling in the urban area. Meteorological data, both long-term records and forecast information, has a role to play in modelling overall climatic impact on the urban area and even micro-climatic data (for example windflows and temperature patterns within 'urban canyons') may be useful. Each of these datasets must be accurately positioned and integrated with topographic and locational information to create a faithful complete VR model.

The currency of this data is likely to be highly variable: weather data can be sampled and supplied in real-time, but global change impacts are only measurable over intervals substantially greater than one year. The integration of some notably contrasting datasets, for example, hourly meteorological readings with periodic electricity consumption meter readings (which may themselves differ in periodicity from domestic to commercial properties), may prove problematic. The topographic base data can also have significant update variability: if still based on paper mapping, there is likely to be significant variation in the actual date of survey across a substantial urban area. Even when digital mapping is used, with its theoretically speedier flowline from data capture to dataset revision, it is possible to discern varying practices by national mapping agencies, some of which offer preferential and more frequent update services in a commercial manner.

The level of processing refers to an equivalent to the spatial data generalisation exhibited by standard topographic base data, but may have different connotations for non-spatial data. In terms of energy, for example, a primary indicator may be actual monitored energy consumption, whilst secondary information such as socio-economic indicators (population density, household size, deprivation index) may be able accurately to predict energy consumption indirectly. The type of data, for example whether it is sampled or continuous, how it is categorised, its nature as nominal, ordinal or ratio data may also be important.

The content of a metadatabase which describes the characteristics of a set of spatial data is potentially enormous. In addition to those described above in the context of an urban energy model, the following may be of relevance to include: the geographical region covered, locational reference system used, a standard to which the data adheres, language or units or data formats or codes employed, overall data transfer format, the nature of objects represented in the data, attributes of each object, associations between objects, restrictions on use, information on ownership and copyright, notes on reliability and logical consistency, and the actual cost of the data. The history of a data set (lineage) may also be described using a host of observations: original source, usage of the dataset, processing or sampling from original dataset, process of updating or enhancement, geometric editing, aggregation or generalisation, combining of data, change of reference system, or change of format.

The establishment of an urban energy model using VR techniques clearly relies on a considerable range of data sources and an ability to successfully integrate such data. With potentially highly variable characteristics, which may be recorded along with the data to differing levels of completeness, the incorporation of all relevant spatial data into the urban VR world creation process may be difficult to achieve. As indicated in the introduction to this chapter, the data model must be firmly defined before any data collection, manipulation and integration is attempted.

Overview of sample applications

The figures accompanying this chapter illustrate in a practical way some of the issues which must be taken into account when engaged in urban VR world-creation, most notably those aspects relating to data capture and integration. Figure 16.4 is extracted from a virtual model of Hartlepool, North East England. This world is presented to museum visitors who embark on a virtual voyage across the harbour of this seaside town, starting within the contemporary confines of a dock alongside a late-1990s retail park built on the quayside, and sailing backwards through time (and through a virtual fog-shrouded harbour) to the originally established headland on which stands the Anglo-Saxon settlement and monastery of AD 700. The illustration, from the first part of the voyage, shows a combination of an architectural feature (the modern quayside building), a faithfully-rendered vertical surface (the sea wall) which required terrestrial imagery to map, and an accurate depiction of the state of the tide, taken with reference to current nautical charts.

Figure 16.4 Multiple data sources for an urban visualization (image courtesy of Universiy of Teeside VR Centre).

The paradigm of the urban computer game simulation, noted in the introductory chapter to this section, will be evident in the planned 'SIMCITY Newcastle', an experimental VR application incorporating a city model of the centre of Newcastle upon Tyne in England with real data and real parameters. In addition to the measured physical fabric of the city, this model will include a broad spectrum of large-scale data sets from map sources, sampled measurements (e.g. of traffic), socio-economic indicators (e.g. of deprivation), environmental knowledge (e.g. noise levels) and public health factors (e.g. disease). Different levels of resolution, significant detail on attributes of the urban physical structure and various socio-economic statistics (all including temporal data) will be required.

In creating a simulation of the city centre for sustainable urban planning, the result will be used to model energy use (as postulated above), transport, pollution impact and socio-economic development, and it is hoped that this will assist in the implementation of a true urban sustainability policy. This use of VR is not solely for the visualization of the urban environment, but is a fundamental attempt to model the operation of a complex system, and simulate the behaviour of various multidimensional spatial indicators of sustainable planning and development. In effect, the model contributes to the realisation of emerging theoretical frameworks for urban planning.

Moving away from specific examples, Figure 16.2 illustrated the complexities of accurate data collection. Here the requirement was to extract three-dimensional information from a single historical photograph, clearly taken with no perceived need for the stereo coverage normally required to enable such measurement. Certain assumptions are made relating to the parallel components within some features, such as top and bottom lines in window and door frames and along the edges of wall faces. By accurate calculation of vanishing points within the image, and subsequent determination of the camera or viewing position, it has been shown that it is possible to extract valuable data from such images. For the task of creating an urban VR world for centuries past, such techniques, applicable to both photographs and older engravings and etchings, have been extremely useful. Some real-world distance is required (for example the frontage of the building), so the feature must still exist or documentary evidence must be sought.

The inclusion of three-dimensional sub-surface objects has also been considered. Utility companies, in particular, can see the potential role of virtual world-creation for modelling, visualization and simulation. Incorporating ground survey and map sources, the utility worker in Figure 16.1 also requires detailed information about sub-surface plant, often difficult to obtain from typically incomplete and inaccurate records. In the UK, there is a requirement of the utilities, in the 1991 New Road and Street Works Act, to hold such data. Its practical implementation involved the planned creation of a street works register: this did not materialise, but

1999 saw a replacement, based on more complete information, including the *National Street Gazetteer*. There may still be problems in actually recording the route of underground facilities, often only approximated by surveying inspection covers, junction boxes and other surface-visible manifestations of the route. However, the technologies outlined above give considerable scope for more accurately locating such objects: ground-penetrating radar in conjunction with packages such as Slicer can create three-dimensional visualizations of sub-surface features; inertial surveying technology, in the form of 'pigs', can assist in gas and water pipe mapping; and 'scientific dowsing' can be used to determine connections of cables and wires.

Conclusion

It is important to note that although the creation of virtual worlds for VR applications relies, of necessity, on measuring and collecting individual objects, the eventual use of these worlds will be holistic, may involve simulation, and will test connections and inter-relationships. It should be appreciated that the virtual world is a seamless continuity. Data collection as described in this chapter may well involve sampling – certainly GPS and land surveying require the discrete acquisition of points in the urban environment. The users and viewers, however, of the VR model will assume an interaction with continuous data. The contribution of continuous imagery, such as photogrammetric data, or even pixellated data, such as satellite imagery, is important in this respect.

In summary, the process of urban VR world-creation may address a range of measurement, metadata, data structuring and data integration issues. To date, the majority of urban VR models have been used purely for visualization purposes. In order to obtain the full benefit from new VR technology and the environmental representations which it can handle, it is necessary to think broadly about its impact on the complexities of the urban world.

References

Clayton, C.R.I., Simons, N.E. and Matthews, M.C. 1995. *Site Investigation.* Oxford: Blackwell Science.

Cohen, F.S. and Cayula, J.P. 1984. 3-D object recognition from a single image. *Intelligent Robots and Computer Vision*, 521, 7–15.

Cooper, J. and Pepper, J. 1997. Mapping for the new millennium. *Proceedings of the 34th British Cartographic Society Annual Symposium.* London: British Cartographic Society, 11–16.

Dorffner, L. and Forkert, G. 1998. Generation and visualization of 3D photo-models using hybrid block adjustment with assumptions on the object shape. *ISPRS Journal of Photogrammetry and Remote Sensing*, 53 6, 369–78.

Edwards, T. 1991. *Virtual worlds technology as an interface to geographic information*, unpublished MSc thesis, University of Washington, 92 pp.

Gruen, A. (1996) Development of digital methodology and systems. In Atkinson, K. (ed.) *Close Range Photogrammetry and Machine Vision*. Latheronwheel, Scotland: Whittles Publishing, pp. 78–104.

Gruen, A. and Wang, X. 1998. CC-Modeler: a topology generator for 3-D city models. ISPRS *Journal of Photogrammetry and Remote Sensing*, 53, 5, 286–95.

van den Heuvel, F.A. 1998. 3D reconstruction from a single image using geometric constraints. *ISPRS Journal of Photogrammetry and Remote Sensing*, 53, 6, 354–68.

Intermap Technologies. 1999. Off-the-shelf SAR Digital Elevation Models. *GIM International*, 13, 1, 70–1.

Lo, C.P., Quattrochi, D.A. and Luvall, J.C. 1997. Application of high-resolution thermal infrared remote sensing and GIS to assess the urban heat island effect. *International Journal of Remote Sensing*, 18, 287–304.

Parsley, S. and Taylor, G. 1998. Methodologies and tools for large scale, real world virtual environments: a prototype to enhance the understanding of historic Newcastle upon Tyne. In Carver, S. (ed.) *Innovations in GIS 5*. Chichester: Taylor & Francis, pp. 78–86.

Stilwell, P., Taylor, G.E. and Parker, D. 1997. Precise three-dimensional measurement and modelling of structures using faces. *Surveying World*, 5, 3, 31–3.

Streilein, A. 1994. Towards automation in architectural photogrammetry: CAD-based 3D-feature extraction. *ISPRS Journal of Photogrammetry and Remote Sensing*, 49, 5, 4–15.

Tempfli, K. 1998. 3D topographic mapping for urban GIS. *ITC Journal, 1998–3/4*, 181–90.

17 Virtual world heritage cities

The ancient Thai city of Ayutthaya reconstructed

Cliff Ogleby

Introduction

There has been a rapid increase of interest recently in generating digital reconstructions of ancient cities and artefacts, both within the academic community and also the general public. Whilst there has always been a fascination with picturing ancient monuments and cities (Piggott, 1978), there seems to be an increased popular awareness. There may be many reasons for this high level of activity; it could be partly as a result of the movie and computer games industry, partly also perhaps because of the advent of computer networking and high-speed computer graphics on the desktop, and probably due also to the interest that many people have for things historic/exotic. The digital reconstructions (or 'virtual heritage', or visualizations, or even virtual reality) exist as CD ROM products, as museum or exhibition displays (for example, several of the Pavilions at Expo '98 in Lisbon featured virtual reality historical experiences), as proprietary research documents or in a variety of formats on the Internet. The visualizations exist as serious research material and as sets for games and entertainment. A large selection of material available is also naturally on the Web, and searches with key words like 'virtual heritage' will bring up many sites presenting the results of reconstructions of cultural monuments and city scapes.

In addition to learned journals and books (Forte, 1997; Novitski, 1998b), general interest computer graphics magazines like *Computer Graphics World* (Novitski, 1989a; Shulman, 1998) and *3d Artist* (Hamilton and Breznau, 1998), as well as television programmes feature articles on the application of visualization technologies in archaeology and history. There is a high level of public interest in the use of this visualization technology when it is applied to historical sites and cultures. A special session at the Virtual Systems and Multi Media conference in Gifu Prefecture, Japan in 1998 was held with the name 'virtual world heritage'; this term is general enough to cover artefacts, buildings, palaces, statues and entire cities so the term will continue to be used in this chapter.

Introduction to the project

Ayutthaya is presently a modern city of around 60,000 population, some 80 km from Bangkok in Thailand. Part of it is a UNESCO World Heritage Site, as it was the capital of the Kingdom of Siam for around 400 years up until 1867 CE, and contains ruins of many of the religious structures from this and previous periods. In its day Ayutthaya (there are many variations in the spelling of this word like Ayuthaya, Ayodhaya and Iudea for example) was reported to be one of the most spectacular cities anywhere in the known world. Today it is a collection of monuments and precincts surrounded by the bustle of a modern regional city in Thailand.

The University of Melbourne in Australia, and Chulalongkorn University in Thailand are conducting a joint research project to re-create the greatness of Ayutthaya in the form of a photo-realistic, three-dimensional virtual world heritage model. In order to achieve this, photogrammetric records of the remaining monuments have been combined with early written records of the city and the assistance and expertise of the Royal Thai Government Fine Arts Department. The basis for the creation of the CAD models is, in many cases, however, the photogrammetric record which will be described later. The models resulting from the research has formed the basis of a series of 'visualizations' and animations of the city, which are under continual development and enhancement.

Many of the dates used in this chapter are those based on the Christian era (denoted CE in Thai literature); the Thai Buddhist dates (BE) can be converted by subtracting 543 years from the dates shown.

A brief history of Siam, Thailand and Ayutthaya

Ayutthaya was a well-established town before it became the capital of Siam, following the fall of the previous 'capital' at Sukhothai. It was supposedly founded by a Prince of Ut'ong (U-Thong) in the year 1350 or 1351. Or, more to the point: 'So he had his troops cross over and establish themselves on Dong Sano Island. . . . In 712, a Year of the Tiger, second of the decade, on Friday, the sixth day of the waxing moon of the fifth month, at three nalika and nine bat after the break of dawn, the Capital City of Ayutthaya was first established' [i.e., Friday, 4 March 1351, shortly after nine o'clock in the morning] (Wyatt, 1984, translation from Cushman). It is named after *Ayodhaya*, the home of Rama in the *Ramayana* epic, which means 'unassailable' in Sanskrit. It is set on an island situated at the confluence of three rivers: the Chao Phrya (also known as the Menam, which flows south to the sea via Bangkok), the Lopburi and the Pasak.

Ayutthaya became one of the wealthiest and greatest cities in Asia, and attracted the interest and awe of visitors. In the sixteenth century, travellers from elsewhere in Asia and Europe were arriving in Thailand, for trade in both goods and Christianity. Visitors, traders and missionaries from Portugal arrived around 1511, from Japan around 1690, from Holland around 1605, England around 1612, Denmark around 1621 and France in 1662. Many of these foreign missions were allowed to settle effectively as 'embassies' and it is from many of these travellers that the historical details of Ayutthaya can be discovered. The glory of the city was reported widely in Europe, and most visitors claimed it to be the most splendid city they had seen.

> We never saw a Fabrick no not in France, where Symmetry is better observed, either for the body of the Building, or the Ornaments about it, than in this Pagod. The Cloister of it is flancked on the outside on each hand with sixteen great solid Piramids, rounded at the top in form of a Dome, above fourty foot high, and above twelve foot square, placed in a Line like a row of great Pillars, in the middle whereof there are larger niches filled with gilt Pagods. We were so long taken up with the sight of these things that we had not time to consider several other Temples close by the Post within the same compass of Walls.
>
> (*sic*, Tachard, 1688, in Smithies, 1995)

The city is associated with a high time in Thai history, with the borders of Siam extending into Burma, Cambodia and Malaysia. It was also a period of many wars, and Ayutthaya was severely damaged on several occasions. In time, the large canons of the Burmese Army once again helped to conquer the Kingdom of Ayutthaya, and on 7 January 1867 CE the resultant fire consumed much of the inner city, destroying some 10,000 houses. Ayutthaya did not rise again, and eventually the capital of Siam moved to Bangkok leaving a ruined and plundered landscape. As a result, all that really remains of the greatness of historical Ayutthaya are the *chedi*, prang and defaced statues of the temples; all of the timber buildings, the palaces and the houses from the period are gone. (There are, however, many examples of Ayutthaya period architecture and sculpture in other locations and as more recent constructions.) In addition, apart from a few selected areas, modern Ayutthaya infects the historical city with roads, shops, houses, condominiums, cars, trucks, dust, mud and exhaust fumes. The city of Ayutthaya was synonymous with Siam the Kingdom in the eighteenth century. After the Second World War, on 11 May 1949, Siam was renamed Prathet Thai, or Thailand as it is known to the world today.

Thai architecture and sculpture

The form and structure of Thai monumental architecture during the periods leading up to the zenith of Ayutthaya was greatly determined by religious belief and tradition, primarily those resulting from Hinduism and Buddhism. If an 'accurate' reconstruction is to be obtained, some understanding of the social traditions that derived the cultural heritage is important.

Buddhism arose in India during the fifth century before the Christian Era, and is based on the teachings and enlightenment of Prince Siddharta Gautama (later to become the Buddha, or the 'enlightened one'). It incorporates much of the religion of Hinduism and shares many of the gods and deities. Buddhism offers, however, an escape from the endless cycle of birth and rebirth in Hinduism, and followers can achieve *nirvana* by generating *karma* through particular actions in their lives. One action that gained merit towards achieving *nirvana* was donations to temples, hence there tends (or at least at Ayutthaya, there tended) to be a proliferation of donated books, images, statues and stupa generating a large collection of visual media. As Buddhism spread across South East Asia it was able to adapt and evolve within the differing indigenous cultures it encountered, ranging from the Javanese, Sri Lankan, Thai, Khmer, Tibetan and Burmese to the Chinese and Japanese. Buddhism was able to achieve this by generally harmonising with earlier religious practices, by claiming a common origin with native or animist gods and emphasising the aspects within Buddhism that most closely paralleled existing traditions, a trait not common with many other religions. The spread of Buddhism was not accompanied by the sword, the burning of libraries, the persecution of scholars or unbelievers, or by the destruction of existing cultures. As a result of the conversion of many other cultures to Buddhism, there developed quite distinct regional styles of Buddhist art where representations of the Buddha often resembled a stylised, idealised form of the local inhabitants. This makes Buddha statues from the historic Ayutthaya stylistically unique. This style would not have been found far from the influence of the kingdom. For the purposes of this project it meant that actual Ayutthaya-period Buddha statues needed for the reconstruction, readily-available three-dimensional models from organisations like Viewpoint Digital (www.viewpoint.com/, an online supplier of 3D models) were unsuitable.

The Buddhist school that established itself in Thailand is the Theravada School, or 'lesser path (Hinayana, as it is often referred to by the Mayayana (highest path) school). This is understood to have been most closely allied with the traditions in Sri Lanka, and has many architectural and sculptural parallels especially with the free-standing majestic *chedi* found at temple sites in Ayutthaya. The sometimes perilous relationships between the Thai people and their neighbours, especially the Khmer and Burmese, also meant periods of occupation by people of differing artistic traditions, resulting in the more geometric Khmer-influenced *prang* also found at Ayutthaya.

There is much symmetry in the form and layout of Buddhist monuments and temples (as there is in the religion), resulting in a limited variation in 'type'. This facilitates the measurement of the relics of Ayut-

(a) (b) (c)

Figure 17.1 The three *chedi* at Wat Sri Sampet before and after physical reconstruction, and during visualization. Photograph (a) supplied by Royal Thai Government Fine Arts Department, Ayutthaya.

thaya in this project, as a minimum number of façades needed to be recorded in order to determine the size and shape of the monument (Figure 17.1).

The chronology of the monuments at Ayutthaya

Recent papers by Krairiksh (1992) have developed a revised dating system for the monuments of Ayutthaya and note that, for most of what remains, much has been reconstructed by a variety of people over different periods. For the epoch chosen in this project, all of the monuments being dealt with were extant at the time, although it cannot be guaranteed that they had the exact form that they have today. Care has been taken to ensure the veracity of the computer-based reconstructions, and any variation would most likely be minimal and would not affect the verisimilitude of the end result.

Krairiksh (1992) also notes that, for much of the analysis, the records of the European visitors have provided the most reliable sources of information. Those few records regarding the built environment that were produced by the Siamese had been destroyed during the sack of Ayutthaya. Much of the recent analysis of the dating of the monuments corrects some of the earlier incorrect assumptions, which would be of interest in this project if the multi-temporal aspect becomes important.

Other architecture

Thai domestic architecture of the period would also be relatively predictable in form, constructed primarily from timber and/or bamboo, with palm thatched roofs and often found in the form of the traditional Thai compound (of which there are illustrated references in the historical documents). The structures would also have been highly inflammable and would decompose very easily in the tropical environment.

Overview of the project

The aim of the overall project is to create, as realistically and as faithfully as possible, a three-dimensional computer model of the ancient city of Ayutthaya as a foundation for an animated visualization of life in the city some 300 years ago. This virtual world is to include architectural models of buildings, structures and decorations; models of trees, models of animals and people; models of boats and carts, and so on. Much of this can be provided by using photogrammetry to record the existing architectural features, and to provide base mapping if this is not available, but other data sources are also required to determine the size, shape and location of missing features. The visualization also incorporates live video sources of traditional dance, and sounds of life typical of the period, and exploits the opportunities of virtual reality and multimedia.

In general the project involves: the creation of CAD models of existing cultural monuments using photogrammetry as the basis; the investigation of early travellers' reports as to the form and decoration of the missing monuments; the reconstruction in the computer of the missing architecture and landscape features; the recreation of the other attributes of Ayutthaya like canals, elephants, barges and carts; sampling of sound and video for use in video production and post-production; compilation of the animations into a video-based product; and development of an interactive computer-based Ayutthaya experience.

The photogrammetric recording process

In order to create models of virtual world heritage, there is a need for data: on the existing form and state of the site and monuments, on their spatial relationship, on their cultural and historical relationship and on the historic or prehistoric form and landscape. In the project being described in this chapter, details are given on where these data were discovered; however, in the light of Chapter 16 by Fairbairn and Taylor, the role of photogrammetry will be discussed in a little detail.

Data collection: photogrammetry and cultural heritage

'Photogrammetry' is a term coined by Albrecht Meydenbauer in the 1880s to describe measurement from photographs, and Meydenbauer is credited not just with coming up with the term but also for having developed the science as a result of his collection of photographs of historic buildings in Germany. The discipline comprises the art and science of making measurements of objects or terrain from photographs of the object, traditionally acquired with purpose-built 'metric' cameras.

Since the development of the science of photogrammetry there have been many applications of the techniques and technology in the recording and documentation of cultural monuments and sites of importance (Ogleby and Rivett, 1985). Whilst there may have been a redirection of effort when aerial mapping boomed following the invention of aircraft, there has been a shift again to the 'other' measurement applications offered by photogrammetry, especially those in architecture and archaeology.

Developments in the sciences of photogrammetry and image processing over the last decade or so have seen an increase in the automation of the data-collection process, ranging from high-precision industrial applications (Beyer *et al.*, 1995; Brown and Dodd, 1995) through to simple solutions for non-traditional users such as *3D Builder* (Patias *et al.*, 1998) and *Photomodeler* (Hanke and Ebrahim, 1997). In addition systems that use imagery from consumer digital and analogue video systems (Streilein and Niederöst, 1998), and sequences of images (Pollefeys *et al.*, 1998), have almost automated the creation of 3D models.

In order to create the digital reconstruction of the city, there needed to be a starting point from which to venture. Because of the well-established three-dimensional recording possibilities offered by photogrammetry it was an obvious choice as a tool to create the initial models of the main remaining architectural elements, and to provide digital maps of the location in preference to field survey. Initially there was the intention to use Small Format Aerial Photography (SFAP) from a military helicopter to provide a map of the area under study, as this approach has been found to be a successful mapping tool (Fraser *et al.*, 1995). The survey control network was established with this in mind, and easily identifiable points on the ground were coordinated during the GPS survey. Fortunately recent digital maps of the precinct were discovered during the investigation so this was not necessary.

The technique for mapping the monuments that was employed at Ayutthaya is relatively straightforward and in many cases very traditional. In general, stereo-pairs were taken of the façades of selected *chedi, prang, stele* and walls so that a 'library' of architectural elements could be derived. Convergent photographs were also taken of many of the features so as to have the possibility of strengthening the control network for the features if necessary by processing through bundle adjustment software. This combination of stereo-pairs and convergent photography has been found to be a successful method of supplying additional control whilst reducing the field time required to coordinate features or targets.

Numbered targets were temporarily placed on the surface of the features being photographed, and coordinates determined for each of these. Three-dimensional coordinates for the targets were provided by theodolite radiation from either an arbitrary control point, or an instrument point coordinated as part of a survey network. A system of controlled instrument points were placed around the main *chedi* area of *Wat* Pra Si Sampet (the most recognisable remains at Ayutthaya), and coordinated onto the

Thai mapping grid using a *Leica* System 300 GPS as part of the preparation for the planned aerial photography. Where control was needed in awkward locations, features on the monuments were coordinated in addition to the targets. At other locations a local coordinate datum was used as only the shape and size of the feature was required; its true position could be determined by other means as required. The theodolite radiation was undertaken using a total station and a laser reflectorless EDM. This meant that distances could be obtained to the targets without the requirement to use either glass prisms or retro-target tape.

Sources of topographic information

In order to re-create the landscape of ancient Ayutthaya, it is necessary to modify the contemporary terrain model in order to accommodate the changes that have occurred over the centuries. The early maps published by the European visitors to Ayutthaya (Figure 17.2) vary in scale, shape and level of detail, so they are subject to some interpretation regarding the true position of the features they show. However, they do indicate that many more canals were present, as well as decorative water features like lakes and ponds. As many of the original canals have been filled for roads, it is not difficult to interpret the historical documentation as to the whereabouts of the missing waterways.

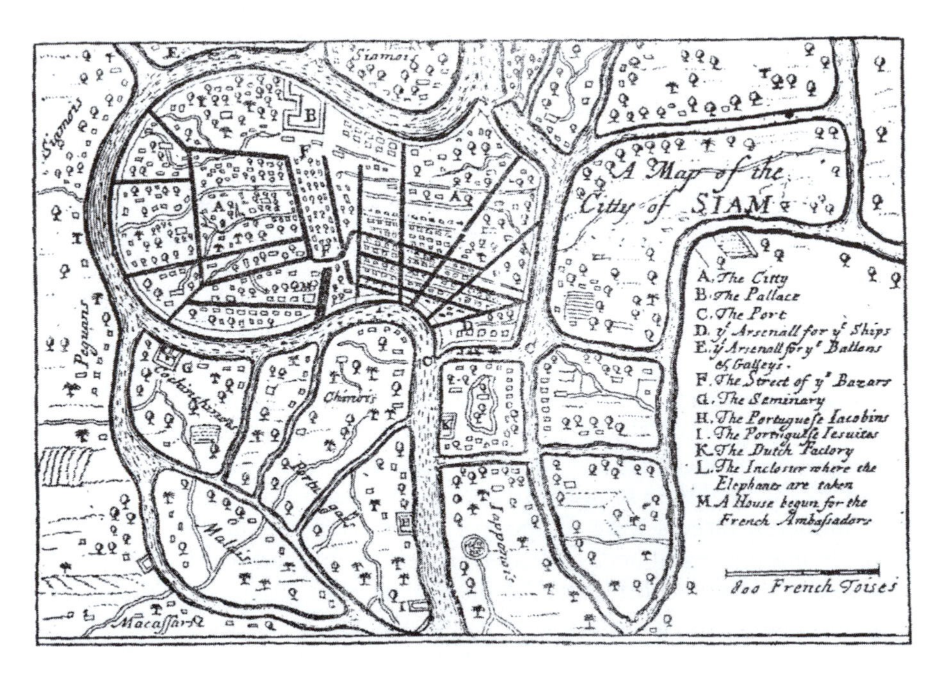

Figure 17.2 A source of topographic information: Map of Ayutthaya, from De La Loubere (1812).

The information regarding the modern topography was acquired from existing maps, with the occasional field survey to augment the data. The recent mapping performed as part of the Royal Thai Government Department of Fine Arts reconstruction project was used to derive the terrain model as it showed contours of 1 m intervals over the project region.

Sources of architectural information

There are several very different sources of information regarding the architecture of the historical city. These come from a wide variety of sources; the drawings made by European visitors of the time, the written descriptions of the city and a variety of other 'reconstructions' including scale models. This project has attempted to combine successfully these into a feasible and accurate vision of the precinct dating from around the 'late Ayutthaya' period.

Ayutthaya was visited by a variety of European travellers and officials, who returned with tales of the glory of the city. Some of these tales were translated from their original Dutch and French into English, and published widely at the time. It is from these publications that much of the architectural description has been obtained.

> About an hundred paces South of the Palace there is a great Park walled in, in the middle whereof stands a vast and high Fabric built cross-ways in the manner of our Churches, having over it five solid gilt domes of Stone or Brick, and of extraordinary Architecture, the dome in the middle is far bigger than the rest, which are on the extremities and at the ends of the Cross. This Building rests upon several Bases or Pedistals, which are raised one over another, tapering and growing narrower towards the top. The way up to it on the four sides is only by narrow and steep Stairs of betwixt thirty and forty steps three hands broad apiece and all covered with gilt Calin or Tin like the Roof ... These Piramides end at the top in a long very slender Cone, extremely well gilt, and supporting a Needle or Arrow of Iron, that pierces through several Christal balls of an unequal bigness. The body of those great Piramides as well as of the rest, is of a kind of Architecture that comes pretty near ours; but it has too much Sculpture upon it, and wanting both the simplicity and proportions of ours, it comes short of its beauty, at least in the eyes of those that are not accustomed to it... All the Fabrick and Piramides are inclosed in a kind of square Cloyster, above six-score common paces in length, about an hundred in breadth, and fifteen foot high. All the Galleries of the Cloyster are open towards the Pagod; the Cieling thereof is not ugly; for it is all painted and gilt after the Moresko way...
>
> (*sic*, Tachard 1688, in Smithies, 1995)

Passages like this tell size, shape, number, decoration, surface finish and construction method. They are often the only clue as to the appearance of the timber buildings in the ancient city. They also express the awe in which the ancient city was held.

In addition to the narrative descriptions there are also 'artists' impressions' of the architecture of Ayutthaya published as engravings to accompany the text. These, too, are useful sources of information regarding location and number; however, they cannot be used as 'exact' drawings of the monuments as they have been exposed to several levels of interpretation between the field sketches and the final publication. The European interpretation of the architectural forms is apparent in many of the illustrations found in the early publications (Figure 17.3).

Modelling considerations

The restitution and observation of the photogrammetry was undertaken solely for the purpose of creating a base 3D model from which the structure could be re-created. It was not the intention to create architectural drawings of the façades, nor to observe everything that was visible in the stereo-model. The most important aspect was to understand the CAD package and how individual graphical elements could be changed into surfaces suitable for rendering.

The images of the monuments were digitised and measured using a modern digital photogrammetric workstation. Stereo-photogrammetric observations were made to all of the main construction elements of the monument in a manner so that they could be used to either create a surface or derive a 'perfect' element in the CAD system (in this case, Bentley System's *MicroStation*). The intention was to create models of the monuments as they appeared before they became damaged; a tilting *chedi* does not best represent the situation 300 years ago. This consisted of observing profiles along decorative surfaces that could be made into a surface of revolution (also known as 'lathing'), lines along the edge of surfaces that could be connected into regular or irregular shapes and so on. Other, more complex shapes like, say, Buddha statues, are observed as a non-uniformly-spaced surface model with break-lines along the main surface features.

Whilst the photogrammetry provides a convenient way of quantifying the architectural forms, and the historical documentation provides some of the attributes of these forms, considerably more is needed to make the visualization succeed. As much of the architecture has disappeared, the other CAD models required are created wholly using the imagination of the operator tempered by some geometric constraints resulting from the archaeological plans. For example, the main palace buildings have been created from a combination of the archaeological plan for the dimensions of the exterior walls, photographs of the model in the Historical Research

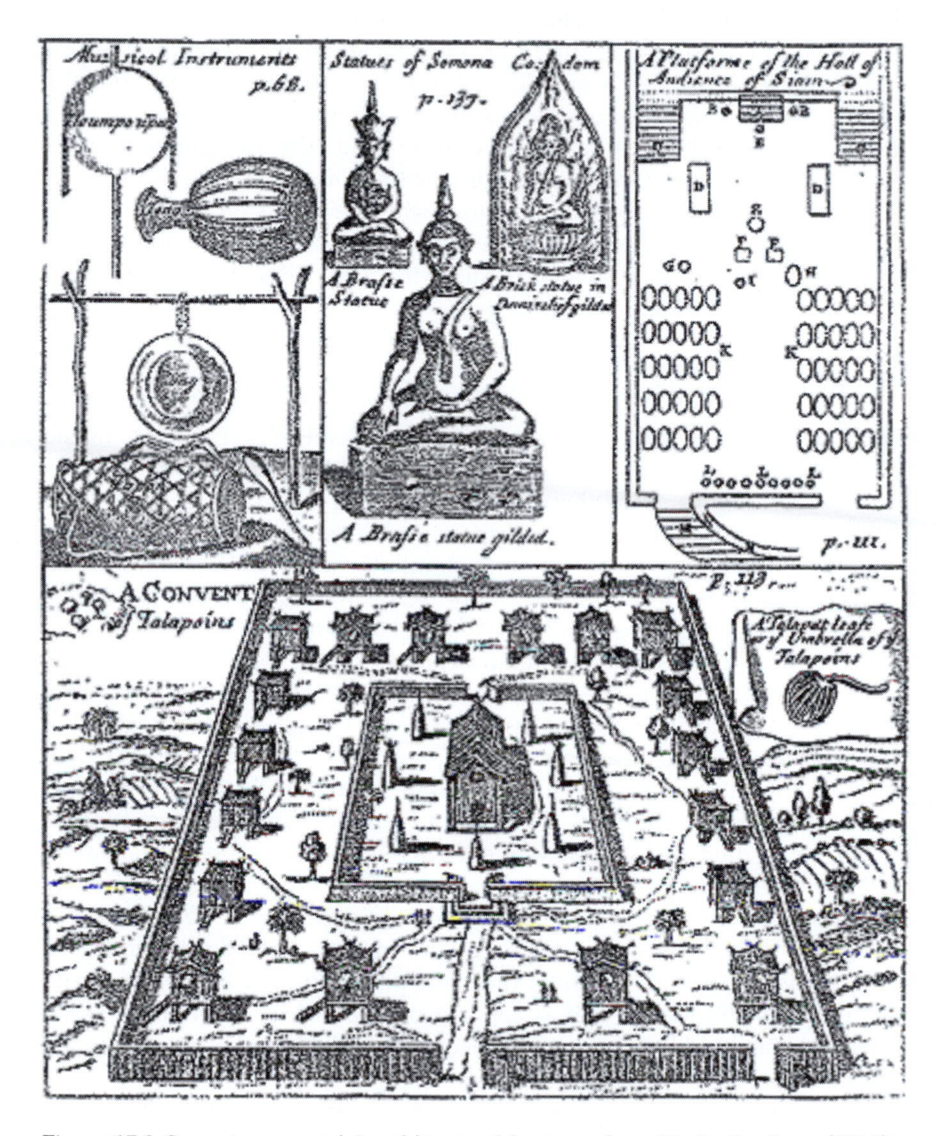

Figure 17.3 Some 'eurocentric' architectural features, from De La Loubere (1812).

Centre, photographs of the half-scale recreation at the Ancient Thailand exhibition centre in *Muang Boran* and photographs of suggested similar architectural styles. There is no building there to record, so the reconstruction is based on whatever material is at hand, as is often the case in archaeological visualization. A similar technique has also been used for other design elements like temple gongs and bells, gateways and other smaller features.

One additional source of information regarding the typical architectural decoration likely to have been used at Ayutthaya are published photographs and photographs acquired at other locations in Thailand. A selection of these have been scanned, and used as a tracing template in the CAD package. The *MicroStation* software allows the display of raster images, allowing these 'ortho-images' to be used as a background so that intricate shapes can be traced from the pictures, and converted into three-dimensional shapes by projection. This technique has been used to create some of the roof and gable decoration, as well as windows and doors. These elements can be re-scaled to suit whatever building model they are being used to decorate.

The visualization process

The last few years or so have witnessed remarkable developments in the speed of computer and graphics processors, and the expansion of mass storage devices. Much of these developments for desktop systems have been a result of the need for high-speed graphics processing in computer games. What this has meant for many computer users and software developers is a move from specialised hardware/software combinations to software operating under the Windows 95/98/NT systems.

The rendered images and animations produced to date were generated using the functions in the CAD package and *MicroStation Modeller*. Originally it was intended to use a Silicon Graphics system and the Alias|Wavefront visualization software, as this platform has an excellent reputation in the visualization community. There were, however, many problems encountered in the exchange of the model data between formats which meant that this option was not followed. The future visualization will migrate to this software as better format translators are now available, although packages like *3D Studio Max* on powerful Windows NT machines offer serious competition to the bigger Unix-based systems. A CAD system was used to model the structures as a high level of integrity was being sought for the size, shape and position of the monuments; with hindsight some level of dimensional accuracy could have been sacrificed in order to use better visualization software.

The visualization process undertaken in this project consists of producing rendered scenes of the city from a variety of viewpoints, creating animations of various 'fly-throughs' and adding live dancers and pedestrians to populate the environment.

Still and animated images

The method of producing rendered scenes of computer models is common to many of the visualization packages. The surfaces of the CAD model are mapped with textures and materials, and the environmental conditions such as the camera parameters, lighting, fog, reflectance, depth cuing and even radiosity are determined. For animation sequences, the camera path and kinematics are also developed. Once selected, the scene or animation is sent for processing, which can take many hours and even days depending on the speed of the graphics processors.

In this project the surface textures and material have been determined from analogy with existing contemporaneous structures, and from the travellers' reports that described the way that some surfaces appeared. Generic materials like gold have been used from the software material libraries, and additional materials have been generated from scanned photographs of real surfaces. A rendered scene is shown in Figure 17.4.

In order to provide a limited amount of user interaction with the virtual world on the CD ROM, several animated sequences have been compiled as QuickTime Virtual Reality (QTVR) panorama and objects. This technology, originally developed by the Apple Corporation, enables a panoramic image to be digitally 'wrapped' on the inside of a cylinder, and displayed so that it is possible to navigate around this image. The images can be connected through a series of 'hotspots', so that it is possible in this project to look around a building, enter through a door, look around the room, and then pass through to another. The panorama phenomenon is not new (see Hyde, 1988), but the migration of the concept to a digital medium does allow interaction without the need to render scenes 'on-the-

Figure 17.4 Postcards from Ayutthaya, *c.* 1650+/−.

fly'. The technique has been used to show the site and monuments of *Wat Si Sampet* as they appear today (from digital photographs acquired for QTVR production) and also various reconstructions (from rendered scenes generated so as to be suitable for QTVR production).

The use of live actors

Many virtual worlds have been criticised for their 'ghost-town' feel; there are few people visible and when they do arrive they often appear as simplified forms or avatars if they have been computer generated. One method used extensively by the special effects industry in television and cinema is to add real people into a virtual set (or more to the point, to add the computer graphics around the actors). The process is based on the generation of a matte based on a selected chroma value (traditionally blue but modern digital systems can use a variety of chroma) and replacing this matte with another image or video sequence.

For the Ayutthaya project a troupe of dancers and musicians were recorded onto digital video in a studio against a pure blue background, so that they could be incorporated into the animations in the audience chamber of the palace. Computer-generated scenes and animation sequences in the virtual world were generated to match the view seen from the real camera, and composited with the video sequences of the dancers in a digital video editing system. Some new software packages like Alias|Wavefront's *Maya Live* allows very precise alignment between the real and virtual worlds and solves for the camera attributes (position, lens, zoom, path) needed to register the two worlds (Alias|Wavefront, 1998). In this project the lower end *Adobe Premiere* was used as the video editing package, along with a Canon DV1 digital video camera and a *Miro3000 Firewire* digital video board from Pinnacle Systems (Figure 17.5).

Figure 17.5 The live dancers merged with the computer reconstruction. (Still frame from video sequence.)

Other results

The work to date has been very successful, and well received. A proof-of-concept CD ROM has been produced to accompany an exhibition as part of the Golden Jubilee celebrations for His Majesty The King of Thailand. The CD ROM contains sequences of rendered images, QTVR animations of the reconstructions and the *in situ* monuments and a sequence with live actors. Currently this is being expanded, and supported by material developed for the World Wide Web.

Recent advances in the software has made the conversion of the CAD files into *Virtual Reality Modelling Language* (VRML) 'worlds' a relatively simple task and several of the models are now available on the Web (see Figure 17.6).

Further results of the project can be found on the Web at http://www.sli.unimelb.edu.au/~cliff/virtual and linked pages.

One example of the VRML output from the project can be found at http://www.sli.unimelb.edu.au/~cliff/ayutvrml/roomnew.wrl. These pages are in a constant state of change and modification as improved results become available.

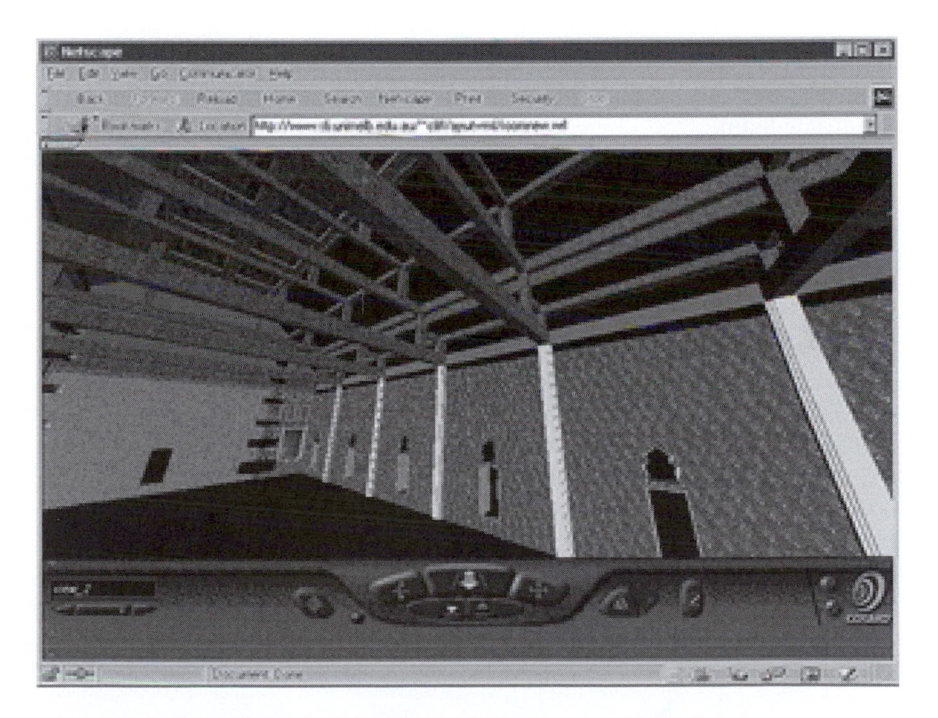

Figure 17.6 The VRML model of the audience room, *c.* 1650+/−.

Conclusions

This project is a joint project between Thailand and Australia that not only uses appropriate digital technology to solve a problem but also ensures the transfer of this technology to where it may be most usefully applied.

As this project hopes to show, with the appropriate use of computer visualization technology, a monument can be reconstructed to its former glory so that an experience can be gained of past greatness, but not at the expense of the antiquity of the remains. What differs between the approach taken here and other 'reconstruction' efforts is the use of photogrammetry to provide the metric base for the visualization of history. It is, however, not an essential component; it merely expedited the field documentation process and facilitated the production of the virtual city.

The combination of state-of-the-art computer visualization technology with measurement science and historical narrative is capable of producing visualizations of historic sites that can serve as research documents as well as be visually pleasing and engaging. What is important in this project is the closeness to the reality of the reconstructions, the verisimilitude of the visualizations. Considerable effort has been made to ensure that all elements of the reconstructions belong to the period, have the correct form and are covered by typical materials. The use of Thai advisers has facilitated this, and any incorrect assumptions are most likely the fault of the expatriate team.

Acknowledgements

The author wishes to acknowledge the financial support of the Australian Government Department of Employment Education Training and Youth Affairs (DEETYA) *Targeted Institutional Links Program*, and the competent team of fellow workers including John Cazanis, Jessica Smith, John Ristevski, Hugh Campbell, Karen Urquhart, Banjerd Phalakarn, Sanphet Chunithipaisan and Chalermchon Satirapod.

Thanks should also be given to those who participated in the video recording session, especially Morakot Sujintahisun and Srirate Yimyoh the two dancers, and Rungroj Phalakarn who acted as assistant director.

In addition the assistance willingly offered by the Royal Thai Government Fine Arts Department is recognised, along with the continuing logistical support offered by the Population and Community Development Association in Bangkok.

Glossary

The following contains definitions of some of the Thai terms used throughout the chapter. In addition, common transliterations of Thai words into English script have been used; however, there may be some variations in spelling between publications.

Stupa The most venerated structure for Buddhists. The *stupa* (or *chedi*) originally enshrined a relic of the Buddha, but over time they also contained kings or holy men, and now they are a religious symbol similar to the Christian cross.

Prang Originally modelled on the corner tower of a Khmer temple, it developed into a new form of Thai *stupa*. More geometric in form than a *chedi*, it generally consists of a tall tower mounted on a large base with three niches on three faces and a staircase leading to a doorway on the fourth face.

Wat The term *wat* generally refers to a collection of religious buildings surrounded by a wall with gateways. A *wat* may contain *stupa*, *chedi*, *prang*, *vihara* and a *bot*, as well as monks' accommodation, open shaded pavilions for relaxing and learning, as well as funerary monuments.

Bot The building within a *wat* containing the principle Buddha image. It generally has a rectangular plan with one nave, often covered with superimposed roofs with decorated gables.

Vihara Also known as *wiharn*, this is generally a building within a *wat* which houses the other Buddha images. The roofs are generally tiled and the walls white stucco.

Klong A canal, used for transport and communications. Unfortunately many have been filled in and turned into roads, others have become very polluted.

Chofa A decorative finial that graces the end of the roof of the *bot*, thought to perhaps represent a horn from an ancient protective mask.

References

Alias|Wavefront. 1998. *Maya Live*. http://www.aw.sgi.com/entertainment/solutions/index.html, 8 May 1999, Alias|Wavefront.

Beyer, H.A., Uffenkamp, V. and van der Vlugt, G. 1995. Quality control in industry with digital photogrammetry. In Gruen, A. and Karara, H. (eds) *Optical 3D Measurement III*. Heidelberg: Wichmann, pp. 29–38.

Brown, J. and Dold, J. 1995. V-STARS: A system for digital industrial photogrammetry. In Gruen, A. and Karara, H. (eds) *Optical 3D Measurement III*. Heidelberg: Wichmann, pp. 12–21.

De La Loubere, S. 1813. *Description du Royaume de Siam. MDCCXIII*. Amsterdam: Gerard Onder De Linden.

Forte, M. (ed.). 1997. *Virtual Archaeology: Re-creating Ancient Worlds*. London: Thames and Hudson.

Fraser, C.S., Ogleby, C.L. and Collier, P.A. 1995. Evaluation of a small format mapping system. *The Australian Surveyor*, 40, 1, 10–15.

Hamilton, C. and Breznau, T. 1998. Designing museum applications. *3D Artist*, 34, 14–16.

Hanke, K. and Ebrahim, M.A.-B. 1997. A low cost 3D measurement tool for architectural and archaeological applications. *International Archives of Photogrammetry and Remote Sensing*, 32, 5C1B, 113–20.

Hyde, R. 1988. *Panoramania!: the art and entertainment of the 'all-embracing' view*. London: Trefoil in association with the Barbican Art Gallery, 1988.

Krairiksh, P. 1992. A revised dating of Ayudhya architecture. *Journal of the Siam Society*, 80, 1, 37–55.

Novitski, B.J. 1998a. Reconstructing lost architecture. *Computer Graphics World*, 21, 12, 24–30.

Novitski, B.J. 1998b. *Rendering Real and Imagined Buildings: The art of computer modelling from the Palace of Kubla Khan to Le Corbusier's Villas.* Rockport, MA: Rockport Publishers.

Ogleby, C.L. and Rivett, L.J. 1985. *Heritage Photogrammetry*. Australian Government Publishing Service. 115 pp.

Patias, P., Stylianidis, E. and Terzitanos, C. 1998. Comparison of simple off-the-shelf and of wide-use 3D modelling software to strict photogrammetric procedures for close range applications. *IAPRS*, 32, 5, 628–32.

Piggott, S. 1978. *Antiquity Depicted: Aspects of Archaeological Illustration*. London: Thames and Hudson.

Pollefeys, M., Koch, R., Vergauwen, M. and Van Gool, L. 1998. Virtualising archaeological sites. In *Future Fusion: Application realities for the virtual age. Proceedings of the 4th International Conference of Virtual Systems and Multimedia*, November 1998, Ogaki, Japan. VSMM, pp. 600–5.

Shulman, S. 1998. Digital antiquities. *Computer Graphics World*, 21, 11, 34–8.

Smithies, M. 1995. *Descriptions of Old Siam*. Kuala Lumpur: Oxford University Press.

Streilein, A. and Niederöst, M. 1998. Reconstruction of the Disentis monastery from high resolution still video imagery with object oriented measurement routines. *IAPRS*, 32, 5, 271–7.

Wyatt, D.K. 1984. *Thailand: A Short History*. New Haven: Yale University Press.

18 Visualizing data components of the urban scene

Kate E. Moore

Introduction

Interactive, three-dimensional models of cities are proliferating as a means of visualizing urban data. Many, however, are either pure imitations of visual scenes or limited to visualization of two-dimensional spatial data. Models are also often closed environments with the only interaction for the user being navigation through the scene. Virtual reality has the potential to become a useful tool for the visualization of more abstract urban data. This is useful within an educational and planning context.

The urban geography background

The mapping of land-use has concerned planners and politicians for centuries as an aid in enabling taxation and development control. The importance of cadastral mapping continued as society moved from predominantly agrarian to metropolitan but, because of limitations of the map production process, little account was taken of the development of high-rise buildings. Only ground-floor activities or predominant use were mapped, higher levels seemed less important in the urban structure. Plans that mapped the central retail areas of many cities showed mainly ground-floor information or, at the most, the number of floors devoted to single major retail outlets. Census data collection, too, was based on two-dimensional spatial analysis. Census enumeration districts (EDs) were allocated according to their footprint on the ground, a high-rise block of apartments would be classed as a single ED. Early navigators and then planners in new world high-rise cities were aware of a third dimension but often these were mainly superficial perspective drawings of the city structure with no other implied meaning or associations. Classic examples of isometric paper maps are the 'optical perspective' plans produced by Bollman Bildkarten of Braunschweig, which culminated in an oblique view of Central Manhattan showing fine architectural detail but no indication of building function.

Deterministic geographers of the 1960s and 1970s who defined urban form and morphology were primarily concerned with the spatial (2D) spread of the city but referred little to the simultaneous growth of the vertical component. Geographic analysis of selected city features such as the Central Business District (CBD) did in some instances take account of the volumetric content of city structures. 'The popular concept of CBD size is two-dimensional, and of course, the district does occupy a certain gross area on the ground. Obviously, the factor of height is neglected . . . volume should be a better measure to use in comparing CBD size than ground area' (Murphy, 1972: 43). However, even this was not satisfactory for the analysis of the CBD as only a subset of land-use activities was considered to be part of the CBD. Therefore a final set of indices were defined that analysed 'business space'. These indices incorporated weighting factors for proportions of land use against total floor area. Other factors used in the delimitation of the CBD were the intensive use of space, higher land values, pedestrian flows and taller buildings. A second interest for geographers was the delineation of shape or form of the CBD, which was often achieved through generalizations and inclusion of subjective factors. Certainly the height factor was not fully considered although Murphy theorizes that the CBD is pyramid shaped and that barriers on the ground force abnormal vertical expansion of CBD characteristics (e.g. Manhattan). CBD land-use mapping solutions were limited to 2D maps or diagrams (Carter, 1981) and occasional aggregated zonal data by floor described in cumulative histograms.

Perhaps the most successful paper-based cartographic representation of 3D urban land use, or actual space use, is a map of Enschede produced by the ITC in 1975 (Figure 18.1). This map uses a graded method of depicting land use by floor level. Buildings up to four storeys high have their plan view coloured by ground floor with small circles superimposed showing levels two to four. Buildings over four storeys are shown in perspective view and the sides are colour coded. This map was produced jointly by cartographers and urbanists and overcomes many of the limitations of depicting three-dimensional attributes of urban space in a two-dimensional medium.

The introduction of 3D technologies

It has taken a change in available technology to trigger a reconsideration of the concepts of mapping urban form and urban phenomena (Batty *et al.*, Chapter 15). New technologies have provided ways of collecting, processing and visualizing three-dimensional data that should overcome previous limitations on mapping the urban environment.

The primary developments in geographical information systems have concentrated on analysis across an x, y plane or, at best, applying surface modelling and visualization methods to terrains and population surfaces (Langford and Unwin, 1994). Models of buildings by floor have been handled as 'layers', i.e. 2D slices across the city. With the advent of VR

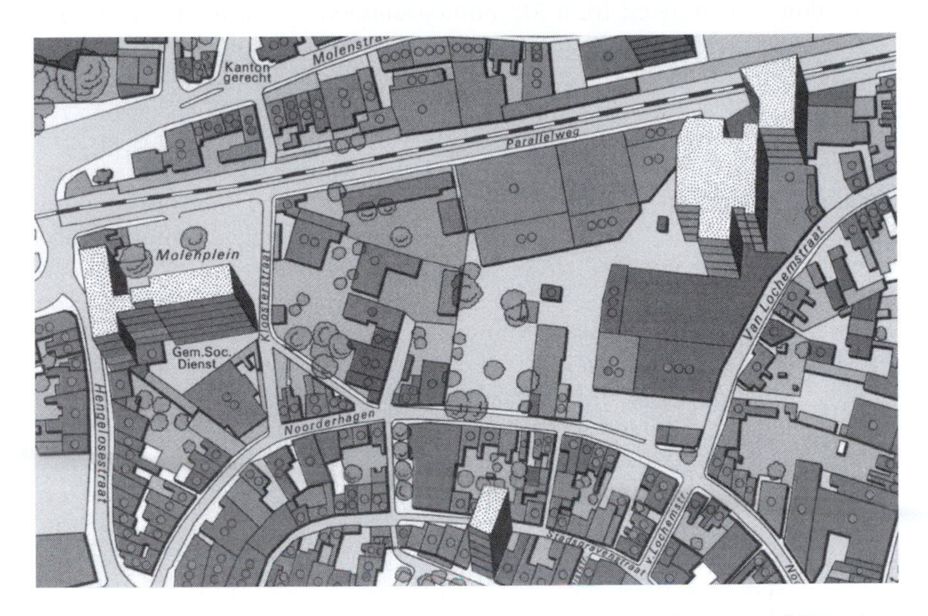

Figure 18.1 Actual space use in Enschede – a semi-perspective map of 3D urban use of space where the cartographic representation is graded by the number of floors in the building. Copyright: International Institute for Aerial Survey and Earth Sciences (ITC), Enschede, the Netherlands, 1975.

add-on components such as Pavan (Smith, 1997) were developed to generate virtual reality models from GIS data but these are in the form of models isolated from the data. Not until the late 1990s did any GIS include any form of interactive virtual reality component from which the database itself may be interrogated. However, little concern has been shown to the 'hidden' geographic relationships in the built environment and mapping truly three-dimensional urban data that varies in three planes. This is due to a lack of previously-collected height-related data which, in turn, was due to the difficulties in collecting such data and also the lack of facilities to display or cartographically describe the data.

A major obstacle in developing 3D thematic urban models has been the lack of 3D urban data. This seems to be the case for both the geometry of buildings and attribute data associated with the geometry. Some of the geometry may be interpolated by remote means, attribute data collection, however, is often more difficult and time consuming. A growing awareness of the need for this data has led to the instigation and development of different forms of data collection.

Automatic methods for building reconstruction on a large scale are being developed to interpolate building height data from aerial photography (The GeoInformation Group, 1999). This data collection methodology is derived from 3D photogrammetry using multiple images (Moons *et al.*, 1999). Using this technique the shapes of blocks of buildings are interpolated. Height information can now be created accurately and is readily available (Jackson, 1999). Fairbairn and Taylor (Chapter 16) discuss automated 3D model generation from photographs. This process may provide structural detail if not available from CAD or architectural drawings. 3D urban modelling has mainly focused on the most readily available data. The modelling of realistic views by applying multiple textures and a high level of detail onto a geometrically-precise model is prevalent. However, the use of VR for visualizing thematic data related to urban environments is far less common.

Attribute data collection is problematic and a major task. Data relating to the constituent floors of buildings may, at best, be recorded in architectural CAD packages, and more typically are collected by manual survey. In the UK the data collected by Experian GOAD for mapping of city retail areas relates primarily to the retail outlets themselves; detail on other uses is less well defined. In the past the UK census has been concerned only with 2D spatial units. The 2001 UK Census (Office for National Statistics, 1999) had a single question related to the lowest floor level of a dwelling, only in Northern Ireland will the number of floors in a dwelling be recorded. The socio-economic implications relating to dwelling height are at present unknown. Environmental concerns, such as air pollution and car emissions, may have dimensional factors relating to their influence on public health but these are completely unknown. Collection and visualization of 3D data may help to spotlight any such relationships.

Recently the trend to a greater awareness of environmental issues concerned with urban design has prompted an interest in modelling in three dimensions. Some pioneering work has been undertaken within the fields of urban planning and urban design to model 3D thematic data in conjunction with temporal (4D) data that has particular relevance to environmental issues. Koninger and Bartel (1998) have developed a preliminary 3D urban GIS that incorporates a 3D database management system and a 3D computer graphics environment to generate an initial 3D GIS. It is envisaged that this will at a later date be integrated into a full 3D GIS with spatial analysis (2D and 3D) functionality. Bourdakis (1997) explores the applications of 3D environments for visualization of urban data as a tool for the design and planning process. Socio-economic and pollution data are mapped onto the 2D and 3D form of the city. However, the data itself is not subdivided into height levels. Instead two approaches are illustrated to '3D' mapping. First, data values for two-dimensional space are symbolized on the 3D form and, second, the height of the building or block is extruded to represent attribute value. In the urban planning arena there are various examples of early attempts to utilize 3D space in the community planning process. Researchers at the University of Illinois (IMLAB,

1999) are developing a VR interface to encourage the participation of the city's poorest communities in planning matters. This example also generalizes land use to block level with no subdivision of functionality by floor or even building.

As the twentieth century closed, space was being re-evaluated by planners and politicians and was becoming a valuable resource. The development of brown field sites is increasingly important to make efficient use of redundant or previously-developed land. This has also encouraged the mixing of land use in the vertical plane. It is UK government policy that, in the decade up to 2006, 60 per cent of new residential space should be allocated within existing urban areas to decrease the pressure being applied to rural areas. Identification of vacant property above existing office and retail space is important in the trend towards urban renewal, and mapping this space equally so.

> Although it is fashionable to play down the 'mapping' role of geography, the identification of spatial patterns and spatial linkages within the city provides a basic source of material both for educationalists and policy makers. Similarly the geographer's traditional concern with the 'distinctiveness of place' and areal differentiation is very relevant to the study of the city...
>
> (Knox, 1995: 4)

Mapping in multiple dimensions may also be used to highlight other urban environmental issues such as decentralization of retail outlets from the traditional CBD to out-of-town sites. Identification and visualization of the dynamics of space utilization in inner-city areas may help plan and control movement of business sectors. Mapping the changes in land use, particularly in the transitional zones at the edge of the CBD, provides an indicator to the decline or rejuvenation of the CBD. This dynamic change in the multi-level functionality of the city has never seriously been explored.

The connection between factors of dimension and our understanding of the environment are therefore important. From cartographic, the analytical and the political perspectives, understanding the third dimension (and indeed the temporal components) of urban spaces must lead to new insights into the way people occupy their environment.

Virtual reality for visualization of urban data

It is worth considering the features of virtual reality that are of benefit to urban designers and planners in a more general context. The role of maps is fundamental in the process but the application of VR in a cartographic context is as yet less well-defined or appreciated.

Urban model building in virtual reality environments has seen a profusion of imitations of reality. This approach limits the full potential of VR for urban mapping. Reality itself need not be limited by what can be seen or experienced directly with the bodily senses. VR offers the ability to create new versions of reality – manufactured intensities (Gillings, 1997) that are based on unseen data. 'VRML is less about making virtual architecture look like physical architecture and more about creating dynamic space that can present information to users in new ways' (Lawton, 1999: 4). This data-driven approach to visualization in VR is, as yet, under-explored, particularly in the area of urban modelling. Virtual reality has been utilized to build many pseudo-real models of the urban environment but less emphasis has been given to the utility of VR for visualizing urban data or the non-tangible elements (Kraak, Chapter 6) of cityscapes.

Virtual reality is being used increasingly for the depiction and visualization of geographic data. It is unique in its ability to simulate the natural experience of being in and moving around spatial locations. However, a greater understanding is required of both the conceptual role of virtual reality for geographic visualization and the practical cartographic methodologies that may be employed. Currently a wide range of empirical investigations is being pursued to fully understand how to apply virtual reality in this context.

There are two key concepts that may add value to virtual environments for visualization of spatial data. First, by understanding that virtual reality is a continuum of reality experiences and second, that as users becomes more integrated into the virtual world they can interact directly with the data capture and input process.

Virtual reality can be seen as a continuum of realities with reality itself, or the re-creation of it, as at a central point on the scale (Figure 18.2). This recreation of reality may be considered as the manufactured deficiency – somehow lacking. The ends of the scale are what have been termed as manufactured intensities. The possible extremes are augmented reality at one end of the scale, where virtual reality adds information to or goes beyond reality, and alternative reality or abstractions of reality at the other. Virtual reality can supply a range of 'virtualities' that exploit the

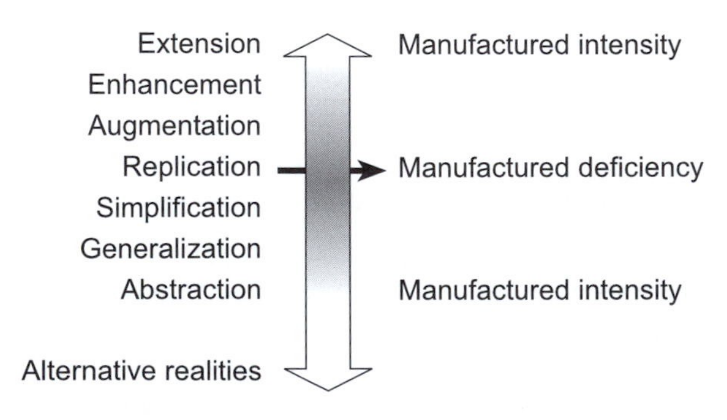

Figure 18.2 Virtual reality as a continuum of realities.

capabilities of virtual reality. The 'abstraction' end of the scale increases the intensity of information much in the way that cartographic design processes can aid cognition of the real world. Simplification, abstraction and generalization selectively remove or alter detail to allow a clearer perception of the general form. Alternative realities communicate information via symbolic or iconographic means. Virtual reality may utilize cartographic precepts to form virtual 'maps' depicting multidimensional geographical data.

Virtual reality need not be purely the navigation of predefined models, but users can interact with the data and input the data directly whilst within the virtual world. Brodlie *et al.* (Chapter 2) have adapted the cartographic model of Tobler (1979) to show how VR can alter the map-making process to integrate the user into the map. This process may be much more profound as the model contracts still further towards the data end of the flow by allowing the user to input and modify data concurrently with experiencing the virtual reality map (Figure 18.3) and, in a planning context, influence the geographical information of the future. Ryan *et al.* (1998) have already demonstrated how computers can be a useful addition to field data collection for subsequent processing. In the foreseeable future the user will be able to collect data, using palm-top computers, have the model, not just map, appear in real-time, and then modify the model to influence projected scenarios of the real world. The virtual reality characteristic of 'immersion', being inside the virtual space, may be extended to allow the user to be 'immersed' in the data. This embodies the concept of

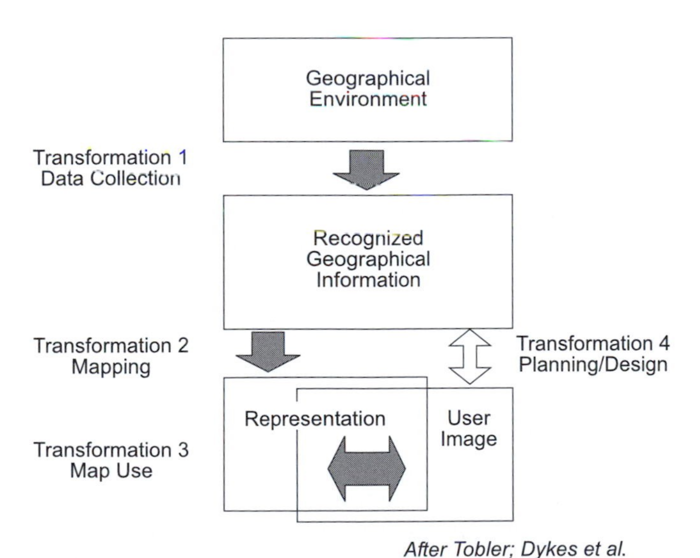

Figure 18.3 Fundamental information transformations in cartography (see Chapter 2; Tobler, 1979) as modified by virtual reality visualization systems.

'exploratory data analysis' (Tukey, 1977), where the user explores data in a free-form interactive manner. The user is immersed in the representation of the data, interacts with the data and creates the information from 'inside'. The data collection process is contracted into the virtual mapping. This process then becomes akin to augmenting reality or the other end of the realities continuum. Virtual reality offers much more than other visualization techniques and gives much greater control to the user who is able to interact directly with the scene and objects in the scene, so creating his/her own view of the model of reality.

The technical ability to portray data using multiple dimensions and dynamics offers a range of possible visualization methods that may be used. It is useful to consider these and their relationships. Many elements in spatial data visualization may be varied to provide a rich, dynamic picture of the structure and relationships between different aspects or components of the scene or map. Bertin (1986) introduced a scheme for defining 'visual variables' and their use in cartography. Many of the standard cartographic symbols can be applied successfully to volumetric models: colour hue, shading intensity, texture, shape or size variables. The third (volumetric) dimension itself may be used to symbolize value but comparative baseline features may be lost and data is obscured by other values.

This range of symbols has subsequently been extended and modified by many others to take account of a wider range of symbolic elements including time components (Sheppard, 1995). Both two- and three-dimensional components of an urban scene may be modified by dynamics to bring in the temporal dimension. The ability of the user to alter viewpoint may be considered as a variant of time within this context.

When combining data themes for visualization in virtual reality, it is possible to mix representations in a way in which two-dimensional spatial themes are still visible in conjunction with 3D themes. However, the number of dimensions used in any visualization should be a product of necessity rather than forcing visualization into an inappropriate form. For instance, traffic flows at street level are essentially two-dimensional data.

For 3D cartography of urban data many of these concepts are applicable and can be used to great effect in the visualization of urban phenomena. New visualization environments may be devised by withdrawing from the neo-realistic interpretations of cities to investigate the cartographic or abstracted models of cities. Modelling urban data as a three-dimensional phenomenon, addition of temporal factors such as pedestrian flows or pollution distribution and the interactive input and concurrent visualization of the data are all important concepts in the design of VR urban environments.

UrbanModeller

UrbanModeller was designed with several objectives. It was, first, an empirical study of virtual reality techniques for the visualization of 3D urban data with consideration of the integration of two-, three- and four-dimensional data types. The system was developed for integration into urban fieldwork and it was to provide an interactive method for building a three-dimensional database and thematic urban models. It was this combination of fieldwork requirements, technological development and cartographic innovation that led to the initial conception of the UrbanModeller program described here.

As a means to enhance student fieldwork, a series of software components, which are essentially spatial interfaces to geographical data, have been developed, linked to a networked geographically-referenced database (Dykes *et al.*, 1999). UrbanModeller fits into this suite of programs to visualize hidden aspects of the urban environment. UrbanModeller initially has facilities to build land-use models of a multidimensional urban scene and to visualize pedestrian counts at sample point locations. Other features are the geo-referencing of the model and dynamic tracking on a 2D map view. It is envisaged that the utility shown here will be extended to visualize other variables such as rents, socio-economic characteristics or epidemiological data that may be appropriate in certain multi-storey urban environments. UrbanModeller has been developed along the same lines as the VFC software component traVelleR (Moore, 1999; Moore *et al.*, 1999). They both utilize the Virtual Reality Modelling Language (VRML) and the facility to extend beyond the modelling capabilities of this language using Java and the External Authoring Interface (EAI). The EAI provides several useful functions for interactive model building. These include the ability to add objects to the scene, a callback method which can alter parameters of any object either defined in the initial file or added later. Some simple geographic examples of these methods are given in Moore (1999).

The UrbanModeller interface is embedded in two frames of an Internet browser window. The left-hand frame is a VRML viewer and the right-hand frame a Java applet consisting of an interface and a 2D map (Figure 18.4). The VRML window initially loads a simple VRML file containing a single root node to which new children nodes may be added by referencing its DEFined name.

The models are generated from ground plans of the buildings based on *x*, *y* coordinates that for polygons for each building plan outline. These are loaded into UrbanModeller in the form of ArcInfo ungenerate polygon format ASCII files. Attributes may either input from associated polygon attribute files or to provide the greater engagement with the model building the VRML models generated interactively by the student in UrbanModeller. Each building is selected in turn from the 2D map interface. A

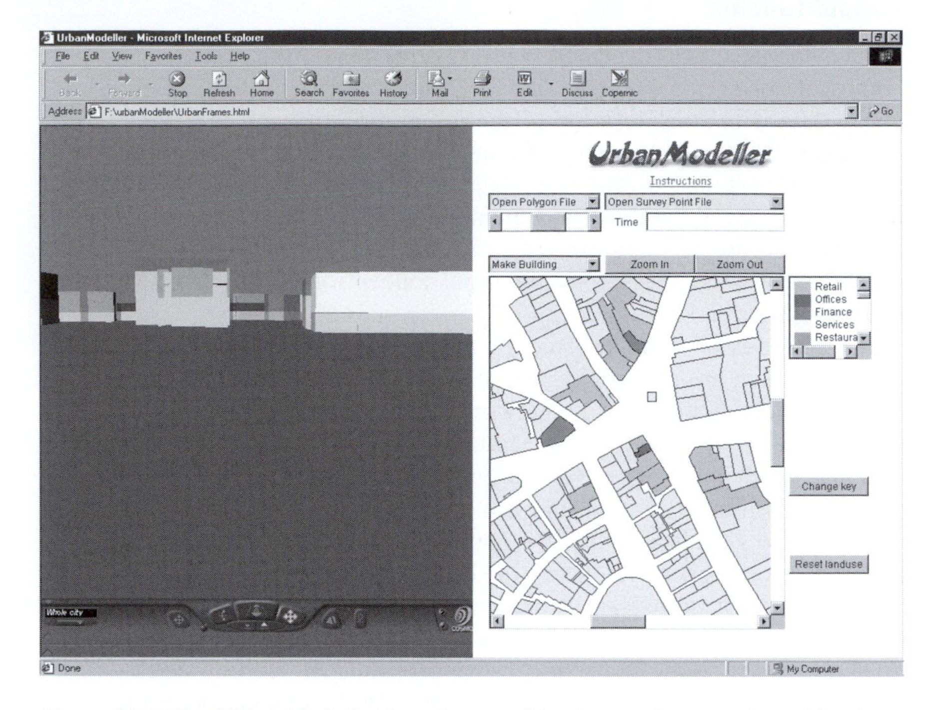

Figure 18.4 The UrbanModeller interface enables interactive entering of land-use categories to the building by floor level. (Derived from Experian Goad Digital Plan; Experian and Crown copyright reserved.)

building dialog then requests number of floors. The land use (or other variable) may then be input for each floor in turn. Colour is used to map the land-use value for each floor. The colour is applied to each floor which is constructed as an extrusion from the building plan. On completion of all floors the building is added to the scene. An associated output VRML file may also be constructed simultaneously. The building number and floor level reference every floor. Additional multimedia data may be added by building, as anchors to that building, to add real views of the buildings to compare to the representational view.

Two methods have been tried for visualizing pedestrian flows (Figure 18.5) and a third is envisaged. In UrbanModeller an empirical use of dynamic bars within the scene was developed primarily to test VRML interface functionality rather than as a visualization methodology. Although an interesting experiment to test the capabilities of VRML technology, this was seen to be inappropriate as a visualization method for these data. First, there was confusion between buildings and symbols, and second, there was no static base line to the data values that also overlapped. A second method reverted to 2D visualization by overlaying a

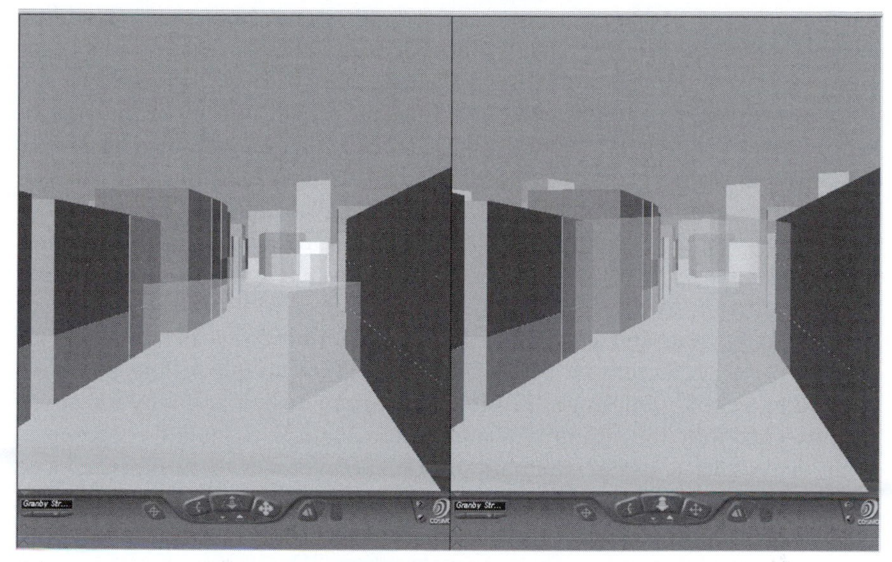

(a)

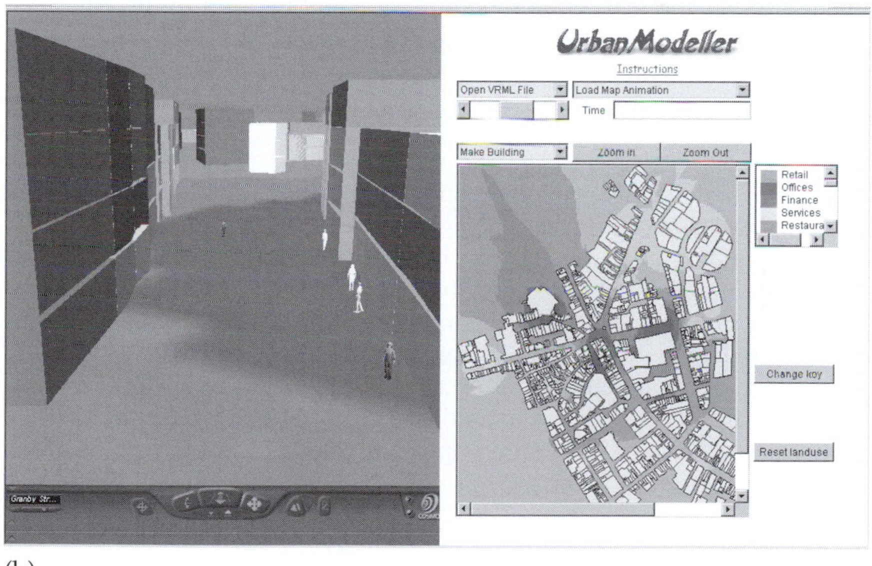

(b)

Figure 18.5 Experiments in visualization of pedestrian flows. (a) animated 3D columns at survey points. (b) interpolated density surface 'draped' on the pavement. (Derived from Experian Goad Digital Plan; Experian and Crown copyright reserved.)

colour 'pavement' on the map and on the model that did help to visualize the quantity of people. A third method, as yet undeveloped, is adding avatars to the scene to represent numbers of people. To develop this latter method into a more interesting solution would be by visualizing the dynamics of people moving through the city scene with some intelligent agent to simulate their position and direction of movement. This would require collection of a greater level of seed data to feed into the simulation. Some work on an intelligent agent in a two-dimensional scene has been done by Batty *et al.* (1999) to emulate the behaviour of pedestrians. Schelhorn *et al.* (1999) in the STREETS model apply an agent-based approach in a two-dimensional street scene. From initial survey of people's behaviour and analysis of the component influences that steer this behaviour, a predictive model is built. This type of modelling could be applied in a three-dimensional context for urban design and analysis applications. Certainly the addition of even static avatars, modelled representations of people, to urban thematic models provides a form of visual cue that aids cognition of an abstract urban scene.

This initial model builder explores the feasibility of integrating data-input and model building into an interactive 3D environment. A more elegant and effective solution would be to use a small palm-top computer on-site to collect data. At base these data could be used to directly build the 3D model and would provide an efficient mechanism for other data collection of this type.

Conclusions

Virtual reality modelling and thematic visualization are useful in the urban planning and design context. It provides a way to monitor change in city areas, provides indicators to the behaviour of people in their urban environment and could potentially be used in the optimization of sites for marketing purposes. With the incorporation of 3D data in the UK 2001 census, modelling 3D socio-economic data becomes feasible. Analysis of these data could potentially offer new insights into some problems in urban areas such as in epidemiological implications of height. Virtual reality in the future will take an important role in many aspects of urban analysis and administration.

References

Batty, M., Jiang, B. and Thurstain-Goodwin, M. 1999. Working paper 4: local movement: agent-based models of pedestrian flows. The Centre for Advanced Spatial Analysis Working Paper Series, http://www.casa.ucl.ac.uk/working_papers.htm.

Bertin, J. 1983. *Semiology of Graphics: Diagrams, Networks, Maps*. Trans. by Berg, W.J. Madison: University of Wisconsin Press.

Bourdakis, B. 1997. Making sense of the city. Paper presented at CAADFutures97, August 1997. http://fos.bath.ac.uk/vas/papers/CAADFutures97.

Carter, H. 1981. *The Study of Urban Geography*. London: Edward Arnold.
Dykes, J., Moore, K. and Wood, J. 1999. Virtual environments for student fieldwork using networked components. *International Journal of Geographical Information Science*, 13, 397–416.
The GeoInformation Group. 1999. Cities Revealed®, http://www.crworld.co.uk/.
Gillings, M. 1997. Engaging place: a framework for the integration and realisation of virtual-reality approaches in archaeology. In Gafney *et al.* (eds) *Computer Applications and Quantitative Methods in Archaeology*. British Archaeological Reports International Series. Oxford: Archaeopress.
IMLAB Imaging Systems Laboratory. 1999. SmartCity, http://www.imlab.uiuc.edu/.
Jackson, M. 1999. A forgotten dimension? *Mapping Awareness*, 13,1, 16.
Knox, P. 1995. *Urban Social Geography*. London: Longman, p. 4.
Koninger, A. and Bartel, S. 1998. 3D-GIS for urban purposes. *GeoInformatica*, 2, 1, 79–103.
Langford, M.L. and Unwin, D.J. 1994. Generating and mapping population surfaces within a geographical information system. *Cartographic Journal*, 31, 1, 21–6.
Lawton, M. 1999. Advancing 3D through VRML on the Web. *IEEE Computer Graphics and Applications*, 19, 2.
Moons, T., Frere, D. and Van Gool, L. 1999. A 3-dimensional multi-view based strategy for remotely sensed image interpretation. In Kanellopoulos, I., Wilkinson, G.G. and Moons, T. (eds) *Machine Visions and Advanced Image Processing in Remote Sensing*. Berlin: Springer Verlag, pp. 148–59.
Murphy, R. 1972. *The Central Business District*. London: Longman.
Office for National Statistics. 1999. http://www.ons.gov.uk/ons_f.htm.
Ryan, N.S., Pascoe, J. and Morse, D.R. 1998. FieldNote: extending a GIS into the field. In Barcelo, J.A., Briz, I. and Vila, A. (eds) *New Techniques for Old Times: Computer Applications in Archaeology*. 1999. BAR International Series number S757. Oxford: Archaeopress, pp. 127–32.
Schelhorn, T., O'Sullivan, D., Haklay, M. and Thurstain-Goodwin, M. 1999. STREETS, an agent-based pedestrian model. *Centre for Advanced Spatial Analysis Working Paper 9*. University College London, http://www.casa.ucl.ac.uk/streets.pdf.
Sheppard, I.D.H. 1995. Putting time on the map: dynamic displays in data visualization and GIS. In Fisher, P. (ed.) *Innovations in GIS 2*. London, Taylor and Francis, pp. 169–87.
Smith, S. 1997. A dimension of sight and sound. *Mapping Awareness*, 11, 18–21.
Tobler, W.R. 1979. A transformational view of cartography. *The American Cartographer*, 6, 101–6.
Tukey, J.W. 1977. *Exploratory Data Analysis*. Reading, MA: Addison-Wesley.

19 Virtuality and cities

Definitions, geographies, designs

Michael Batty and Andy Smith

Introduction: converging technologies

Computer graphics came of age in the late 1970s when the classic fly-through of Chicago was developed by the architectural firm of Skidmore, Owings and Merrill (Deken, 1983). Until then, visualizations of large-scale artifacts were esoteric, static affairs that simply provided snapshots of real or imagined structures with little or no user interaction other than through offline preparation of data and designs. Although computer aided design (CAD) began in earnest soon after microcomputers were invented, software such as *AutoCad* remained at the level of the advanced drafting package until machines became fast enough to allow users to interact with and change designs 'on-the-fly'. Simultaneously, the idea that users could 'immerse' themselves within the computer environment used to visualize such designs – so-called virtual reality (VR) – was being fast developed with the first headmounted displays and interactive gloves emerging at much the same time as the first primitive CAD packages (Rheingold, 1991). After almost twenty years of intense development, both CAD and VR are now largely interchangeable but their convergence has been somewhat narrow. This has been mainly restricted to single user interaction with an emphasis on better and more realistic rendering of scenes rather than the development of any specific functionality that makes CAD and VR special purpose tools for better analysis, decision making or design.

In one sense, progress in strict CAD and VR has been a little disappointing, for the focus has been on technical rather than scientific developments; but, more recently, these areas have begun to be transformed (Kalawsky, 1993). Their technologies are rapidly converging with quite different conceptions of the digital world based on communication and participation, on visualization in its widest sense, and on simulation and modelling. Notwithstanding the development of VR and CAD in contexts wider than the desktop – the theatre, the CAVE, and such-like forums which clearly widen participation – the most profound developments have been in software and networking. Different kinds of visualization where data has dimensionality other than three – for example, in temporal forecasting, in geographic information systems, in image processing and so on – are having a significant influence. The ability to interact with software and data across networks employing a variety of hardware is broadening communication and use of these technologies as well as enabling new modes of participatory science and design.

Virtual reality systems are no longer restricted to single users immersed in their own worlds but are now accessible to many users simultaneously. Ways of building environments which do not depend exclusively on 3D rendition but embrace data structures appropriate to analytic functions and models, enable a complete spectrum of applications. These range from the real to the fictional and the surreal at different levels of abstraction and involve users across the continuum from passive observation to active participation. Consequently the term 'virtual' has become ever more ambiguous (Heim, 1997; Woolley, 1992). VR now embraces a variety of systems from the totally immersive, centralized, single-user tools with which it began, to the entirely decentralized, remote and anonymous technologies spawned by the Net. Into this milieu has come the 'city' metaphor which likens a virtual world or virtual reality to an urban complex, often represented in 2D or 3D form, as maps and scenes in which users explore and interact with the objects of their interest from buildings to themselves, from clocks to clouds. In the very recent past the term 'virtual cities' seemed encompassing and focused enough to capture the idea of VR in this domain but it no longer suffices to define that set of technologies and their uses which all have some impact on, and draw their inspiration from the city metaphor. Instead, in this chapter we will simply refer to such technologies as embodying the property of virtuality, implying some kind of digital abstraction with respect to the use of computer technologies in the context of the city.

Accordingly we need to be clear about the scope of our inquiry into 'virtual cities'. We will begin by some specific discussion of the idea of virtuality, reality and abstraction in the digital domain and more specifically within the context of geography. The idea of VR as simulating 'real' domains in their physical form drives the idea of virtual geography and we will illustrate how far 2D and 3D representations of cities have been developed with state-of-the-art digital technologies. This emphasis on representation has a parallel in analysis and communication, and although we will point to analysis and what VR holds out for this traditional focus of geographic inquiry, we will then turn to ways in which the 'real' might be communicated using VR. In particular we will describe the rapid transformation of what might have appeared to be esoteric technologies in the very recent past to the public arena. We will illustrate these notions with some recent work showing how such technologies can help non-professional audiences concerned with urban issues in large cities. There is another side to VR which is evolving quickly and this concerns the representation and simulation of non-physical digital domains, which are surreal rather than real. We will briefly chart the emergence of these virtual worlds, focusing on ways in which we might transfer their insights to physical situations where we have a scientific or professional interest. We will conclude by defining a set of themes which we consider to be of relevance to geographic inquiry and which are informed by the emergence of virtuality in these domains.

Definitions: the 'virtual' and the 'real'

Defining terms such as the 'virtual' and the 'real' can become an elaborate pastime with no agreed consensus ever in sight. The relativist position would have it that there is no difference; or at least any difference has little to do with the digital or non-material domain per se. In a recent movie, *The Matrix* (Warner Brothers Pictures, 1999), one of the heroes says: 'What is real? How do you define real? If you are talking about what you can feel, what you can smell, and you can taste and see, then the real is simply electrical symbols interpreted by your brain', with the unspoken implication that the virtual reality in which this is being said is the actual reality. Yet in this context, we do require a working definition to make progress, for we need to provide some form of classification of the wide array of digital environments which are appearing under the guise of virtual cities.

A good dictionary definition of virtual is 'almost real', containing some of the essence of the reality but this does not go far enough, for it does not separate the digital from the non-digital. In one sense, a better definition equates virtual with digital abstraction in that everything that has hitherto been referenced as a virtual reality, in the digital sense, is an abstraction of various properties of cities. But this too is changing; what is assumed to be reality in and of itself is becoming digital with accessibility to this virtual world depending intrinsically on the technology of VR. As the city gets itself wired, and as the population begins to interact online, then many of the tools of VR are being used to enable users to manipulate the information and the information-processing activities that constitute the most important aspects of interaction in the real economy (Negroponte, 1995). There is still abstraction at play here but our model is one where the process of abstraction is nested into itself (Figure 19.1)

A reality that is abstracted can become another reality that is further abstracted and so on, into infinite regress. This nightmare of circularity is never far from the surface when virtuality is at issue as we will clearly see in the various examples that we present within this chapter.

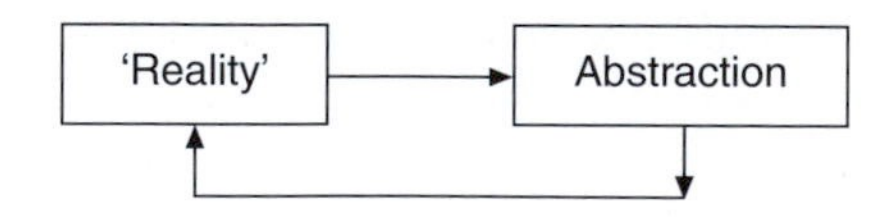

Figure 19.1 The nesting of abstraction into itself.

The hallmark of virtual reality, then, is that it must be digital in some non-human and/or non-material sense, but even this is too broad, for it would include all computer endeavour. The traditional and narrow definition of VR in fact emphasizes two elements: *interactivity* which now almost uniformly characterizes computer use but also the *intensity of the experience* of the user which conventionally has been biased to the visual but also includes smell, touch, taste, and so on (Steuer, 1992, quoted in Borgmann, 1999). This is still too ambiguous. Many experiences with digital phenomena heighten normal intensities of experience; indeed it might be argued that this is the very point of computer use, and thus to refine our working definition we need to consider VR as being the *heightening of experiences beyond the current norm* of existing computer use. This implies that we are dealing with a moving target, that the boundary between virtual reality and other forms of digital abstraction is continually shifting which has been the case ever since the term was invented. Yesterday, virtual cities were those like the Chicago fly-through: individual users simply flew around in order to understand and perhaps change through design, heightening their experience of cities by navigating entire areas rather quickly and gaining insights through unconventional, often impossible views. Today, VR is participatory, stretches across the Net and is deeply embedded within it, drawing on many tools that traditionally come from the VR of yesterday. Tomorrow, this kind of VR will be commonplace and, if VR still exists as a distinct digital domain, then it will be different and beyond whatever we are able to see today.

Such a shifting boundary implies a very high degree of fuzziness concerning what is and what is not VR. An increasing number of applications contain bits of virtuality. In the past, because the heightening of experience has been entirely associated with the technology of interactivity – headsets and gloves very largely – then VR projects have been quite distinct, embodying little other than technologically-driven applications. However, all this is changing and technologies are increasingly embracing the Web, opening access, and thereby changing the way experiences can be heightened. Three elements seem best to characterize the current processes affecting VR: simulating, experiencing (in physical terms) and communicating. None of these are mutually exclusive or distinct categories in any sense for they all overlap and VR usually involves all of these in some way. However, we can use these quite fruitfully to identify applications with respect to virtuality and cities. If we array these against passive and active applications, defining passive as those in which users 'normally' interact with their VR individually, and active as those in which users interact collectively in some participatory sense, then we form a matrix in which we can specify typical and predominant applications (Table 19.1).

In terms of techniques, many of which are illustrated throughout this book, CAD, VR and multimedia represent the heartland of virtual cities, all being used across the six categories defined in this matrix. However, it

Table 19.1 Typical and predominant applications of VR

	Passive	*Active*
Simulating	Modelling: e.g. simulation on the desktop	Modelling and changing: e.g. immersive VR models of cities
Experiencing	Observing: e.g. exploring maps, designs	Engaging and changing: e.g. immersive & Web-based worlds
Communicating	Displaying: e.g. reading websites	Delivering: e.g. making decisions about services

is in developing active experiences in immersive, theatre or Net-based environments that their most significant use has developed. Frequently such applications are referred to as models, although the use of this term is as 'iconic' model rather than the more frequent use of simulation model which is invariably based on more abstract, formal ideas. CAD representations of cities reflect 'iconic' digital models in contrast to GIS where the focus is on more 'abstract' or 'aggregate' representations and simulations of urban structure and process. In this sense, GIS embodies tools more likely to inform passive simulation, experience and communication, although increasingly this too is evolving into Net-based, participatory contexts (Plewe, 1997).

Although our chart provides a convenient way of grounding examples, we will not attempt to be comprehensive. We will 'sample' the space in a way that at least engages and illustrates these concepts. This is because, almost everywhere you turn, you now see aspects of the city being involved in applications which embody virtuality. Four years ago, when one of us (Batty, 1997) wrote about virtual cities, the range was much narrower, as was our definition of the real and the virtual. In the rest of this chapter we will organize our presentation into three sections, beginning with those applications where the focus is on representation and experience such as 2D maps and 3D physical models, then turning to the way such abstractions of physical realities can be communicated, and finally illustrating how non-physical environments are emerging as a result of participatory interaction across the Net. These may embody reality in some measure but are usually based on virtual communities that have no material, geographic extent.

There is an area of virtuality and cities that we will exclude from our discussion, or at least only refer to indirectly. Real cities are increasingly becoming digital as the kinds of materials and communications infrastructure that they are composed of is used to move information. These 'wired cities' blur almost imperceptibly into 'virtual cities' and although we will detail many features of virtuality, particularly Net cities or cities whose physical presence is communicated through the Web, we will not present examples of how cities are being wired or how socio-economic activities are adjusting in response to this new digitality. This more properly belongs to the new economic geography and the geography of telecommunications (Graham and Marvin, 1996), although we recognize that measuring and modelling cyberspace is forever merging with the development and use of virtual cities for analysis and design.

Cities in 2D, 3D and in traditional VR

Conventional CAD which forms the basis of physical models of the city such as the early Chicago fly-through has reached a fine art in the current generation. The model of the town of Bath, UK, is a quintessential example. These models are geometrically-accurate portrayals of buildings, and to a lesser extent, urban landscapes, built from a range of plan scales, some of which, such as the Bath model, use digital street and terrain data associated with GIS. These models originally appeared for single users on the desktop, but were then ported to single-user immersive environments where the experience of flying along, over and through buildings was heightened, thence to the VR theatre, and now they exist on the Web. But in a sense, the improvements since the primitive versions appeared fifteen or more years ago, have all been in the detail of rendering. These visualizations look so clinical and clean that most of the effort has been to make them more irregular. The first step to this, not surprisingly, has been to add detail to the buildings in terms of form and façade. Fifteen years ago, there was a strong 'Ghee Whizz' element to them, generated by the surprise that such animations elicited. With each subsequent tweak in the technology and increase in rendering detail, new audiences have been found to continue the praise.

A scene from the Bath model which is a complete rendition of the town at the level of individual buildings is shown in Figure 19.2. You can fly through this structure at will and even take a virtual tour within a sightseeing bus along a preprogrammed route but that is all. In so far as the model is useful for design, it is to show the visual impact of new buildings. But in terms of an enhanced understanding, everything is left to the user's imagination. This model and its like have minimal functionality, although a large number have been constructed. A recent survey (Batty *et al.*, 2000) reveals that more than sixty serious models for large cities with greater than one million population exist, with some cities such as New York having several competing applications from different databases. However, few of these applications have anything other than visual interest, often designed simply to show that this kind of virtuality is possible, or for 'infotainment'. There is much debate as to how useful these models are for analytical purposes, especially given the cost of construction, although this too is falling rapidly with the advent of new techniques for the remote sensing of the relevant data.

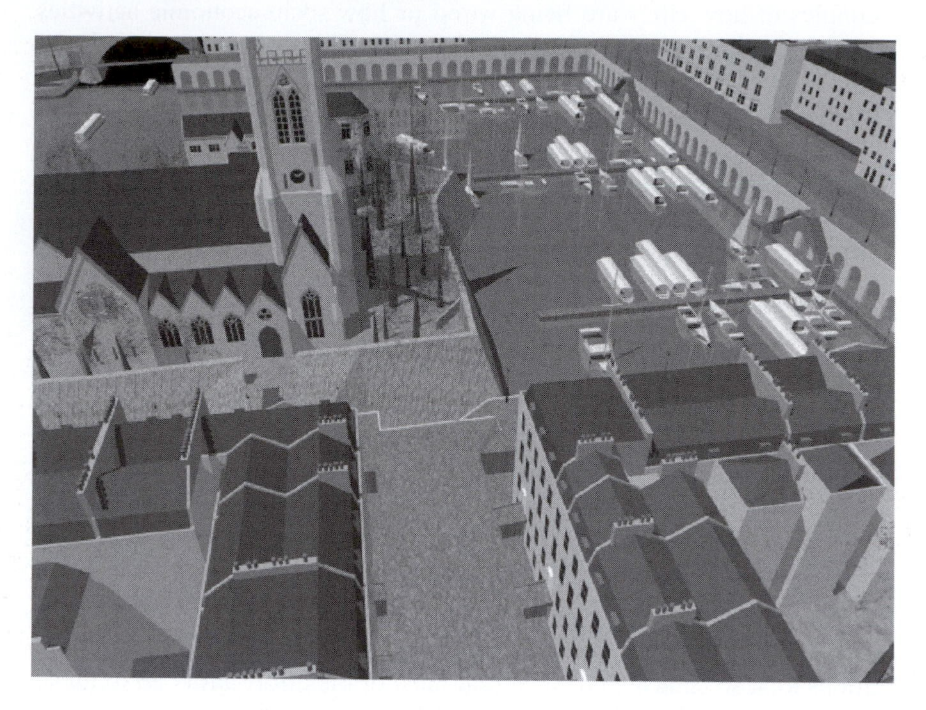

Figure 19.2 Typical CAD in a VR context: a scene from the Bath model (http://www.bath.ac.uk/).

This kind of CAD-VR, of course, has followed a tradition from architecture but not from architectural design, more from drafting. All that a CAD package can do is visualize and, although these now have some limited capability to deal with the physics of the objects being visualized, their origins are in the mechanical, routine functions of drawing. There is in fact, as we will argue below, enormous potential for a new science of the 3D urban environment which is consistent with, and a natural extension to, the 2D. But this will embody a much more analytic perspective than that associated with the entire line of 3D urban models so far, where the emphasis has not been on simulation, but on limited insights about the visual urban world through experiences generated by novel navigation. Communicating designs to non-expert audiences whose abilities to appreciate urban change and urban form through immediate visual imagery will always be the main focus of these virtual city models.

In stark contrast, a more abstract move from 2D to 3D is rapidly coming from GIS which, in many respects, represents one of the main traditions of many examples in this book. In the Introduction to this section, we showed an example for Westminster in central London and such examples can be massively scaled as we show below for the entire area of inner London. The data inputs that drive these types of virtual city are much more abstract than those above. Any attribute associated with any locational referent can be visualized in 3D in most desktop GIS, thus enabling data to be 'sculpted' into volumes, lines and surfaces associated with the third dimension. The emphasis is on spatial distribution, not on visual morphology per se. Insights emerge from seeing patterns which may not have been evident using non-visual analysis or even just 2D. In Figure 19.3, we show various renditions of Manhattan which have been developed in 3D GIS, where the buildings are coloured according to use. Such examples are no more than visual extrusions of 2D building outlines and, as such, do not add much beyond the basic thematic map, other than imparting a sense of place and perhaps complexity through the third dimension. But this particular form of analysis does have enormous promise.

It is only when these kinds of renditions are developed at a more aggregate level that new insights begin to emerge, and of course, this then pushes applications towards more analytic uses. In Figure 19.4(a), we show population density – 1991 population normalized by the areal extent of each enumeration district (block group) which is the basis of collection in the UK Population Census. At this scale, the pattern of density is quite irregular although when we project it into 3D, as in Figure 19.4(b), a semblance of the typical decay pattern (with respect to distance from the CBD) appears. In fact what is so valuable about this kind of visualization from the point of view of urban theory is the way open space affects density. The form of the fractured and irregular pattern which exists within the overall decay envelope is something that is lost in more aggregative, non-visual analyses. The implication, of course, is that there is some underlying surface behind this distribution although once again such visualization shows that this, too, is a convenient fiction. Visualization and VR thus become important tools in exploration of the aggregation problem.

The issue here is not so much identifying patterns within single spatial data layers which 3D undoubtedly aids but in examining the association of patterns between data layers. In 2D, this is very well worked out. Indeed an entire battery of spatial statistics can be brought to bear on such associations and GIS software is increasingly embracing appropriate functions either directly or through plug-ins and extensions. There is nothing remotely like this in 3D. The nearest one can get is through visual associations where different layers can be visualized within the same 3D projection. But this is tricky and not immediately obvious for layering different distributions within the same 3D needs great care. A simple example of this is shown in Figure 19.5 where we have interpolated a density surface from measurements of air pollution (NO_2) at many sites in central London. This surface can be overlaid in its translucent form for any other distribution of interest – in this case it is simply the extent of the London boroughs on a 2D map surface – and in this form, the most interesting

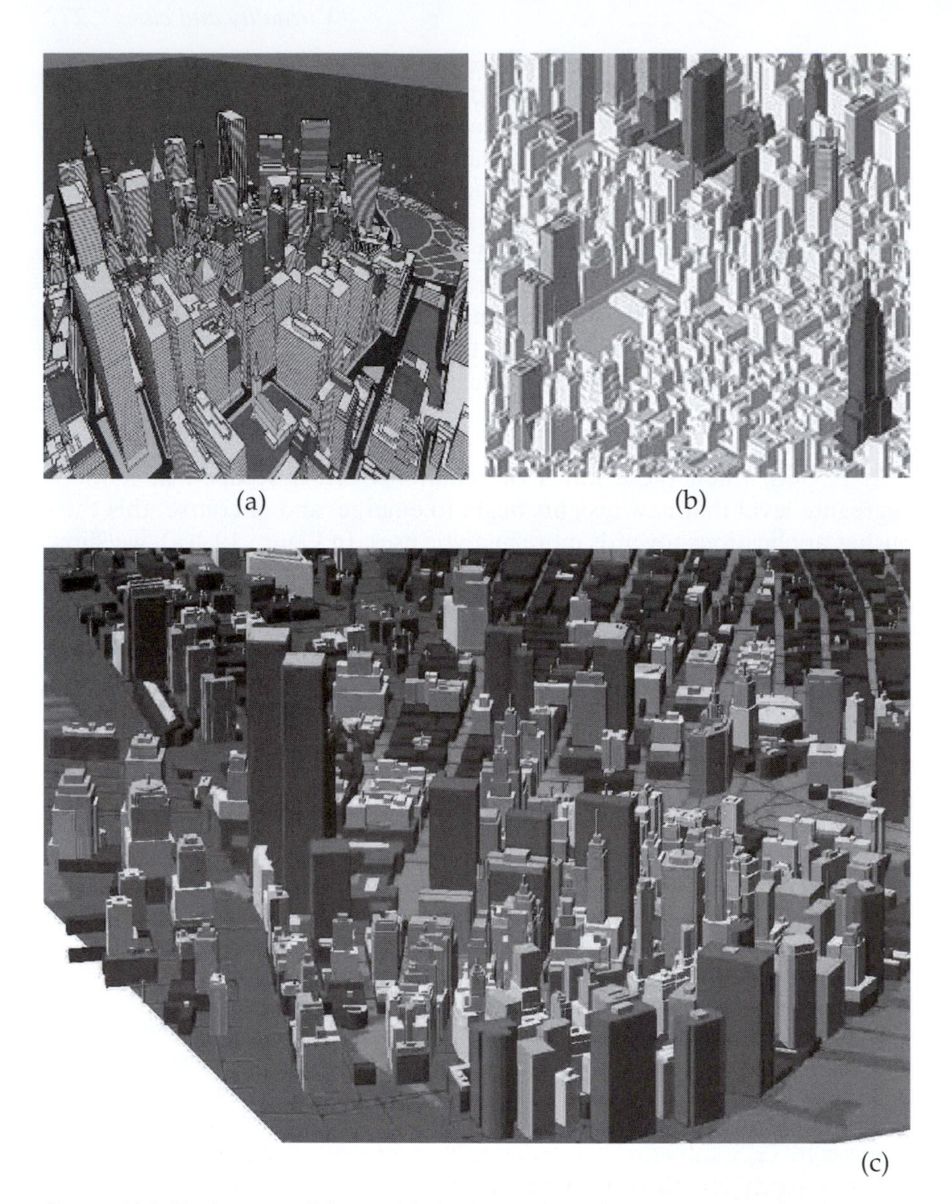

Figure 19.3 Various renditions of building outlines in Manhattan using 3D GIS: focusing on land-use type, floorspace volume and quality of use: (a) Slabs (in dark grey) greater than 8,000 sq. ft floorspace from http://www.simcenter.org/. (b) Building Attributes Data Base from http://www.u-data.com/. (c) A Photogrammetrically Accurate Buildings Database from 3d Metric at http://www.3dmetric.com/. (Source: Urban Data Solutions, New York, NY.)

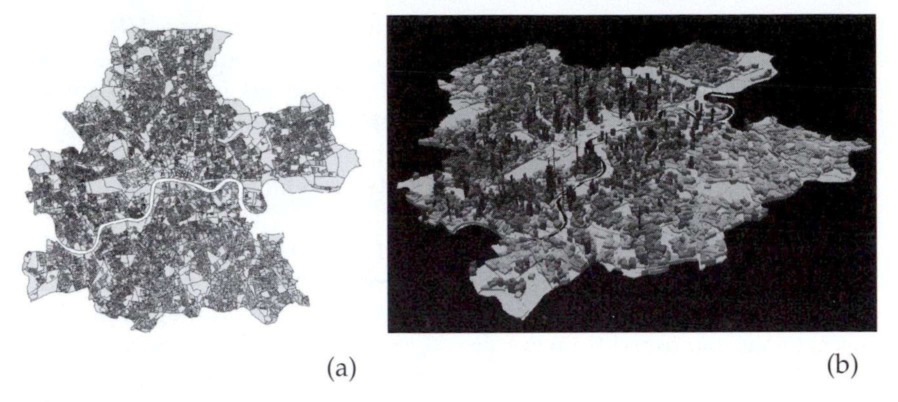

Figure 19.4 Visualizing Inner London in desktop GIS: (a) Population density in 2D, and (b) in 3D.

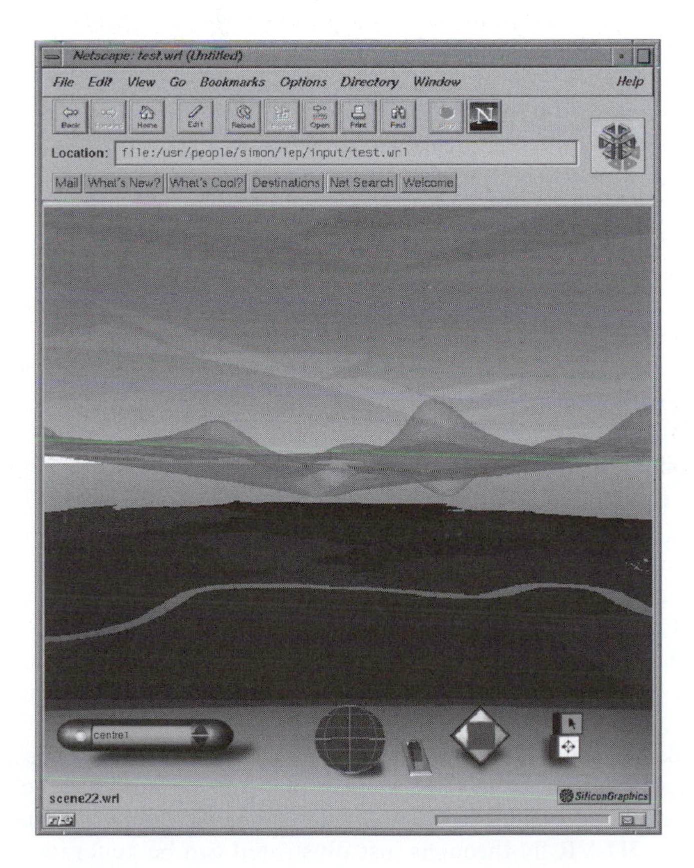

Figure 19.5 Flying through an urban pollution surface.

visualizations are formed by flying through the surface using VR technology (Batty *et al.*, 1998). From quite minimal and often suspect data points, a fly-through around this surface in a VR theatre has generated remarkable discussion concerning the pollution problem, the quality of the data, and its association with other variables. But the lack of functionality in 3D other than simple query, limits our ability to really explore the deeper impacts of this spatial variation.

Although there are some examples of moving back from CAD-VR representations to the logic of GIS, the ideology and focus of CAD is quite different from the urban analytical tradition. To date, nothing of any note has been accomplished from this perspective, whereas the move to 3D GIS from 2D is much more promising. Indeed the possibility of consistently handling and exploring the third dimension as an extension of the second, raises the prospect of this kind of IT forcing a new kind of urban geography. In the past, our understanding of cities has largely ignored the third dimension but, with the existence of 3D GIS, then a new world of relationships awaits. There has been no real urgency to grapple with the analytic modelling of location in 3D because most western cities have a simple 3D over most of their area with no more than two-storey building. This is not the case, however, in most of the world's cities where high-rise living is more the norm. Nevertheless, even there, the third dimension is treated in somewhat uniform terms. It is the CBD that is the really complex element in most examples where there is need for an explicit geography of the third dimension.

In desktop GIS, thematic mapping of point and areal locations, areas, the interpolation of surfaces, and the spatial analysis of different aspects of spatial distributions might all find extensions and parallels in the third dimension. But doubtless there are more involving geometry such as interior lighting, energy usage, tenure and use, amongst many other features which dictate a rather different kind of urban geography from that which is restricted to the 2D. And then there is the transition from 2D to 3D and analysis that takes place across both. This world has barely been considered, but all of it will involve embedding new functions into 3D GIS or, rather, evolving a true 3D GIS with the same kind of requisite functionality that exists for the 2D. An example of how this might be progressed is illustrated in Figure 19.6 for the centre of Swindon. Building layers have been visualized using *ArcView's 3D Analyst*, where the accessibility to the exterior of each block has been represented in 3D isochrones, providing one of the first examples showing how standard 2D spatial analytic functionality might be extended to 3D.

Our last tradition which develops aspects of virtuality can be generally classified as using multimedia to engage in virtual realizations of the city, the archetypal example being the 'virtual tour'. Visualizations as equally impressive as the 3D-VR fly-throughs just illustrated can be generated by interposing maps, movies, and panoramic images around which one can

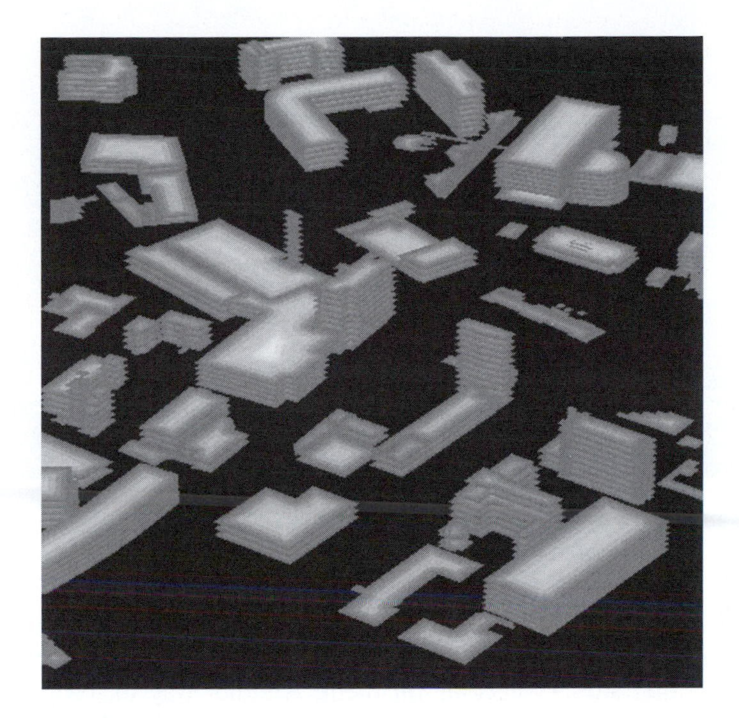

Figure 19.6 Internal accessibility surfaces based on depth within building blocks in 3D GIS for the centre of a small English town (Swindon).

rotate. Some of these systems have been extended into CAD, as, for example, in the rendering software *Canoma* which enables 3D scenes to be constructed from photos taken from different viewpoints. For example, the CAD model in Figure 19.7 which shows that part of Docklands centred on Canary Wharf, London's second CBD, was constructed from two oblique aerial photographs, the software making assumptions about the façades of buildings not seen and extracting the wire frames from the induced perspectives. Simpler but analogous techniques are used extensively in generating panoramas such as those used in virtual tours, field courses and other multimedia. Although existing on the desktop, increasingly such tours are embedded into Web browsers largely because their visual demands on storage and software are modest but also because the real advantage of such navigation is to link points of interest to other Web pages.

An example is Wired Whitehall developed by one of us (Smith) to show part of the UK capital's heritage to visitors. When a user accesses the page, up comes a map of Whitehall with key sites marked – at Parliament Square, Trafalgar Square, Whitehall itself, Buckingham Palace, Leicester Square and Covent Garden. Users can hop from site to site by clicking on a site within the map. This loads a panorama of the site captured by video

Figure 19.7 A 3D CAD building scale model of Docklands (left) alongside a more aggregate 3D population density frame (right) for the same area.

around which the user can pan. As the user pans around the site, key landmarks are identified. You can get nearer to them by zooming in or you can move to other sites by clicking on them. Another way to navigate is by moving from panorama to panorama directly from within the scene using hotlinks. However, what Wired Whitehall really illustrates is how you can link to remote sites from such an interface. When you get into Whitehall, you can point towards Downing Street and although you cannot move virtually into Downing Street (as you cannot move in reality either for security reasons) you can click on the 10 Downing Street website. An example of this is shown in Figure 19.8. When you click on the relevant link from within the scene, you are transported to the remote site from where you can navigate elsewhere. The power of this kind of virtuality, of course, only comes into its own when the connections both inside the virtual city and outside are sufficiently rich and clear that they provide really useful information for learning about the real city. This will, of course, depend upon how many of us are online and how well the sites are designed.

Communicating geometric virtuality to the public at large

There is little doubt that, apart from some of the more abstract GIS noted above, most of the virtual city models we have illustrated so far emphasize simple geometric qualities where the focus is on exploring intricate geometries and appreciating the impact of differentially-rendered scenes in aesthetic terms. As we have implied, the most frequent aims of many of these applications is 'infotainment' targeted at a diverse variety of audiences. However, to explore ways in which such virtuality might be communicated,

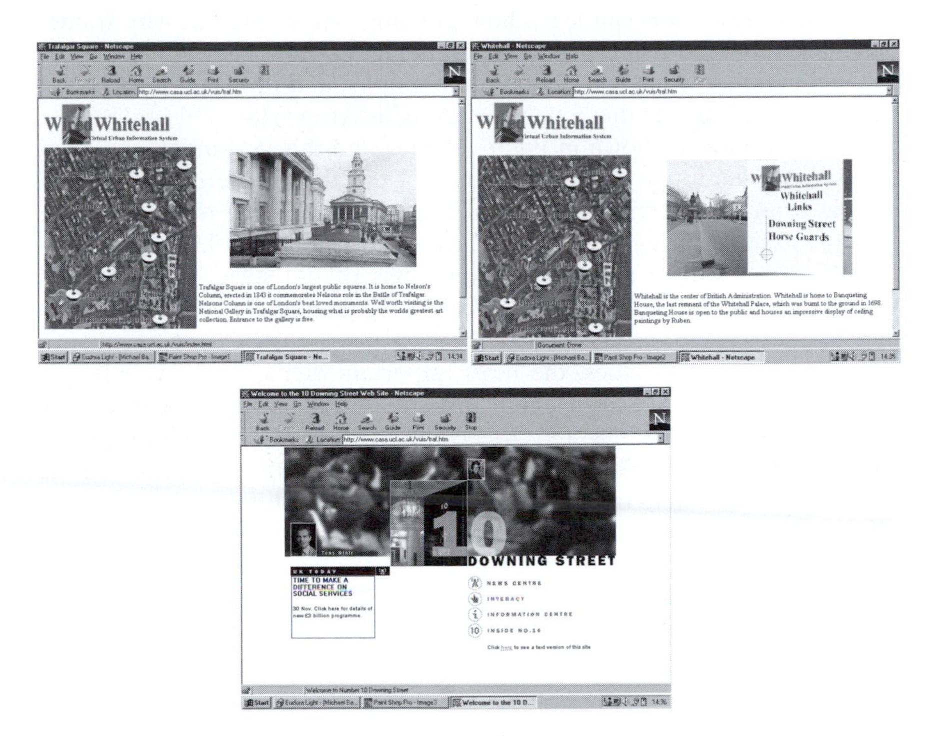

Figure 19.8 Wired Whitehall: virtual cities as virtual tours.

we have developed a simple pedagogic tool which is aimed at non-professional audiences. Moreover, this tool illustrates that what was literally state-of-the-art yesterday, is common practice today. The Hackney Building Exploratory – Interactive! is a Web-based interface which demonstrates how 2D and 3D visualizations of cities can be developed and used in a straightforward, routine manner (http://www.casa.ucl.ac.uk/hackney/). The Hackney Building Exploratory itself is a non-profit-making centre located in Hackney – a deprived borough in inner London – which is focused on raising environmental awareness within the local community, particularly amongst school children.

The software that has been developed (Smith *et al.*, 1999) embraces the three approaches to virtual cities already demonstrated for professional concerns earlier in this chapter. The site is organized around three themes – a chronology of housing in the development of the borough represented using CAD, panoramic scenes of typical streets and squares within the borough based on the multimedia software used in the construction of Wired Whitehall, and display of geographic variations in the social and economic quality of life in the borough using 3D GIS. Each of these techniques and themes, illustrated through examples in Hackney, is backed by explanations of how the software and data create the appropriate digital representations. Users can learn how buildings are created as wire frames and then rendered using photographic imagery, how panoramas can be stitched together using still digital photographs, and how 3D surfaces can be developed from 2D thematic maps using desktop GIS.

The homepage is shown in Figure 19.9(a) and the card-index-like identifiers along the top give access to each of the three different themes. A sample from each theme is shown in Figures 19.9(b)–(d) which illustrate how the user can bring up different house types (Figure 19.9(b)), panoramas (Figure 19.9(c)), and 3D block models of the social geography of Greater London in which Hackney can be identified (Figure 19.9(d)). In fact, although these themes indicate how virtuality can be used to enhance our understanding of cities, the most important element of this software involves letting users 'design' their own city. 'Drag and Drop Town' is an interface which enables a user to position different key building types which are found in Hackney in different locations on the map. Along the edge of the street map taken from part of Hackney, different house types, a pub, church, supermarket, tree and fast food outlet as well as a London

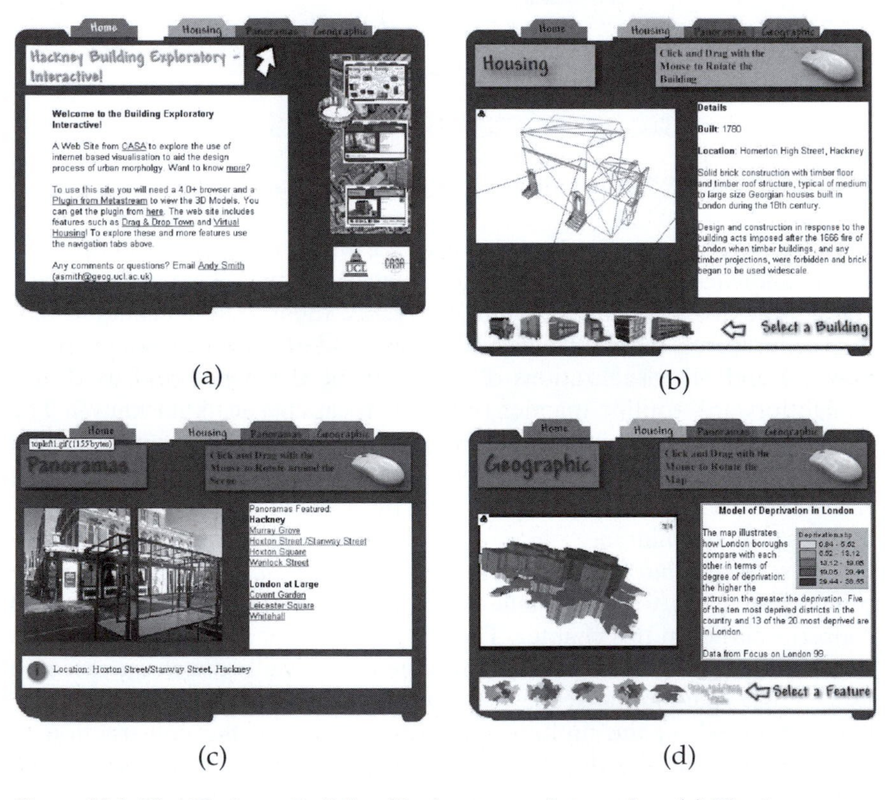

Figure 19.9 The Hackney Building Exploratory – Interactive: (a) The homepage, (b) A house type in wire frame mode, (c) A street panorama, and (d) A block model of social deprivation in London boroughs.

bus can be positioned anywhere within the scene by dragging, then dropping. Users thus gain some idea that cities are composed of different elements that sometimes go together, sometimes don't, and under gentle supervision, such 'drags' and 'drops' can be supported by issues that heighten awareness about the conflicts endemic in the design of cities. This interface is clearly designed for school children but, in a wider context, it begins to illustrate how Web interfaces might ultimately be the media through which many kinds of visual design take place. In Figure 19.10(a), we show the basic Drag and Drop Town interface and in Figure 19.10(b), an example of a design.

In the near future, the Web will become the main medium for visual representation as all kinds of software traditionally on the desktop migrate. VRML – the Virtual Reality Modelling/Markup Language – and similar graphics-based Web representations are beginning to become the preferred medium for CAD but much simpler forms of Web-based interaction are now possible as in the 'Drag and Drop Town'. In short, the fact that the Web now has over one billion pages (January, 2000) and is still growing dramatically, is forcing software developers to port everything to this medium in the quest to provide the easiest and most elaborate graphics interfaces for the most basic computer user. There are now countless digital or virtual cities, so-called, which are simply Web pages of various elaborations communicating a variety of place-specific information targeted at tourism, service delivery, or other aspects of e-commerce using the metaphor of the city. Municipalities and city governments have exploited this technology as quickly as anyone else. Digital advertising of cities and what they have to serve their citizens, new businesses searching for new locations, and those wishing to visit as tourists – in fact, the whole plethora of what cities and the activities that compose them, have on offer – is now widespread. This has also become a way of publicizing plans for their future as well as proposals for change, such as applications for planning permits.

The simplest sites are those which simply print their city brochures

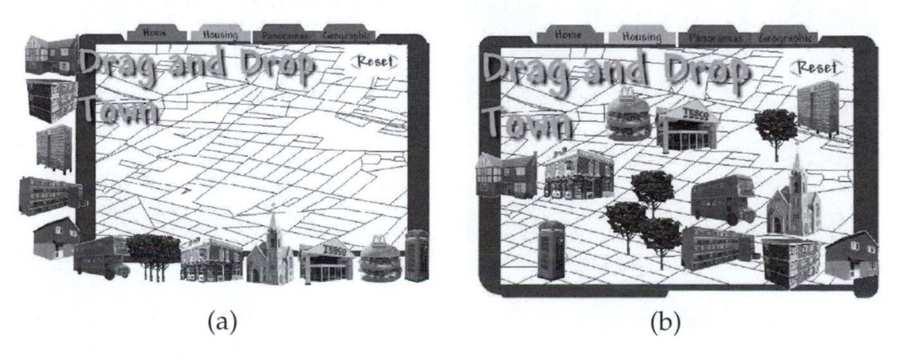

(a) (b)

Figure 19.10 Drag and Drop Town: (a) The original canvas, (b) A 'design'.

digitally, but all such sites have some structure in guiding the user to other pages of information which on the Web provide a total picture of what is on offer. Bristol in England is as good an example as any; here one might find information about the city, about its plans for a high-tech future and about various local interests and how they might be indulged through access to Web-based information. A page from its various websites illustrates the typical form and this is shown in Figure 19.11(a). Sometimes the information is structured somewhat more intelligently around an idealized map of the city or a series of city-like icons which direct entry to areas of the city such as City Hall, the schools system and so on. The Apple Computer Company used such imagery to provide access to its e-mail system some five years or more ago, and the same kinds of icons have been used in many places. An early example was in Singapore but one that has lasted longest is in the Italian city of Bologna whose homepage is shown in Figure 19.11(b).

When you point your mouse and then click on an appropriate icon, you are transported to the next Web page in the tree of pages which deals with the information implied by the icon. From the Bologna page, clicking on shopping takes you to the pages dealing with lists of shops in the city, each in turn being linked to the main sites of the stores and companies that are distributing their products in the town. Clicking on education takes you to information about the various schools, their locations, their sizes and specialisms. The transport icons move you to information about the bus system, the railways, taxi services, and so on. The city government icon takes you to information about your taxes, as well as the social services on offer in the city. The list is endless. The best websites like Bologna are well structured and clear, but one point to note is that this emerging digital world is so new that sites appear and disappear with amazing alacrity. The fact that cities can be represented virtually means that they can be changed immediately which is perhaps an inevitable feature of the digital

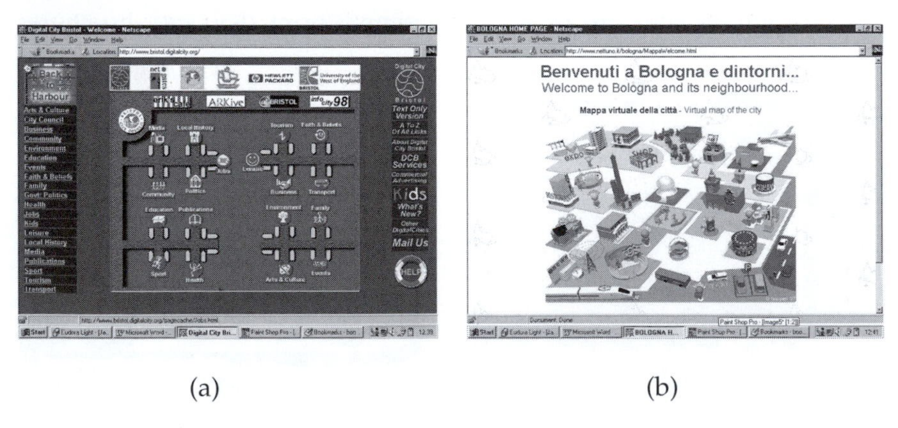

(a) (b)

Figure 19.11 Web-based digital cities: (a) Bristol, England, (b) Bologna, Italy.

world that we will all have to get used to. By the time this book reaches print, many of these sites might be gone.

The virtual future: cities in cyberspace and cybertime

So far we have presented many ways in which virtuality is embodied within digital images, text and other forms of representation reflecting different ways of simulating, experiencing and communicating ideas about cities. Increasingly routine services which are being communicated or even delivered across the Web are adding visual content in which the virtual city metaphor is writ-large. In short, the latest visual software is being used for purposes other than the professional communication and analysis of city problems and plans. An added twist to this story is the emergence of virtual worlds such as those described by Dodge (Chapter 21) in this book, based on Web servers whose purpose is to serve common information arranged around a fictional place through which users can communicate in diverse ways. These are truly virtual cities in that they involve users colonizing land, building structures, even designing, in the context of online communities which have the visual shape of real cities. AlphaWorld is the best known of these and, as Dodge (Chapter 21) shows, the structure of such a world is quite city-like in plan form as well as in the kinds of activities that have developed as more users have joined and developed land. These worlds are largely extensions of gaming in a passive way where users might communicate but where the rules of the game are no more (or less) than those which any real community might evolve. There is no long-term competitive reward structure which mobilizes and focuses users who interact within such environments.

Virtual worlds such as these, however, are increasingly being used not for infotainment but for professional use such as urban design. Like CAD models, these worlds may be as real in intent as one might imagine, for, although ethereal in form, real-world imagery and data can be used to manufacture scenes within which users can interact in many ways. In fact, basic VR came from similar origins where a single user was able to fly through CAD models of cities, experiencing such environments not only visually but through other senses. Networked versions of the same allow communication amongst participants, and thus enable possibilities for collective action. In this context, collaborative design and decision making is the quest. Cities themselves provide a remarkably rich metaphor for various types of concerted action concerning decisions in environments which might be analogous to cities. But here decision making about cities within virtual cities is a direct application of this notion, and this is where virtual worlds have enormous potential.

One of us (Dodge and Smith, 1998) has been involved in developing such worlds for urban design through the medium of the Collaborative Virtual Design Studio (CVDS). As a prelude to this, an experiment in asking users to build and communicate in such a virtual world was devised, and initially for a trial period of thirty days, users – anyone, anywhere, at any time, and from any place – were invited to interact. The city in question was set up as a small plot, about the size of London's Soho, some five square kilometres in size, and, once the experiment began, users began to colonize this world. During the thirty days, eighty-four members were registered and these users constructed 30,000 distinct objects. It is not the place here to describe the social mechanics of this experiment, the trials and tribulations of the 'owner' of the world, and the way the community fractured and evolved during its short turbulent life. But the critical point is that this experiment demonstrated that virtual cities, like real cities, not only have a spatial structure but also a temporal one. To date, virtual reality methods have rarely dealt with temporal dynamics other than the dynamics of motion posed by navigation, yet as these virtual worlds come to be used ever more extensively (and intensively), they take on a life of their own with their own internal dynamics that is much more than the routine.

In Figure 19.12, we show how the city or space evolved up to day twenty-nine, as users began to colonize land, improve their constructions, and interact with each other within the limited space available. If you track the detail of these pictures, then every spatial decision made, as recorded by the end of the day, is embodied in these images. The space fills quickly but once it has been developed, decisions become more subtle as land uses are adapted and detailed shapes changed. The real point of this illustration is to raise the notion that, besides a virtual geography based around digital space, cyberspace, there is a virtual time to these creations which we have hardly begun to broach. At the beginning of this era of digital geography, what we urgently require are good theories that tell us how to handle not only virtual space – information space – but also new chronologies and new streams of time which are likely to be as important as our current concern for the morphology of cyberspace (Kitchin, 1998).

The methods introduced in this chapter have largely emphasized the visual at the expense of dealing with issues of social and economic structure. Geometric and aesthetic considerations dominate most of these developments and there is a clear need for a more analytic focus if these methods and models are to move much beyond the superficial. It is likely that CAD models, which still constitute the majority of applications, will be rendered obsolete by developments in remote sensing. Within a generation, perhaps within no more than ten years, large-scale 3D geometric models will be appearing automatically as data is routinely sensed. This will, of course, depend upon demand. If 3D GIS develops with new functionality and new theory about urban structure in the third dimension, then demand might grow substantially from the professional communities. But if such development is stymied by lack of good theory and lack of

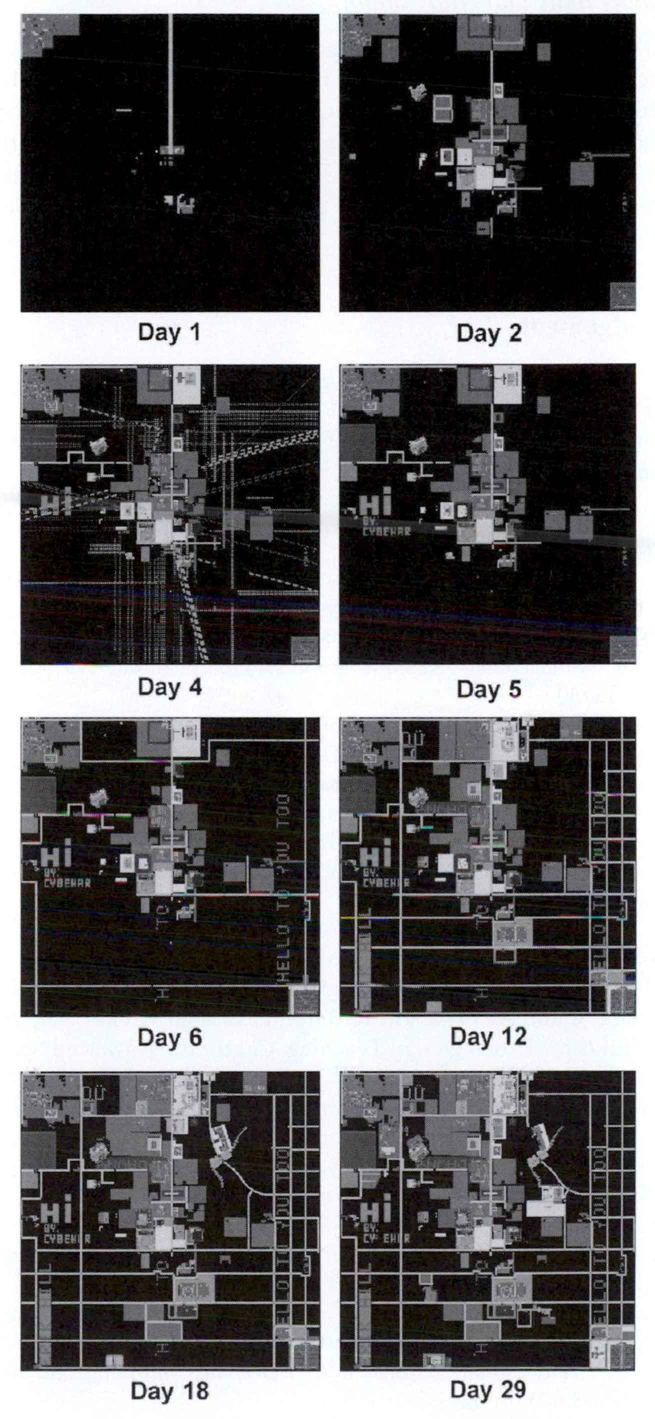

Figure 19.12 The evolution of a virtual world: charting cybertime during 30 days.

interest, then the demand for such models will remain largely in the infotainment sector and will become increasingly routinized as the technology develops. What is urgently required, then, is the development of spatial analysis which embraces not only the third but also the fourth dimension (time). In the last analysis, for these concepts of virtuality to be useful for analysis, a new theory of cities must evolve to embrace these concerns.

Acknowledgements

This project was partly financed by the Office of Science and Technology-EPSRC Foresight Challenge Grant for the VR Centre for the Built Environment (GR/L54950), and EPSRC's Partnerships for the Public Understanding of Science (P007172).

References

Batty, M. 1997b. Digital planning: preparing for a fully wired world. In Sikdar, P.K., Dhingra, S.L. and Krishna Rao, K.V. (eds) *Computers in Urban Planning and Urban Management: Keynote Papers*. Delhi, India: Narosa Publishing House, pp. 13–30.

Batty, M., Dodge, M., Doyle, S. and Smith, A. 1998. Modelling virtual environments. In Longley, P., Brooks, S., McDonnell, R. and Macmillan, B. (eds) *Geocomputation: A Primer.* Chichester, UK: John Wiley and Sons, pp. 139–61.

Batty, M., Chapman, D., Evans, S., Hacklay, M., Shiode, N., Smith, A. and Torrens, P. 2000. A Review of Visualization Methods for Urban Management, A Report to the City of London, abstract at http://www.casa.ucl.ac.uk/londoncity/.

Borgmann, A. 1999. *Holding onto Reality: The Nature of Information at the Turn of the Millennium.* Chicago, Illinois: University of Chicago Press.

Deken, J. 1983. *Computer Images: State of the Art.* London: Thames and Hudson.

Dodge, M. and Smith, A. 1998. Virtual Internet Design Arenas: The Potential of Virtual Worlds for Urban Design Teaching, Centre for Advanced Spatial Analysis, University College London, UK.

Graham, S. and Marvin, S. 1996. *Telecommunications and the City: Electronic Spaces, Urban Places.* London, UK: Routledge.

Heim, M. 1997. The art of virtual reality. In Droege, P. (ed.) *Intelligent Environments: Spatial Aspects of the Information Revolution.* Amsterdam: North Holland Publishing Company, pp. 421–37.

Kalawsky, R. 1993. *The Science of Virtual Reality and Virtual Environments.* Wokingham, UK: Addison-Wesley Publishing Company.

Kitchin, R. 1998. *Cyberspace: The World in Wires.* Chichester, UK: John Wiley.

Negroponte, N. 1995. *Being Digital.* New York: Alfred A. Knopf.

Plewe, B. 1997. *GIS Online: Information, Retrieval, Mapping, and the Internet.* Santa Fe, NM: OnWord Press.

Rheingold, H. 1991. *Virtual Reality.* New York: Touchstone Books.

Smith, A., Sheppard, S. and Batty, M. 1999. *Hackney Building Exploratory – Inter-*

active! Centre for Advanced Spatial Analysis, London: University College London, http://www.casa.ucl.ac.uk/hackney/.

Steuer, J. 1992. Defining virtual reality. *Journal of Communication*, 42, 79–90.

Warner Brothers Pictures. 1999. *The Matrix*. Hollywood, CA.

Woolley, B. 1992. *Virtual Worlds*. Harmondsworth, Middlesex, UK: Penguin Books.

Part IV

'Other' worlds

Part IV

'Other' worlds

20 'Other' worlds

Augmented, comprehensible, non-material spaces

Jo Cheesman, Martin Dodge, Francis Harvey, R. Daniel Jacobson and Rob Kitchin

Introduction

In this section of the book we are concerned with spaces which do not seek to *represent* geographic space, but which utilise geographic concepts to depict other types of information, extend the portrayal of geographic phenomena by non-visual VR supported methods, or render non-material characteristics in virtual reality environments. These spaces do have a spatiality, with qualities often equivalent to geographic space, and yet they are entirely socially produced/constructed and exist as parallel (non-related) or symbiotic (augmenting) spaces. As such, the discussion within the chapters is significant, at both philosophical and practical levels, because they utilise VR spaces implicitly to challenge traditional, physically-bound conceptions of space and place, and explore geographic visualizations that do not necessarily seek to be representations of 'real world' environments. Whilst these digital constructions using hardware and software spaces are spatially organised, and metaphorical spaces and imagined spaces are nothing new, VR spaces are qualitatively different in that they actually enable 'geographic-style' (spatial) behaviour – they are spaces you can be immersed in, you can move through, and where a new form of interaction with other people and objects can take place. They create a sense of place that has meaning to people.

Situating VR spaces

It is worth at this juncture to briefly consider conceptions of space and the representation of space in order to situate VR spaces. There is a long historical precedent, tracing back to Ancient Greece, of philosophising the nature of space. Curry (1998: 24) suggests that conceptions of space lie on a continuum, 'where at one end are those wherein the relationship among objects is strictly contingent, and where at the other are conceptions where the objects in space have very strong, even necessary and intrinsic relations with one another and the space in which they are located'. In other words, space is viewed as either absolute, a container filled with objects

(Aristotelian, Newtonian), or relational, which is either the consequence of interrelationships between objects (Leibnizian, Kantian) or is 'constituted through social relations and material social practices' (Lebevrian; Massey, 1994: 254), or as a combination of absolute and relational aspects (Sayer, 1985; Urry, 1985). Moreover, Curry (1998) details that the position of the viewer varies within these conceptions. In some the viewer is intrinsically connected with the viewed and/or has a privileged position, whereas in others the viewer is disconnected and/or occupies an unprivileged position. The conceptualisation of space is the foundation for its representation.

As we will discuss, VR spaces (both those that seek to represent geographic space and those that do not) seem to us to offer challenges to current conceptualisations. VR spaces offer 'geographic-style' interaction, and yet the spaces are *not* essential (given) or absolutist. Instead they are purely relational (both spatially and socially) and yet, unlike geographic space, they possess a number of other qualities which set them apart. As discussed in the chapter by Kitchin and Dodge (Chapter 23), VR spaces are constructions – productions of their designers and, in a few cases (Dodge, Chapter 21) by active users. VR spaces adopt the formal qualities of 'geographic' (Euclidean) space only if explicitly programmed to do so. Spaces are often purely visual, objects have no weight or mass. VR spaces have spatial and architectural forms that are dematerialised, dynamic and potentially devoid of the laws of physics; spaces that are not physically tangible, in that they can be explored by the mind, yet metaphorically relate to bodily experience. It is only now that we are starting to wrestle with wider questions concerning the spatiality of VR spaces and the nature of interaction within them. For example, Virilio has recently come to question whether objects in VR are simulations or substitutions:

> this is a real glass, this is no simulation. When I hold a virtual glass with a data glove, this is no simulation, but substitution. . . . As I see it, new technologies are substituting a virtual reality for an actual reality. . . . there will be two realities: the actual and the virtual. . . . This is no simulation but the coexistence of two separate worlds.

This, in turn, raises questions about representation. Geographers and cartographers have long wrestled with notions of how to represent geographic space. For example, for centuries cartographers have been experimenting with ways to map and visualize spatial relations across the Earth's surface. Here, those creating representations are interested in issues such as accuracy, precision, verisimilitude (having the appearance of truth, realistic depiction), and mimesis (imitation, mimicry), and the degree to which a representation is separate from that which it seeks to represent. These issues still have pertinence for VR (see Part I) along with related issues concerning hyper-reality. The spaces described below, however, do not seek to represent geographic space. They are spaces that have no tangible geographic counterpart – creations rather than representations. In some cases (e.g. information spaces) the geographic metaphor and territory become synonymous (one and the same). Here, the use of a geographic metaphor to structure the data becomes the means by which this new territory is navigated. For example, a VRML Web page is both the territory and the means in which to navigate this territory. Moreover, unlike representations such as maps, VR-based representations can be viewed and navigated in ways analogous to the ways people habitually navigate geographic space (e.g. wayfinding along streets to visit the local bank), plus in addition by utilising ways such as exterior viewpoints and teleporting. The separation of territory and map, the continuous nature of that being mapped, and a qualitatively different form of interaction (e.g. reading a map), are central to traditional geographic representations such as mapping. Clearly these tenets are being arbitrarily violated in VR spaces.

The spaces identified

In the chapters that follow, five types of 'Other' worlds (imaginary, information, sensory, statistical and social spaces) are identified and discussed. Before discussing these five spaces it is necessary to highlight their distinguishing traits. In the first instance, these spaces differ in the ways in which they transform traits of a wide range of phenomena – geographic and non-geographic – to spatial relations in a VR environment. While these transformations rely on geographic and topological concepts (e.g. proximity), they do not inherently make phenomena geographic. Frequently, geographic concepts are used here to aid the presentation of complex relationships or even construct new relationships that have never existed, and to open new analytical and didactic vistas. These concepts can also extend human creativity into new social spaces which currently do not physically exist, but provide a substantial touchstone for a wide range of human activities. Given these arguments we identify three different types of virtual realities, which the chapters in this section explore:

- *Augmented* – additional information to provide a multi-modal VR experience that extends the representation of the virtual and geographic environments by alternative sensory means enriching the participant's experience. For example, a textual MUD can be augmented by the inclusion of visual indicators (icons) or interaction with the urban environment by visually-impaired people can be augmented by virtual auditory overlay to provide external frames of reference.
- *Comprehensible* – application of geographic concepts, such as proximity, to provide cues to aid the navigation of large amounts of data that would otherwise be difficult to understand. For example hypertext documents can be restructured into a browsable site map which utilises human ability to understand spatial relations.

- *Non-material* – VR representation of physically non-existent phenomena, figments of the imagination, or characteristics of physically-present, non-material phenomena. For example, the social virtual environments of AlphaWorld are a dematerialised, dynamic, disembodied VR space inhabited by real people through their avatars. The VR representation of non-material phenomena, such as heat, gravity and the representation of indirect measurements (i.e. precipitation, population density) is yet another example.

These types of virtual realities are instantiated in following five *metaphorical* spaces. These types of spaces are not mutually exclusive and, in fact, combine different aspects of augmented, comprehensible and non-material virtual realities. Indeed, these aspects alone do not fully differentiate these spaces, and as the examples discussed in this section show, other characteristics – such as purpose and intent – need to be considered as well.

- *Imaginary spaces* are pure creations with no geographical referent, no geographical materiality. They only exist as virtual spaces, although they may use geographic-style spaces to foster interaction and adopt many of the characteristics of real-world places. They do not, however, represent a geographic space (Dodge, Chapter 21).
- *Information spaces* are conceptual, metaphorical spaces created by utilising spatial metaphors such as proximity/density/connectivity to make metadata more comprehensible and navigable (e.g. site maps to aid comprehensibility of Web pages) (Harvey, Chapter 22).
- *Statistical/graphical spaces*, similarly to information spaces, use spatial metaphors to add understanding. However, they are qualitatively different in that they provide a spatial structure to immaterial data in the geographic world (e.g. gravity, heat) (Kitchin and Dodge, Chapter 23; Cheesman and Perkins, Chapter 24).
- *Sensory/perceptual spaces* do not represent geographic space but provide augmentation to such spaces to provide a qualitatively-enriched experience of a geographic space. This augmentation might be through the provision of immaterial information, such as labels or directions, using sound (Jacobson *et al.*, Chapter 25).
- *Social spaces* are spaces in which several users can interact through some form of communication (Dodge, Chapter 21).

These spaces are constructed by a wide range of individuals institutions, and large corporations who set out to develop and refine them as means of developing new spaces for social interaction, understanding or VR spaces to enhance existing information technology environments.

These types of spaces already exist in a number of forms, but in terms of cyberspace are new, qualitatively different, and change socio-spatial relations in the 'real-world' supported by digital computers and networks. The changes in people's awareness of social spaces through the telephone, for example, is an earlier technology that extends social forms of interaction and our comprehension of geographic spaces we inhabit. VR-based spaces go yet further in offering distributive immersive interaction. Whereas the telephone only extends our auditory interaction with a limited number of participants at a given moment, VR offers a profusion of enhanced means for relating material and non-material phenomena through the persistent enrolment of a spatialisations as references for interaction. Indeed, these spaces illustrate the various ways in which VR concepts can go beyond the approximate rendering of geographic phenomena and enable creative interactive and immersive environments to support new and enhanced human learning and social interaction. We are now just beginning to understand the spatialisations that make these spaces qualitatively different and the chapters of this section discuss some of the more salient issues.

Concerns/issues

The promise of these VR spaces does not go without concerns. The chapters in this section voice some of these concerns, but we feel it is most appropriate in this introduction to point to a wide range of issues we feel is, and will remain, important in the development and critical examination of the development of VR technologies.

Power of VR

As Harley (1989) and Woods (1993), among others, have argued in relation to maps, and Pickles (1995) in terms of GIS, VR is not an objective, neutral space (Markley, 1996). VR spaces are imbued with the values and judgements of those who construct both the technology and the medium, and are situated within broader historical contexts – VR has not arisen in a vacuum, but has been guided by historical context (namely military's desire for flight simulators, funding bodies, capital venturists, etc.). Moreover, VR spaces are not egalitarian. Access to them is limited to those who can pay, and they are regulated by the institutions that build or maintain them. As such, they are situated in a nest of power-relations that influence social relations. This in turn raises ethical and social questions. We can clearly foresee that the financial ability to obtain the necessary technical infrastructure and maintain it will be a primary means of including and excluding groups in society. The augmented, comprehensible and non-material spaces we discuss are thus not neutral spaces and should not be accepted as such. Scholars such as Steve Graham (1995) and Michael Curry (1998) point to this in their work on surveillance technology.

Embodiment/disembodiment

For many commentators, VR technologies offer the possibilities of new, real-world social relations. For example, Stone (1991) suggests that VR spaces help to blur the relationships between the social and technological, biology and machine, natural and artificial, with the resultant mergers forming the keystones for the new social space. This blurring is leading some theorists to hypothesise that we are adopting new forms of embodiment, becoming nations of cyborgs. Here, human and machine merge, with the machine replacing or supplementing the flesh. We are being reconfigured in new ways that challenge traditional identities. Computers, headsets, bodysuits (as well as cosmetic surgery, biotechnology and genetic engineering) extend our bodies into new spaces and allow us to explore what it means to be human. Feminist commentators (Lupton, 1994; Plant, 1996), for example, suggest that cyborging technology represents 'liberation from the confines of gender and other stereotypes, by rendering cultural categories indeterminate and fluid' (Lupton, 1994: 101). These and other analysts (Morse, 1994; Tomas, 1991) extend these debates concerning new forms of embodiment to suggest that VR technologies are, potentially at least, spaces of disembodiment – spaces that can be explored free of the body – new social spaces where the physical and material are transcended. It will be interesting to see whether gender reinforcement or gender crossing become more predominant in cyberspace; the possibilities for impaired individuals also present interesting questions regarding their involvement. The issues indicated here lead to the question of whether we are now embarking on the way to establishing alternative worlds which interact with our physical world in yet unknown ways and open new vistas for social relationships. If this is the case, what are the implications of such reconfigurations?

Real/virtual

Philosophers have long been troubled over the question of reality. Recent commentators have started to argue that it is becoming increasingly difficult to tell the genuine from the fake.

VR spaces add a new dimension to these arguments by providing spaces which blur further the distinction between real and virtual. For example, Benedikt (1991) argues that cyberspace causes 'warpage, tunnelling and lesioning of the fabric of reality'. Cyberspace rapidly increases the blurring of reality and virtuality first started with the printed word, and further developed by radio, television and film. Each of these media provide us with a representation of the real; a copy of the original, simulations which Baudrillard terms 'hyper-reality' (more 'real' than reality). For Baudrillard, much of our postmodern world is an illusion, full of objects and buildings masquerading as the real – much like VR. For commentators like Slouka (1996) the danger is that many of us are now willing to accept the copy as original, and put our trust in those that re-represent the world to us – to accept simulation as substitution. We are too willing to accept the virtual as real. Many VR technologies are specifically designed to immerse users in a parallel, artificial world that mimics the real. Their appeal is that actions in virtual space seemingly do not have material consequences. Slouka, however, fears that these are self-indulgent technologies that will make it increasingly difficult to separate real life from virtual existence. Some of these issues are extremely pertinent when VR spaces are non-material and non-representational.

Place/placelessness

To many users, VR social spaces (e.g. visual MUDs like Alphaworld, see Dodge Chapter 21) imbue a 'sense of place'. Spatial behaviour within these spaces largely mirrors real-world interaction (with some notable exceptions such as teleporting). Similarly, social interaction mirrors real world interactions, spaces are divided into territories, and there is spatial governance. Does this then make them places or are these spaces placeless (as touted by analysts such as Mitchell (1995) and Rheingold (1994)? If they are places, and given definitions by commentators such as Jess and Massey (1995) (places are characterised by providing a setting for everyday activities, having, and being characterised by linkages to other locations, and providing a 'sense of place'), are they simulations of real places, substitutions or something else (especially when they are immaterial locations)? Moreover, what are the implications for places in the real world? Social scientists have started to direct their attention to these questions (Castells, 1996; Kitchin, 1998). In particular, they have focused on social and cultural issues such as identity, self, community and belonging, and on political issues such as democracy, territory, ownership, regulation, ethics and privacy. VR spaces *do* have implications concerning the social, cultural, political and economic spheres of daily life because they provide alternative arenas/places in which they can occur. Geographers have only just started to think through notions of place and how place as a concept may be changed by VR.

Public/private

The spaces being created through virtual reality techniques have, and will increasingly have, a role in how social groups interact. The mitigation and arbitration of space will continue, even if VR spaces do not abide consistently by the rules we are used to in the 'real world'. Not since the European colonisation of the Americas began have 'territories' this large been opened to new 'settlement'. Obviously VR spaces are different to the physical space humans inhabit, yet, even in its still relatively primitive spatiality, new geographies are being created (see Dodge, Chapter 21). Discourses of colonisation in cyberspace present a 'new frontier' for development. The distinct difference from past colonisation is that cyberspace colonies can theoretically be actively engaged and influenced by anybody from anywhere on the globe. These are not places without history, but with many histories in a flux of space–time that is not bound by the physicality of the earth's environment. A historiographic perspective is therefore beneficial and can be an aid in developing the governance and democratic societies espoused by information technologists from a broad community that ranges from Alvin Toffler to Al Gore.

The ideas of community developed under modern nation–state regimes cannot be simply extended to VR spaces. Expectations on civil society, constitutional premises, and democratic ideals are not simply extendible to places created on computer hardware and software by private companies. In particular, VR spaces, whilst proclaiming to be public spaces, are privately owned and regulated. In ways that perhaps the privatisation of the Internet forebode, governance of cyberspace could become a very contentious issue. Whereas recent European colonisation took place under the guise of the nation-state, VR spaces are being colonised under the flag of private corporations. There are no nations in VR nor are there constitutional laws that guarantee people basic human rights. The law of the corporation is the potential law of an autocrat, and as cyberspace becomes a meaningful part of public life and civil society, fundamental legal questions about free-speech and civil rights in cyberspace will become more prominent.

VR spaces are starting to be colonised. Academics, trained in post-structuralist and post-colonialist critique and analysis stand in good position to make unique contributions to the articulation of these cyber-societies and construction of VR. Euclidean spatial geometries are just one possible spatial organisational form that embodies our generic understanding of the physical space we inhabit. Is the Euclidean structuring of sharp boundaries, precise angular and distance measures and clear abstraction of experienced space best suited for cyberspaces? Other geometries and other spatial forms are not only conceivable, but become experientially accessible through virtual reality. There is an infinite number of spaces with individual rules and special constructions of new places for humans to develop.

Concluding remarks

Trying to think through these issues has not been an easy task and, as such, our discussion and the following chapters should be viewed as initial grapplings rather than polished thoughts. Virtual reality, in our opinion, raises a number of philosophical questions (relating to space and representation) and theoretical questions (social, cultural, political and economic implications), that extend way beyond its use as a geographical tool. There is a tendency as a new technology unfolds to concentrate on technical details and the promises such technology offers. It is only at a later date that we usually examine critically the implications of technological development. Whilst we engage with the technicalities of using VR technologies, designing VR spaces, and implementing practical uses, we should not lose sight of wider questions. Such questions open new avenues of thought and provide vantage points from which to survey developments. We would urge the reader to consider some of the issues we have raised in this Introduction in reading the following chapters (and the book in general), to reflect upon their current and possible, theoretical and practical implications, and influence the development of cyberspace.

References

Benedikt, M. 1991. Introduction. In Benedikt, M. (ed.) *Cyberspace: First Steps*. Cambridge, MA: MIT Press.

Castells, M. 1996. *The Rise of the Network Society*. Oxford: Blackwell.

Curry, M. 1998. *Digital Places Living with Geographic Information Technologies*. London: Routledge.

Graham, S. 1995. *Towns on the Television Closed Circuit TV Surveillance in British Towns and Cities*. Dept. of Town and Country Planning, University of Newcastle upon Tyne.

Harley, J.B. 1989. Deconstructing the map. *Cartographica*, 26, 1–20.

Jess, P. and Massey, D. 1995. The conceptualization of place. In Massey, D. and Jess, P. (eds) *A Place in the World? Places, Cultures and Globalization*. Oxford: University Press.

Kitchin, R.M. 1998. *Cyberspace The World in the Wires*. Chichester: John Wiley.

Lupton, D. 1995. The embodied computer/user. In Featherstone, M. and Burrows, R. (eds) *Cyberspace, Cyberbodies and Cyberpunk Cultures of Technological Embodiment*. London: Sage, pp. 97–112.

Markley, R. (ed.) 1995. *Virtual Realities and their Discontents*. Baltimore: John Hopkins University Press.

Massey, D. 1994. *Space, Place and Gender*. Cambridge: Polity.

Mitchell, W.J. 1995. *City of Bits. Space, Place and the Infobahn*. Cambridge, MA: MIT Press.

Morse, M. 1994. What do cyborgs eat? Oral logic in an information society. In Bender, G. and Druckery, T. (eds) *Culture on the Brink. Ideologies of Technology*. Seattle: Bay Press, pp. 157–89.

Pickles, J. (ed.) 1995. *Ground Truth*. New York: Guilford.

Plant, S. 1996. On the matrix. Cyberfeminist simulations. In Shields, R. (ed.) *Cultures of InternetVirtual Spaces, Real Histories and Living Bodies*. London: Sage, pp. 170–84.

Rheingold, H. 1994. *The Virtual Community. Surfing the Internet*. London: Minerva.

Sayer, A. 1985. The difference that space makes. In Gregory, D. and Urry, J. (eds) *Social Relations and Spatial Structures*. New York: St Martin's Press, pp. 49–66.

Slouka, M. 1996. *War of the Worlds. The Assault on Reality*. London: Abacus.

Stone, A.R. 1991. Will the real body please stand up? Boundary stories about virtual cultures. In Benedikt, M. (ed.) *Cyberspace. First Steps.* Cambridge, MA: MIT Press, pp. 81–118.

Tomas, D. 1991. Old rituals for new space. Rites de passage and William Gibson's cultural model of cyberspace. In Benedikt, M. (ed.) *Cyberspace. First Steps.* Cambridge, MA: MIT Press, pp. 31–48.

Urry, J. 1985. Social relations, space and time. In Gregory, D. and Urry, J. (eds) *Social Relations and Spatial Structures*. New York: St Martin's Press, pp. 20–48.

Wilson, L. 1994. *Cyberwar, God and Television.* Interview with Paul Virilio. CTHEORY. http://www.ctheory.com/a-cyberwar_god.html.

Woods, D. 1993. *The Power of Maps.* London: Routledge.

21 Explorations in AlphaWorld

The geography of 3D virtual worlds on the Internet

Martin Dodge

Introduction

In this chapter I am not so much concerned with the use of virtual reality (VR) *in* geography, rather I want to explore the social and urban geography *of* a social VR space that is inhabited by thousands of people from around the world. This has received little attention from academic geography (some notable exceptions include the papers by Batty *et al.*, 1998; Hillis, 1996; Taylor, 1997). I have chosen to focus on one particular VR space known as AlphaWorld. I hope that exploring the social and physical geography of AlphaWorld will be revealing for wider notions of the geography of virtual reality.

AlphaWorld is one of a number of commercially-developed systems that are publicly available on the Internet for social interaction. The aim of these systems is to create a graphical environment that can be shared by groups of people for real-time social interaction (known colloquially as 'chatting'). They are variously described as multi-user worlds, networked virtual reality (Schroeder, 1997), metaworlds (Rossney, 1996), avatar worlds (Damer, 1997), many-participant online virtual environments (Morningstar and Farmer, 1991), inhabited digital space (Damer, 1996), shared worlds (Roehl, 1997). In this chapter I refer to these commercial systems simply as *virtual worlds*. They are expressly designed as social spaces that are in some senses fun and easy to use. Also, they are an accessible form of VR in that they can be used on ordinary home PCs and phone lines without the high-powered equipment often associated with 'proper' VR.

We will begin by defining virtual worlds, their unique characteristics and antecedence. The main part of the chapter examines, in some depth, the particular social and geographic characteristics of one of these – AlphaWorld. Those interested in the other virtual worlds may like to consult *Avatars! Exploring and Building Virtual Worlds on the Internet* (Damer, 1997), written in a populist's travel-guide style. To give an idea of the type of imagery and metaphors being employed, the names of some of the other competing systems are – Community Place, V-Chat, InterSpace,

Worlds Chat, WorldsAway, The Palace, Deuxième Monde and CyberGate.

Defining virtual worlds

In many respects virtual worlds can be seen as an obvious enhancement to the popular text-based Internet chat services and MUDs (Multi-User Dungeons/Dimensions) by providing a visual and tangible 'physical' interface for social interaction using 2D, 2.5D or fully 3D graphical environments. Much as Web browsers provided an attractive, easy to use graphical interface to the largely textual information of the Internet, so it could be argued virtual worlds are just visual interfaces to the popular Internet activity of real-time chat. Taking this type of definition, virtual worlds could easily be dismissed as they share 'the features that make chat rooms so paralyzingly banal . . . and wed them to the empty eye-candy of the videogame' (Rossney, 1996). However, I would argue that the virtual world phenomena is greater than the sum of its parts (chatting plus eye-candy) for a number of reasons.

First, the social spaces that the virtual world provides are being actively used and, to varying degrees, modified by the participants themselves. This sense that they are environments that can be shaped by user's activities and interactions is crucial. The participants can define the parameters of their environment and experience (Morningstar and Farmer, 1991). In this fashion I would argue that cold, inhumane, digital spaces of VR are rendered into *places* that have meaning and the virtual world users become inhabitants, owners, citizens, developing a very real sense of belonging. This is particularly so with AlphaWorld where users are able to 'own' land and build homesteads, thereby constructing their own places for social interaction. This 'homesteading' facility in AlphaWorld was a conscious part of the design of the software and is unique amongst competing commercial Internet virtual worlds. It has been enthusiastically grasped by many thousands of people since AlphaWorld was opened.

The graphical environments themselves offer more than just an interface. They provide a genuine sense of geographic space and feelings of physicality to the user, with distinct ground and sky, the freedom to move in different directions. Bruce Damer, a virtual worlds evangelist, comments that the development of these system is significant because at last 'there was "space" in Cyberspace and "visiting a place on the Internet" began to have real meaning' (Damer, 1996: 1). The fact that virtual worlds are trying to provide visual environments with many of the characteristics of geographic space should make them of greater research interest to the academic disciplines that have a focus on use and configuration of space – geographers, planners and architects.

The other significant element of virtual worlds is the way the participants are represented and made visible to each other. A graphical object known as an 'avatar' is used as a virtual totem, projecting a tangible sense of self into the space. Virtual world avatars are able to move, manipulate objects, talk to each other and make gestures. The word 'avatar' comes from the Sanskrit language and can be translated as 'God's appearance on Earth' and it was first used in the context of virtual worlds in the pioneering Habitat system of the late 1980s (Morningstar and Farmer, 1991) and popularised by Neal Stephenson's 1992 science-fiction novel *Snow Crash*. Reid describes them thus, 'Avatars, the colonists of the virtual frontier ... A "real" person's proxy, puppet or delegate to an online environment, an avatar can be a faithful representation of its master or (more commonly) a fanciful one' (Reid, 1997: 197). The avatar is important for representation and identity formation in virtual worlds (Jeffrey and Mark, 1998; Suler, 1997) and will be discussed in more detail later.

I like to think of virtual worlds as hybrid places, not truly real like our material surroundings and, yet, also not completely virtual like other digital space (e.g. e-mail, Usenet, the Web). Virtual worlds are in-between, being virtual as they only exist in the digital form in computer memory and files on disk but trying to provide the visual structures of real places in the familiar forms like houses, gardens, ground and sky. This virtual 'physical' environment, with varying degrees of realism, attempts to simulate characteristics of real-world places in the hope of making the online experience less virtual and more naturalistic, therefore more enjoyable and fulfilling. And yet, because of technical limitations or conscious design decisions, the virtual world simulation of real places is still markedly different from the material world. For one thing, the realism of the graphics rendering has to be limited because of the Internet's capacity and social interaction is constrained by the keyboard and screen. Of particular interest are the conscious decisions of the programmers and designers to warp and even transcend the conventions of the real world. This is most evident in terms of movement, with participants being able to fly, teleport or walk through walls.

Virtual worlds, much like their text-based MUD forebears, can appear to the outside observer as strange, ephemeral and unreal places of fantasy. However, upon closer examination it can be seen that participants in the virtual worlds are engaging in very real communication, albeit strange in form and context. They are re-forming existing, and constructing new, modes of interaction and personal identity to suit the 'reality' of the places available in virtual worlds (Schroeder, 1997). From the outside it may appear ephemeral and unreal, but to the participants it imbues a powerful reality and sense of belonging. From a geographical point of view the study of these systems is interesting because of the how the social environment and the built environment combine to create a virtual place. Research is needed to consider how the design of the built-form (both of the space and the software interface), the technical operation of the system and the social management shape the activities and communities that form within the virtual worlds (Donath, 1997; Huxor, 1997; Schroeder, 1997).

What were the antecedents of these virtual world systems? There are two major antecedents – text-based, synchronous conferencing systems and computer games. Text-based systems, encompassing MUDs, MOOs (MUDs Object Oriented), IRC (Internet Relay Chat) and chat-rooms, have a long and varied history; the first computer MUD was created by Roy Trubshaw and Richard Bartle when they were students at Essex University at the end of the 1970s. The characteristics and usage of this textual virtual reality have been well studied by social scientists (see for example Curtis, 1996; Rafaeli *et al.*, 1998; Reid, 1991, 1995; Rheingold, 1993; Turkle, 1995). The crucial element of all these systems has been that they enable genuine social interaction between real people in real-time using ordinary PCs. This interaction in turn leads to the formation of so-called virtual communities (Garton *et al.*, 1997; Rheingold, 1993). Although the interfaces to these systems are limited to text communication, this does not seem to have hindered too greatly people's ability to reach out and interact with others through this medium.

The first virtual world system was called Habitat and it fused together a graphical interface with avatars and a game-style MUD environment (Morningstar and Farmer, 1991). It was a genuine pioneering effort, developed with the limited home computers of the late 1980s (the Commodore 64), and is really the progenitor of today's Internet virtual worlds. The creators described their system as:

> A far cry from many laboratory research efforts based on sophisticated interface hardware and tens of thousands of dollars per user of dedicated computing power, Habitat is built on top of an ordinary commercial on-line service and uses an inexpensive – some would say 'toy' – home computer to support user interaction. In spite of these somewhat plebeian underpinnings, Habitat is ambitious in its scope.
>
> (Morningstar and Farmer, 1991: 273)

In many ways this description could apply to the Internet virtual world systems today, in that they aim to be accessible to the average Internet user, yet have the ambitious goal to provide the space and conditions for communities to grow. Significantly, I would argue that virtual worlds draw much less from the experience of the mainstream virtual reality field (what I like to think of as the 'SGI and goggles' brigade). The key goals in the development and deployment of virtual worlds have been sociability and entertainment, rather than rendering algorithms or realistic visualization.

The other key antecedent of virtual worlds has been computer games. The rapid development in computer games technology in terms of the quality of their graphical interfaces and the immersiveness of their gameplay in the last ten to fifteen years has been considerable (Haddon, 1993; Herz, 1997). The computer games market is driven by fierce commercial pressures to deliver exciting new products that run on affordable home computers. Some of the most influential games are the simulation and strategy ones like *SimCity* (Macmillan, 1996), adventure games such as *Myst* and *Ultima* (Carroll, 1997; Kim, 1998) and, in particular, the 3D first-person action games with *DOOM* (Riddell, 1997) and *Quake* (Breeze, 1997; Gifford, 1996) being the best exemplars. In many respects the virtual spaces of the best of these networked games are more sophisticated than virtual worlds, and they are certainly deserving of serious study in their own right by geographers and other social scientists.

AlphaWorld as a case study

I now focus on one particular Internet virtual world called AlphaWorld. It is a massive virtual realm that has been visited by over 250,000 people since its inception in the summer of 1995, and many of them have actually constructed a large, complex city at its centre. It has a realistic three-dimensional environment and Figure 21.1 shows typical views of this, with

Figure 21.1 Various views of AlphaWorld.

buildings, trees and other users, represented by their avatars, walking around. AlphaWorld is a first-person-perspective virtual world in that you see through the eyes of your avatar; as you move your avatar, so your view of the world changes (although it is also possible to view yourself in third person as well).

AlphaWorld is owned and managed by Activeworlds.com, Inc. a small firm based in Newburyport, MA, USA and their homepage is at http://www.activeworlds.com/. Their software product is called Active Worlds, and it a comprises a 'universe' of around 300 different worlds operated by themselves or licensed to other individuals, universities and companies. Each world is really just a piece of server software with a unique name. While they are all based on the same server system they each have subtly different characteristics and visual styles. Some of the worlds are used as virtual versions of real places like Yellowstone National park or the planet Mars, while others are for marketing purposes, for example promoting major movies (such as *Godzilla*, *X-Files*). There are also worlds for different languages and religions (Schroeder, 1997; Schroeder *et al.*, 1998). Russian and Scandinavian language worlds are particularly popular. The company's figures claim that from its launch the total Active Worlds 'universe' has 680,000 unique users (as of 1 March 1999). AlphaWorld is one of the worlds in this Active Worlds 'universe'. However, it is special because it was the first and it is still the largest, most developed and most populated. It is this world that I will describe in detail in this chapter.

There are a number of reasons why AlphaWorld makes a good case study:

- it is one of the most technically-sophisticated virtual worlds, with a realistic 3D environment and avatars,
- it has many small, but active communities,
- it has many social activities, including the 'classic' example of a virtual wedding,[1] as well as classes in building, contests for the best homestead, prayer meetings (Schroeder *et al.*, 1998) and guided tours,
- it is probably the most popular virtual world system (although reliable usage figures are difficult to gather),
- it enables users to design and build the 3D environment themselves,

[1] The wedding took place in May 1996 and Damer recounts, 'Citizens floated their avatars down the aisle, crowded the altar to witness the words "I do" from both the bride and groom, and then floated in around the couple to wish them well. . . . When the bride tried to toss her bouquet, she discovered that it was permanently glued to her avatar. Immediately after the wedding, the groom drove 3,100 miles from San Antonio, Texas to Tacoma, Washington to kiss his bride' (Damer, 1997: 134).

- it has a detailed recorded history.[2]
- its urban geography has been mapped in detail (see Figures 21.2 and 21.3),
- a wide array of community information is available, for example on the Web and via active newsgroup discussions,
- you can try it out for free (although the software only runs on Microsoft Windows and you need to pay a citizen registration fee to gain access to all facilities),
- it has been studied by several social scientists using participant observation (Jeffrey and Mark, 1998; Schroeder, 1997; Schroeder *et al.*, 1998).

In technical terms, the system operates on a client–server basis. An Active Worlds server stores all the details on the world, such as the characteristics and location of all built objects likes houses and trees, and acts as a central hub to share real-time information on what the users are doing in terms of moving their avatars and talking. Users access the world with the client software called the Active Worlds browser which can be downloaded free. The top right picture in Figure 21.1 shows the interface of the browser, with its three key components being the large 3D viewing window, the text-chat interface at the bottom and the control panels on the left-hand side. The browser provides the essential elements that enable people to inhabit the virtual world – the 3D view, the tools to move the avatar and the ability to conduct text 'conversations' with other people. As you move around the world, the browser downloads, from the server, the 3D environment and avatar activity in your immediate neighbourhood. The 3D environment is made of individual building blocks (like wall panels, doors, trees, chunks of road, etc.) which are streamed as you move and then cached by the client. The blocks themselves are in Renderware format, rather than the more standard VRML. The use of streaming, caching and object compression make for a realistic and usable 3D environment that runs well on average home PCs and typical network speeds.

[2] For example, there is the AlphaWorld Historical Society, with a museum <http://www.awcommunity.org/awhs/>. There is also a 'national' newspaper called the *New World Times* (or *NWT* for short), run by AlphaWorld citizens. The first edition of the *NWT* was published only a couple of months after AlphaWorld opened, with regular editions since then. An archive of all editions is available on the Web providing valuable historical documents (*NWT*, 1998).

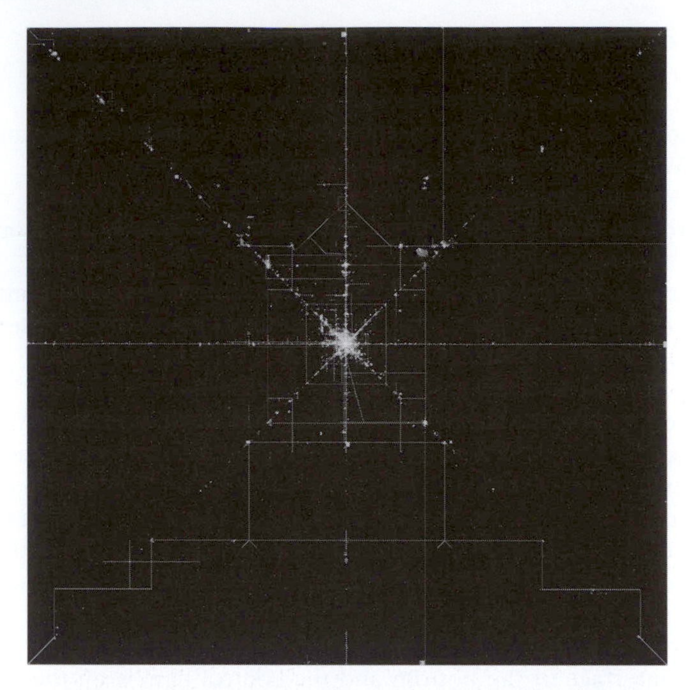

Figure 21.2 (a) Density of building in AlphaWorld on 23 February 1998, produced by Vevo (Source: http://awmap.vevo.com/densmap.html).

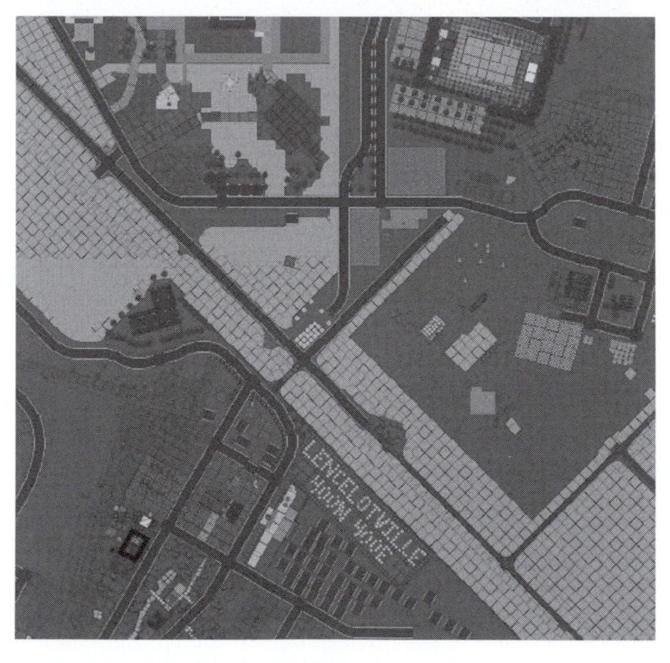

Figure 21.2 (b) Detailed 'aerial-photo' style map of AlphaWorld (Source: http://awmap.vevo.com/).

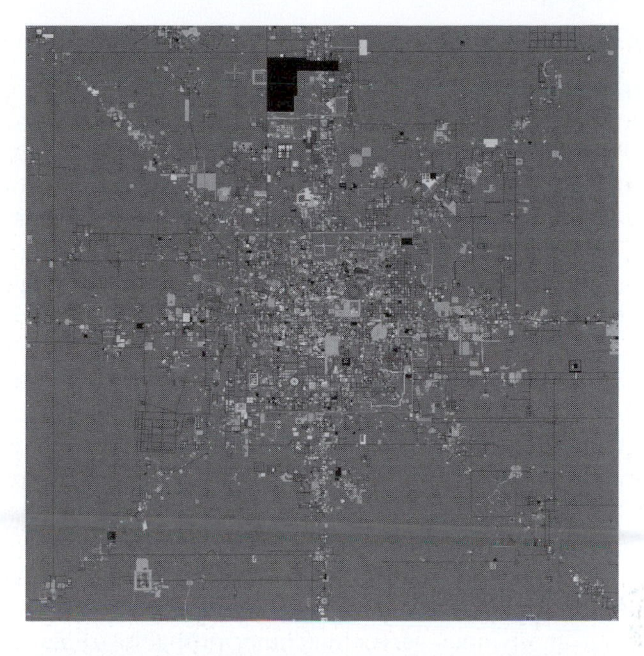

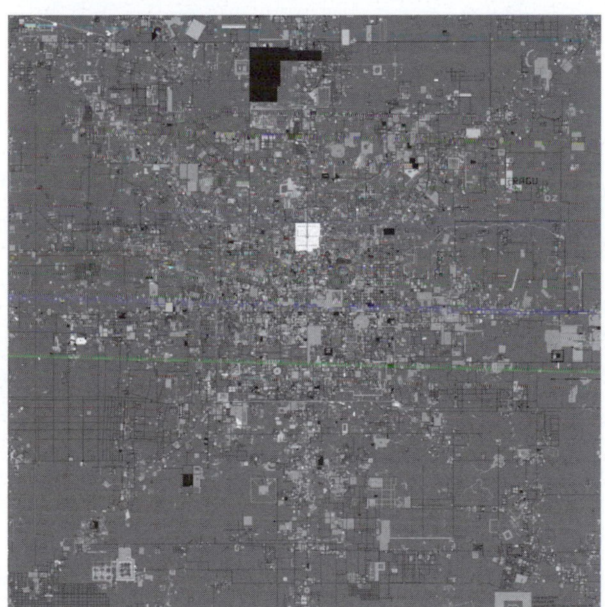

Figure 21.3 AlphaWorld 'land-use' maps created by Roland Vilett from December 1996 (top) and February 1998 (bottom) (Source: http://www.activeworlds.com/events/satellite.html).

The space of AlphaWorld

AlphaWorld comprises a massive virtual space which is undergoing concerted urban development. It was opened to the Internet general public on 28 June 1995. This date marks the beginning of sustained colonisation, with tens of thousands of people leading a virtual land-rush to claim the *terra nullius* of AlphaWorld. The virtual space came into existence as a flat, featureless plain stretching for hundreds of virtual kilometres in every direction, coloured a uniform shade of green to signify it as virgin territory waiting to be claimed. There are no 'natural' features, no mountains or rivers, just a perfect green plain sheltering under an unceasing bright blue sky. There is no weather or wildlife. For geographers, an obvious comparison can be drawn between space of AlphaWorld and the isotropic plain used in early models of settlement patterns. This world would also be the true dream of the flat earth advocates of old! Everything that now exists in AlphaWorld, over 25.2 million objects as of June 1998, has been placed there by the human inhabitants (Vevo, 1998). Of course, the reality is that the virtual expanse of AlphaWorld, with all its urban development, is simply a large server database, around 1.4 gigabytes in size in June 1998, running on an anonymous server machine, probably somewhere in the Boston area of the USA. It only exists in cyberspace, but can be accessed from anywhere in the world with an Internet connection.

So, how big is AlphaWorld? The total area of the flat plain is exactly 429,038 square kilometres, which is some 43 per cent larger than the United Kingdom. Unlike the UK, the borders of AlphaWorld are dead straight, forming an exact square of land 655 kilometres on each side. A Cartesian coordinate system is used to delineate space in AlphaWorld with an origin point in the dead centre of the world at 0,0. This centre point is known as Ground Zero (GZ) to the locals and is the focal point for the world because when people enter this is the location at which they arrive. Consequently, the area around Ground Zero is always the most densely populated. When people give addresses in AlphaWorld they use coordinates such as *67N, 42W* which translates to 670 metres north and 420 metres west of GZ. It is interesting that people know and use coordinates of their homesteads, rather than house numbers and street names. In the real world the only people who are likely to know the location of their house in terms of coordinates are surveyors.

The inhabitants of AlphaWorld have been busy claiming land and building all manner of structures from modest suburban-style homes to grand castles (I discuss how it is possible to build later). Since the world opened in 1995, over half of a million people have visited and about 30,000 of these have built something. However, the activity of these 30,000 has made little impact on the vast expanse of AlphaWorld, despite their best efforts. Figure 21.2(a) is a map of the whole of AlphaWorld showing the density of urban development as of late February 1998. The most heavily built-up areas on the map are represented by the brightest pixels. The end result looks very much like the satellite photographs of the Earth taken at night where the major cities and conurbations are identified as bright areas caused by all the light escaping into space. From Figure 21.2(a) it is clear that the most developed area of AlphaWorld is the densely-built city around GZ in the middle of the image, which sprawls out in all directions for about fifteen kilometres. Ribbons of urban growth project out from this city along the principle compass axes to form a distinctive star shape. Towns and other small settlements lie along these axes, looking like bright beads strung along a necklace. The spatial structure of urban development is largely the result of the power of the coordinate system as a form of addressing in AlphaWorld. Human nature means people like to choose regular and memorable coordinates, such as 50N, 50W or 1555E, 1555S as the location for their homestead. Once a pioneer has started building, other citizens will build alongside either by invitation or just to be close to other people. The fact that AlphaWorld has a single entry point – GZ – (unlike real-world cities), has led to the great concentration of development in the centre of the world.

It is evident from Figure 21.2(a) that a large amount of AlphaWorld's expansive green plain remains undeveloped, with no glow of human activity. Only a tiny percentage of the AlphaWorld's land contains any building. There is clearly still plenty of room for expansion, although the much sought-after land in the centre of the world, as close as possible to GZ, is now heavily developed. With so much space available, AlphaWorld is very under-populated compared to most real-world countries. There are currently about 30,000 different people who have built something in the world; taking this as the resident population of AlphaWorld, this gives a density of just 0.07 people per square kilometre.

Figure 21.2(b) shows a much more detailed map of a small area of the GZ city. At this resolution, the map is like an aerial photograph where it is possible to discern roads and individual buildings. The maps in Figure 21.2 were produced by the Vevo project, which has developed a sophisticated mapping system, using a quadtree structure, that is capable of producing maps of AlphaWorld at twelve different resolutions using the server database that holds data on the location of all objects in the world. The multi-resolution maps can be interactively browsed from their Website at http://awmap.vevo.com/ and the highest resolution map can actually be used as a powerful teleportation tool. By clicking on a desired location on the map, you will then be instantly transported there in the ActiveWorlds browser.

What does the whole of the city at GZ look like? When you walk around at street level it is hard to get a sense of how big it is or how it is spatially structured. So, Roland Vilett, one of the AlphaWorld programmers, has produced two gloriously detailed colour 'satellite' maps of the city (Vilett, 1998). Figure 21.3 shows these for two snaps-shots in time

(December 1996 and February 1998). They vividly reveal the organic complexity of the urbanisation caused by the unplanned action of thousands of real users. Steven Johnson described Vilett's maps as follows:

> You can't help but be startled looking at these images the first time. They have a kind of orderly disorder.... There's a clear pattern to the shape, but it has the blurriness, the granularity of real-world cities seen from above.
>
> (Johnson, 1999)

It can be clearly discerned that the most intensive urban development has taken place in the centre of the maps, which is around Ground Zero. The extent of urban development, spreading out from GZ, in just over a year is apparent by comparing the two maps. Both maps cover the same area of AlphaWorld from 1000N, 1000W to 1000S, 1000E, a 400 square kilometre tract of land. This represents a mere 0.3 per cent of the total extent of the world, although it does contain a large proportion of the building. In the maps, particularly the first created in December 1996, the star-shaped urban development along the compass axes is clearly evident. A year later the star shape is dissolving as fill-in development in the desirable land around GZ continues apace. The dark green areas on these maps is unclaimed land.

Immigrating to AlphaWorld

When you enter AlphaWorld for the first time you go through an immigration procedure, but Damer reassures, 'Don't panic, immigration into this virtual world is much easier than crossing national boundaries!' (Damer, 1997: 107). You can choose to enter the world on a tourist visa or you can apply for citizenship, which grants you several important rights. To gain AlphaWorld citizenship requires the payment of a fee (about $20 a year) to the owners, Activeworlds.com, Inc. It is interesting that the people who run AlphaWorld use real-world metaphors of immigration and citizenship in relation to access to their virtual world. The introduction of differential rights between tourists and citizens, when the fee was instigated in September 1997, has given rise to a two-tier social structure of 'insiders' and 'outsiders' which has impacts on the community in AlphaWorld (Schroeder, 1997). Before the fee was introduced, everything was free and everyone was essentially equal. Now, AlphaWorld citizenship buys you the right to choose your avatar from a wide range of available ones, to own land and build on it and also send telegrams to other users. As a tourist you are free to wander around and engage in conversation, but you are stuck with the default 'tourist' avatars and you cannot build permanent structures. Tourists can be treated differently and unfavourably by citizens, although it is difficult to determine how widespread this is. Just like

Figure 21.4 Typical AlphaWorld avatars, the one on the right is a tourist one.

in the real world, the 'locals' can be unfriendly to tourists and, as Schroeder notes, 'tourists are immediately identifiable; they have single "standard issue" avatars – with a camera hanging from their shoulders!' (Schroeder, 1997, para 5.5). Figure 21.4 shows examples of citizen and standard tourist avatars. There are also more subtle differentiations between citizens with the age of your citizenship being indicated by how low an ID number you have. The lower the number, the more of a pioneer you were, arguably conferring a degree of kudos and status in the world. Social stratification is also played out in spatial terms, with 'newbies' tending to cluster at GZ whereas the 'regulars' are more wide ranging, exploring more of the territory and holding meetings and events at specific locations (Schroeder, 1997). This is due to their greater familiarity with the system and what is available in the world – they know the good places to go. Also, regulars have often built homesteads that they can invite people to visit, a facility denied to tourists.

When you immigrate to AlphaWorld you are required to choose a nickname that is unique in AlphaWorld. This contributes to the construction of a virtual identity which can often be quite different from your real-world persona. The selection of an avatar also adds to this, as it is perfectly possible to choose a body form of a different race or gender. The nicknames of people present in AlphaWorld at any given time clearly show that many are taken from favourite characters in literature, TV, films and games. The construction of virtual identities has been widely noted by other researchers looking at various computer-mediated communication systems, particularly text-based MUDs (see for example the work of Curtis, 1996; Donath, 1998; Turkle, 1995). The allocation of citizenship also requires you to provide a valid e-mail address. This is for practical communication purposes, but it also, arguably, instils some degree of social responsibility on users, knowing that they are not totally anonymous (Damer, 1997).

To what degree are the users of AlphaWorld really citizens in a new world with inalienable rights? The constitutional position of citizens can best be described as vague and I would argue they are really just consumers, despite all the immigration rhetoric. When you immigrate to AlphaWorld you are really signing up to a consumer contract with the world owners, Activeworlds.com, Inc. To use their software you must agree with their licence which entitles you to certain activities. There is evidence that some users are unhappy with the nature of their rights under this agreement and how the world is managed. There have been accusations levelled against the management of arbitrary use of their powers in regard to ejecting and banning people from AlphaWorld. These rumblings of discontent are reported in various online newsgroups for discussion of AlphaWorld. It could be that some of the AlphaWorld old-timers have become frustrated by the changes they see being imposed on their world, for example the introduction of the citizen fee and increasing commercialisation that they have no control over.

The use of powerful real-world metaphors of immigration, citizenship and homesteading mask the reality that AlphaWorld is a privately-owned themepark to which you can buy a ticket (citizenship) that allows you to build new rides and chat to the other customers, but the management of the park have the right and power to refuse entry, throw you out and ban you when they feel you have broken their rules. You have no means independently to challenge the management's actions. In this sense, AlphaWorld is not a wild-west frontier, rather it is more like a Disneyesque FrontierLand. Internet virtual worlds like AlphaWorld can been seen as another example of privately-owned and operated semi-public spaces designed for consumption, just like shopping malls and themeparks in the real world (Graham and Aurigi, 1997; Sorkin, 1992). The true nature of AlphaWorld as a semi-public space was eloquently spelt out by the following message posted on the AlphaWorld community newsgroup by user 'retsmah' under the subject heading 'its a corporation not a country':

> i think what most of the people are forgetting is the cof [Circle of Fire Studios, the previous owners] is a corporation not a country . . . therefore they can run it any way they want . . . and they dont have to listen to the consumer unless they choose to . . . for some reason people seem to think that it is a country, and a free country at that . . . well hate to burst the bubble but it is not . . . aw [AlphaWorld] is a software program owned by a company . . . if you dont like how they run the company dont use the software . . . i mean i hate coca cola . . . but i dont try to tell them how to run the company hahahaha.
>
> (Retsmah, 1998)

Just like themeparks and shopping malls, AlphaWorld has its own private security guards, called Peacekeepers, their job being to maintain order and prevent disruption to other customers. To achieve this they have the power instantly to eject people and then ban them for varying lengths of time from returning to the world.

Building in AlphaWorld

Undoubtedly, one of the main attractions of AlphaWorld compared to competing virtual world systems has been the ability of its users to construct things in the world. Building in AlphaWorld is much like using a Lego construction set. The world was conceived with the means to allow users to build and this has proved to be popular. As I have shown, the citizens have built a huge, sprawling city in the centre of the world, along with many smaller settlements and isolated homesteads out in AlphaWorld's expansive prairie.

The first step in building is to locate a plot of empty land that is not owned by anyone else. This can be difficult if you want to build anywhere near GZ due to the density of existing urban development; however, there is still plenty of land further out. Once you have found some suitable vacant land you can claim this territory for your homestead by simply building on the ground. There is no limit to how much land you can claim. However, it is important to cover every acre you own with buildings or gardens, otherwise others can build in your backyard. This gives rise to building disputes, one of the major sources of conflict in AlphaWorld.

Building is undertaken with predefined objects, much like virtual Lego bricks, such as road sections, wall panels, doors, windows, flowers and furniture. In total there are over 1,000 different objects available and you put them together piece by piece to create larger structures. The Active Worlds browser provides rudimentary tools to select and manipulate the objects, putting them in a desired position. Construction of large buildings, using hundreds of individual objects, requires a considerable amount of skill and effort. It appears that AlphaWorld is providing a powerful new medium of personal self-expression which is denied to most people in the real world. How many people have the time, money and skills to build their dream home in the real world? Well, many thousands have been able to become architects and builders in AlphaWorld and Damer claims that 'Within two years ... home users had built more three-dimensional virtual space than all the laboratory and university virtual reality environments combined' (Damer, 1997). The views in Figure 21.1 and the detailed map in Figure 21.2(b) shows typical AlphaWorld construction.

As with a Lego set, you can only build with the pieces provided. It is not possible to create your own objects in AlphaWorld. This means the built environment of AlphaWorld has a somewhat homogeneous appearance as everyone has to use the same materials. Despite this limitation, the individual creativity of the citizens has flowered, with all manner of interesting structures having been constructed. Some are well-designed and aesthetically pleasing, but there are also equal measures of ugly and half-finished structures. The homesteads that people build in AlphaWorld are analogous to the homepages that people create on the Web, rather than private homes of the real world. Both homesteads and homepages are tangible expressions of presence in the online world, serving as fixed points of reference in an ever-changing landscape. They are both made publicly available, to be examined without your express knowledge or invitation. Jeffrey and Mark comment, 'Building a home provides an opportunity to showcase one's craftsmanship, and create a feeling of ownership as the home is a territorial marker for a virtual habitat' (Jeffrey and Mark, 1998). They both require an investment of time and effort to build properly and maintain. AlphaWorld citizens list the coordinate location of their homesteads in the same way people give the Web address of their homepage. Finally, both forms of virtual expression suffer the same problems of poor design and 'build and abandon'.

From my informal observation of the structures which users have built in AlphaWorld it is clear that they are firmly rooted in people's everyday experience of real-world places. Many homestead designs match familiar architectural forms and layouts despite the complete freedom of the virtual world to stretch and warp the conventional architectural notions of the material world. For example, it is perfectly possible to build abstract structures floating in mid-air and other architectural designs that would be impossible with real-world building materials and the force of gravity. Yet, most people seem to stick to building conventional homesteads, with walls, a roof and a frontage onto the road. Of course this is in part enforced by the types of building blocks provided, but I think it is also due to a powerful need to create places that are in some senses familiar and 'real'. People also have difficulties navigating structures in three dimensions that disobey many of the conventions of real-world architecture. A more detailed and systematic study of the architecture of AlphaWorld could be revealing about how people are adapting virtual space to create places that have meaning in their lives.

The local urban morphology of the city around GZ is, not surprisingly, chaotic and disorganised because it has grown over time from the efforts of lots of individuals with no central coordination. In AlphaWorld there are no building controls or planning zones. I would argue that AlphaWorld's towns can be viewed as similar to the informal squatter settlements that surround many rapidly-urbanising cities in the Third World. These settlements are unplanned and built by the residents themselves from whatever materials they have to hand.

There have been several attempts to form specific communities in AlphaWorld by formally planning and building an actual township. The most well-documented of these has been the Sherwood Forest community project run by the Contact Consortium (Contact Consortium, 1998; Damer, 1997). The project commenced in early 1996 and had a formal charter and a town plan (Figure 21.5). In many ways the community activists of the Sherwood Forest project were aiming to re-create a utopian, Californian-style, suburban township (Kling and Lamb, 1996). Another good example is Pink Village, a gay and lesbian community, which has bars, cafés, a night club, a town hall, remembrance gardens, a museum, galleries, as well as 'private' homes (Pink Village, 1998). The community is active, with a local newspaper, a calendar of social events including a pride festival, and an elected village council.

The unique facility of AlphaWorld to allow uncontrolled and unplanned building by users also provides a fascinating new avenue for aberrant behaviour in the virtual world in the form of virtual graffiti and

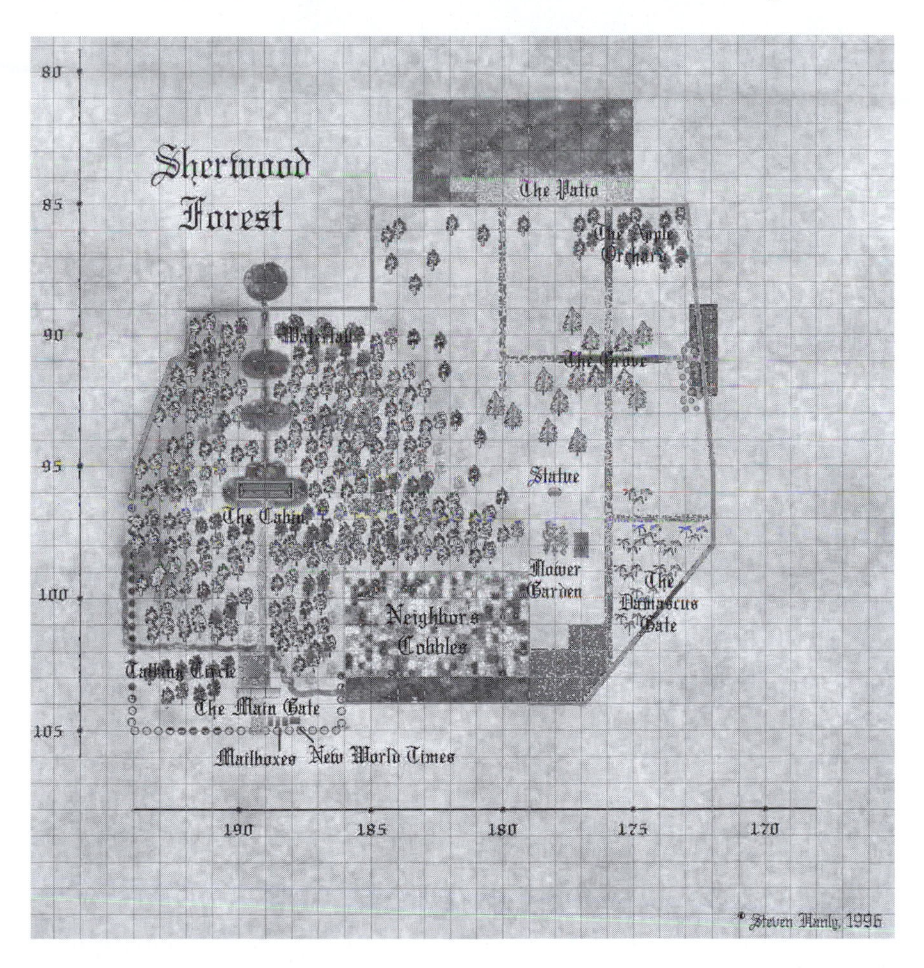

Figure 21.5 Sherwood Forest community project town plan (Source: http://www.ccon.org/events/sherwood.html).

Figure 21.6 An example of vandalism.

vandalism. Even though the software prevents anyone but the owner of the land changing a building, vandalism is possible by deliberately placing objects as close as possible to other people's homesteads. A small number of users appear to take pleasure from this, using annoying objects, like flames, bogus teleports, and even large billboards with offensive pornographic pictures on them, placed right in front of the entrance to people's homesteads. Figure 21.6 shows an example of virtual vandalism, where a large animated flame has been maliciously placed inside the fenced property of someone else (Damer, 1997). The owner of the land cannot move the flame, so has responded in the only way possible by placing a large placard requesting the perpetrator to remove the offending object. Interestingly, the flame object appears to be used as a 'physical' equivalent of a flame e-mail. This type of vandalism is viewed as the serious 'crime' in AlphaWorld because of the importance citizens attach to their homesteads and directly led to early community action in 1995 with the formation of the AlphaWorld Police Department to counter this anti-social behaviour (Damer, 1997). However, there is little people can do as it is very difficult to get vandalism removed. It is not clear how widespread it is or whether they are random acts or a more concerted campaign against certain properties and individuals.

Travel and time in AlphaWorld

The physical movement of avatars in AlphaWorld is greatly enhanced compared to how humans can move their bodies, unaided by machines, in the material world. The nature of space and time is being warped in the virtual world by the superhuman powers of the avatar. By default, when you enter AlphaWorld your avatar walks along the ground. However, it is just as easy to fly up in the air as it is to walk on terra firma. Flying requires no greater effort or special equipment, you simply press the + key and up you go. In fact, it is often easier to travel by flying because you can see better where you are going, so for many users flying is the preferred means of local movement. It is also perfectly possible and acceptable to hold a conversation while floating in mid-air.

To travel any distance by walking or flying quickly becomes tedious. As there are no cars, trains or planes in AlphaWorld citizens use teleportation to get around. Teleportation in AlphaWorld works just like it does in sci-fi films. Your avatar is instantaneously transported to a specified location, with the accompaniment of a suitable 'beaming' sound-effect. Teleportation has seriously warped the concept of distance and geographical accessibility as any location in the 429,000 square kilometre expanse of AlphaWorld can be reached instantaneously from any other point, at no cost in terms of time or money. Consequently, every point in AlphaWorld is equally accessible. This is truly the 'death of distance' (Cairncross, 1997; Couclelis, 1996). The ability to teleport is a powerful feature; however, it was not available at the beginning of AlphaWorld's history. It has only been progressively introduced for fear of its affects on the world. As the AlphaWorld newspaper reported in November 1995:

> Teleportation! Yes Teleportation! The one most common request of AlphaWorld citizens has been teleportation ... With teleportation more of AlphaWorld will become readily accessible. There is still some concern that teleportation will ruin the simulation of reality in AlphaWorld.
>
> (*NWT*, 4: 2)

Teleportation does cause problems in terms of navigation because once it is introduced people become more and more dependent on it and they lose their understanding of the geography of the larger world, how one place relates to another (Anders, 1998). Teleporting can be thought of like riding the subway under a city. It gets you from A to B quickly but you have no sense of the structure of the city you are passing under. Teleportation also has a negative impact on the social life of the virtual world as it reduces the opportunities for chance encounters and discoveries (Anders, 1998). AlphaWorld citizens can teleport directly to their homesteads without any risk of bumping into other people. In a similar manner to the car in real cities, teleportation has the tendency to diminish social activity.

The tyranny of distance may be rendered obsolete by teleportation, but the importance of geographic location is alive and well. When people are choosing a location to visit or, more importantly, a place to build their homestead they want a *good* location. In the real world, the factors that determine a good location vary from place to place and person to person (e.g. a good view, away from the noisy airport or near a good school). In the context of AlphaWorld a good location is determined by two main factors: first, being as close as possible to Ground Zero, the centre of the world. Second, having a location with memorable coordinates, for example the Pink Village is located at 2222S and 2222E.

Navigation in AlphaWorld is made extremely difficult because visibility is limited by the browser so it is only possible to see a maximum of 120 metres in any direction. (The default setting is half this distance.) So, despite the vast expanse of space in AlphaWorld, you never really get a sense of this scale as you can never see distant landmarks, for example. The constrained visibility is a practical necessity to restrict the amount of 3D environment that has to be rendered so the software can maintain a usable frame rate on average PC hardware. However, the effect on your perception of geographic space is really quite unnerving; it is like walking around in an opaque bubble 120 metres across and it takes some getting used to. The problem is that streets and buildings appear to end, with a sharp cut-off line at the edge of the visibility bubble. This impacts on local accessibility because it is hard to orientate and navigate with no fixed landmarks and distant vistas. When this visibility limitation is combined with the use of teleportation to exact locations, the effect is to prevent you building up a mental map of the city as you do not learn the geographic context of features and the spatial relations between them. The practical impact is to make it very hard to find buildings and features of interest unless you know their exact x and y coordinates.

As well as warping space and distance, AlphaWorld also exists in its own time-zone called AlphaWorld Standard Time. The need for a special time-zone arose because of the difficulty of scheduling meetings and events with people from all around the world; confusion often occurred as people tried to agree a mutually convenient time and then convert it into their local time. Consequently, a group of AlphaWorld activists designated Greenwich Mean Time minus two hours as the standard for AlphaWorld in November 1996. This time-zone, known as mid-Atlantic, is not used by any countries in the real world. Subsequently, this has been adopted by other virtual communities who communicate in real-time, with the time-zone recently being renamed Virtual Reality Time (VRT for short). In the latest version of the ActiveWorlds browser software, VRT is displayed on the status bar, so citizens need never be confused about the time in Alpha-

World. The existence of virtual worlds in their own time-zone helps to make them more like real places.

Avatars and identity

The role of the avatar, in providing a tangible, representative form of the user, is a crucial element that distinguishes virtual worlds from other computer-mediated social spaces such as Usenet and IRC (Rossney, 1996). The avatar as a bodily presence in virtual space provides a focus for conversation and social interaction. Avatars can talk to each other, which has the effect of enabling a kind of face-to-face communication between users.

The selection of the avatar is also important in the formation of the virtual persona, just like the physical body is at the core of our real-world identities (Donath, 1998; Suler, 1997). The development of online virtual identities that differ from real-world identities has been a topic of considerable interest in social science research. In addition, the role of the body in cyberspace, the potential for intellectual disembodiment and transcendence of flesh and bone has also interested researchers (Featherstone and Burrows, 1995; Stone 1991).

In AlphaWorld, registered citizens can choose from a list of thirty-odd available avatars. Figure 21.4 shows several typical AlphaWorld avatars. Unlike other virtual world systems, such as the Palace, you are not able to create your own unique avatar (Suler, 1997), which does limit this avenue of personal expression. All the AlphaWorld avatars are based on conventional human body shapes unlike many other virtual worlds which have avatars more obviously aimed at the entertainment market using cartoon-style characters or inanimate objects. All AlphaWorld's avatars have a similar 'look' and I think they resemble virtual versions of Barbie and Ken. They attempt to model some bodily movement with articulated arms, legs and head, along with some limited facial expressions. Avatars move in a simulated walking motion and a number of pre-programmed actions are possible such as 'dance', 'wave', 'anger' (a wave of the fist). Some have argued that avatar gesturing will bring some of the depth and nuances of non-verbal communication that is so vital in the real world to the textual conversations in virtual worlds (Rossney, 1996). However, this does not seem to be the case in AlphaWorld as the gestures are not widely used (Jeffrey and Mark, 1998). The only time gestures are really used is to simulate avatar dancing at parties (Damer, 1997). Instead feelings and emotions are conveyed through short text messages used for conversation.

Conversing in AlphaWorld is achieved by typing short sentences in a panel at the bottom of the ActiveWorlds browser. These messages then appear in sequence in the chat-box, scrolling up as other people 'speak' and also floating above your avatar's head. Like voice conversations in the real world, you can only 'hear' people talking in your immediate vicinity and if too many people try to speak at once it can be difficult to follow the conversation. Conversing using short-typed messages is of necessity direct and 'chatty' in nature. Indeed, it has been observed that this has given rise to new conventions of language unique to real-time text-chat, with prevalence of abbreviations and acronyms (like 'LOL' for laugh out loud or 'BRB' for 'be right back') to represent frequently-used phrases, to minimise the amount of repetitive typing and help maintain the speed of conversation (Menges, 1996; Suler, 1997). In addition, these abbreviations act as a shared dialect which helps to define the particular character and distinctiveness of the virtual community. Other distinctive features of this mode of communication include the use of emoticons (e.g. :-)), action phrases ('Martin smiles'), heavy punctuation ('....????'), capitalisation ('SCREAM!'), and onomatopoeia ('hehehe') to express feelings and emotions that are normally conveyed by body language and tone of voice in spoken conversation. This can be combined with avatar movement and gesturing to create some of the depth of real conversation.

Selection of the avatar is from a menu listing with each one having a unique name. Citizens can change their avatars at any time simply by clicking on the name of their choice; they instantly assume that avatar shape in the world. It goes without saying that the avatar someone has chosen may not have any relation to their real-world bodily appearance. The ease and freedom of avatar selection, arguably, encourages well-known identity deceptions that occur in online social interaction such as gender-swapping, race shifting and exaggeration of physical characteristics (Donath, 1998). One needs to be aware that avatars are powerful means of *mis*representation as well as representation in virtual worlds.

It is apparent that users in virtual worlds both consciously and unconsciously use their avatars as they would a physical body, the best example being the convention of facing your avatar to 'look' at the avatar you are talking to; the need to make eye-contact is obviously important in virtual conversation although it is not technically required by the software. This is largely because the world is presented in first-person perspective through the eyes of the avatar. When people talk in groups they tend to arrange their avatars in a loose circle, all facing each other. The avatar seems to exhibit the same sense of personal space that bodies do in the real world. From their observations, Jeffrey and Mark state, 'Although physically possible to pass through avatars, it was seen as rude and impolite and this behaviour was not observed very frequently' (Jeffrey and Mark, 1998). So people tend to walk their avatars around others, rather than go straight through. Indeed, the sanctity of personal space around your avatar means unwarranted and deliberate attempts to invade it can feel threatening and are known as avabuse (Damer, 1997). The potential for many kinds of anti-social behaviour exists in computer-mediated communication from annoying pranks to deeply offensive verbal abuse and threats (Dibbell, 1996). This is often encouraged by the veil of anonymity that the computer provides. The extended environment of virtual worlds, particularly the avatar representation, gives people new opportunities and avenues for abusive behaviour that go beyond the verbal to physical blocking, shadowing and stalking (Suler, 1997).

To try and counter the social problems of verbal abuse, avatar assaults and virtual vandalism, an organised system of policing by volunteer 'security guards' under the direction of the world owners was instigated. Their concern, running the world as a commercial venture, is that offensive behaviour discourages casual visitors from registering as citizens. One class of control is provided by so-called Gatekeepers who have the power of ejection, although their primary role is to welcome new users and provide assistance. However, their guidelines also state that they 'have the right and power to maintain the levels of decency that you and I would expect in the real world', whatever the level may be. The 'proper' AlphaWorld police are called Peacekeepers and their role is more wide ranging, being able to patrol the world and try to intervene to prevent verbal abuse, investigate stalking and incidents of vandalism and they have the powers of ejection and banning (see http://www.activeworlds.com/Peace.htm). They are organised with a duty roster to provide continuous police cover. Some users have expressed serious concerns over how the Peacekeeper role is executed, with accusations of heavy-handed policing with summary expulsions, and an inadequate appeals systems.

The avatar plays an important role in providing the 'human' scale in AlphaWorld with objects like windows, doors, stairs and furniture being appropriately sized. Buildings and other structures are scaled to the avatar height, just like much of the man-made environment of the real world is built for the scale of the human body. Avatars can also dispense with the hassle of the real-world convention of doors as it is possible to walk through walls by simply holding down the shift key. AlphaWorld encourages the construction of a built environment with solid walls using the metaphors of the material world; however, it also provides Superman-like powers to shatter the illusion and allow avatars to effortlessly glide through structures, and many people do use this feature. Avatars in AlphaWorld do truly provide god-like powers to mortal users.

Conclusions

Virtual worlds systems, like AlphaWorld, are providing fascinating new social spaces that exist only in cyberspace. Geographers can make valuable contributions to understanding the spatial structure, development and use of these spaces. One area which seems especially ripe for investigation by geographers is how the individual actions and the social interactions of real people can transform cold, computer-generated virtual spaces into meaningful, humane virtual *places*. In AlphaWorld, the activity of claiming land and building homesteads to your own designs that are persistent is vital to this transformation.

Some of the key themes that I have explored in relation to AlphaWorld in this chapter are the geographic structure of the virtual space, the growth of urban development, how users build and the meaning of their homesteads, the important role that the avatar plays in virtual worlds, the problems of anti-social behaviour, the warping of the notions of geographic distance and accessibility by the nature of movement and travel.

In this chapter I have taken a fairly positive view of AlphaWorld. However, there are problems and weaknesses with virtual worlds, including AlphaWorld. From my exploration I believe the key problem is the very severe under-population. There are simply not enough people using most of the virtual worlds to make them interesting, engaging and self-sustaining social environments. All too often you enter a virtual world and find yourself alone. After wandering through empty streets and buildings for a while you usually give up as there is nothing to do! These spaces only come alive when there is a real throng of people around to talk to. This population problem is also suffered by the Active Worlds system where the vast majority of the worlds are devoid of people and therefore effectively dead. Even AlphaWorld, the most popular and active, is lacking in people when you get beyond the immediate surroundings of Ground Zero and it is easy to wander through ghost-towns with not a soul around. Virtual Worlds are still a real minority activity on the Internet compared to other real-time social spaces like chat-rooms or games.

For me the key question is whether the apparent failure of virtual worlds to attract a large user population is due to a fundamental weakness with the virtual world concept or simply due to technical failures by the companies implementing and marketing the current systems. In some senses, this question could apply to the larger VRML and VR 'projects' which have, so far, failed to make the dramatic impact on the way people use computers and the Net that was predicted by some in the early 1990s. I am not sure what the answer is, but I think the spatial sciences have an important contribution to make in investigating the question. Therefore, by way of conclusion, I would like to encourage you to take a little time to explore some of these virtual worlds, especially AlphaWorld, for yourself, talk to some of the residents and visitors, experiment with flying and teleporting, and maybe have a go at building. Let me know what you think of the geography of this virtual world.

Acknowledgements

I would like to thank Andy Smith for generously sharing his knowledge and insight on the virtual worlds. I am also grateful to Francis Harvey and Rob Kitchin for comments on earlier drafts of this chapter. I am also grateful to Roland Vilett (Circle of Fire Studios), and Greg Roelofs and Pieter van der Meulen (Vevo project) for permission to use their maps of AlphaWorld.

References

Anders, P. 1998. *Envisioning Cyberspace: Designing 3D Electronic Space*. New York: McGraw Hill.

Batty, M., Dodge, M., Doyle, S. and Smith, A. 1998. Modelling virtual environments. In Longley, P.A., Brooks, S.M., McDonnell, R. and Macmillan, B. (eds) *Geocomputation: A Primer*. Chichester, UK: Wiley, pp. 139–61.

Breeze, M. 1997. Quake-ing in my boots: <examining> clan community construction in an online gamer population. *Cybersociology Magazine*, 2, November 1997. http://members.aol.com/Cybersoc/is2breeze.html.

Cairncross, F. 1997. *The Death of Distance: How the Communications Revolution Will Change Our Lives*. Boston: Harvard Business School Press.

Carroll, J. 1997. (D)RIVEN. *Wired*, September 1997, 5.09, 120ff.

Contact Consortium. 1998. *Sherwood Forest Community Project*. <http://www.ccon.org/events/sherwood.html>.

Couclelis, H. 1996. The death of distance. *Environment and Planning B: Planning and Design*, 23, 387–9.

Curtis, P. 1996. Mudding: social phenomena in text-based virtual realities. In Stefik, M. (ed.) *Internet Dreams: Archetypes, Myths, and Metaphors*. Cambridge, MA: MIT Press, pp. 265–91.

Damer, B. 1996. Inhabited virtual worlds, *ACM Interactions*, Sept–Oct. 1996, 27.

Damer, B. 1997. *Avatars! Exploring and Building Virtual Worlds on the Internet*. Berkeley, CA: Peachpit Press.

Dibbell, J. 1996. A rape in cyberspace: how an evil clown, a Haitian trickster spirit, two wizards, and a cast of dozens turned a database into a society. In Stefik, M. (ed.) *Internet Dreams: Archetypes, Myths, and Metaphors*. Cambridge, MA: MIT Press, pp. 293–313.

Donath, J.S. 1997. *Inhabiting the Virtual City: The Design of Social Environments for Electronic Communities*. Unpublished PhD thesis, MIT, February 1997. http://judith.www.media.mit.edu/Thesis/.

Donath, J.S. 1998. Identity and deception in the virtual community. In Kollock, P. and Smith, M. (eds) *Communities in Cyberspace*. London: Routledge, pp. 29–59.

Featherstone, M. and Burrows, R. 1995. *Cyberspace/Cyberbodies/Cyberpunk: Cultures of Technological Embodiment*. London: Sage Publications.

Garton, L., Haythornthwaite, C. and Wellman, B. 1997. Studying online social networks. *Journal of Computer Mediated-Communication*, 3, 1, http://www.ascusc.org/jcmc/vol3/issue1/garton.htmlhttp://www.ascusc.org/jcmc/vol3/issue1/garton.html.

Gifford, J.J. 1996. Quake tectonics. *FEED Magazine*, September 28th 1996. http://www.feedmag.com/96.09gifford/96.09gifford.html.

Graham, S. and Aurigi, A. 1997. Virtual cities, social polarization, and the crisis in urban public space. *Journal of Urban Technology*, 4, 1, 19–52.

Haddon, L. 1993. Interactive games. In Hayward, P. and Wollen, T. (eds) *Future Visions: New Technologies of the Screen*. London: British Film Institute, pp. 123–47.

Herz, J.C. 1997. *Joystick Nation: How Videogames Ate Our Quarters, Won Our Hearts, and Rewired Our Minds*. New York: Little Brown and Company.

Hillis, K. 1996. A geography of the eye: The technologies of virtual reality. In

Shields, R. (ed.) *Cultures of the Internet: Virtual Spaces, Real Histories, Living Bodies*. London: Sage Publications, pp. 70–98.

Huxor, A. 1997. The role of virtual world design in collaborative working. *Information Visualization '97 Conference*, July 1997, London.

Jeffrey, P. and Mark, G. 1998. Constructing social spaces in virtual environments: A study of navigation and interaction. In Höök, K., Munro, A. and Benyon, D. (eds) *Workshop on Personalised and Social Navigation in Information Space*. Stockholm: Swedish Institute of Computer Science, pp. 24–38.

Johnson, S. 1999. Maps & legends. *FEED Magazine*, March 1999. http://www.feedmag.com/column/interface/ci190lofi.html.

Kim, A.J. 1998. Killers have more fun: games like Ultima online are grand social experiments in community building. *Wired*, May 1998, 6.05, 94ff.

Kling, R. and Lamb, R. 1996. Bits of cities: utopian visions and social power in placed-based and electronic communities. http://www-slis.lib.indiana.edu/kling/pubs/bitsofcities.html.

Macmillan, B. 1996. Fun and games: Serious toys for city modelling in a GIS environment. In Batty, M. and Longley, P. (eds) *Spatial Analysis: Modelling in a GIS Environment*. Cambridge, UK: GeoInformation International, pp. 153–65.

Menges, J. 1996. Feeling between the lines. *Computer Mediated Communications Magazine*, 3, 10, http://www.december.com/cmc/mag/1996/oct/mengall.html.

Morningstar, C. and Farmer, R. 1991. The lessons of Lucasfilm's habitat. In Benedikt, M. (ed.) *Cyberspace: First Steps*. Cambridge, MA: MIT Press, pp. 273–301.

NWT. 1998. *New World Times,* online archive. <http://vrnews.synergycorp.com/nwt/>.

Pink Village. 1998. *Pink Village Info Pages*. http://www.geocities.com/WestHollywood/8382/.

Rafaeli, S., Sudweeks, F. and McLaughlin, M. 1998. *Network and Netplay: Virtual Groups on the Internet*. Cambridge, MA: MIT Press.

Reid, E. 1991. *Electropolis: Communication and Community on Internet Relay Chat*. Unpublished honours thesis, University of Melbourne. http://people.we.mediaone.net/elizrs/electropolis.html.

Reid, E. 1995. Virtual worlds: culture and imagination. In Jones, S.G. (ed.) *Cybersociety: Computer Mediated Communication and Community*. London: Sage Publications, pp. 164–83.

Reid, R.H. 1997. *Architects of the Web: 1,000 Days that Built the Future of Business*. New York: John Wiley & Sons.

Retsmah, 1998. Its a corporation, not a country. Message posted on the *Active Worlds Community newsgroup*, on 10 May 1998, 13:08:04 GMT. news://news.activeworlds.com/awcommunity/.

Rheingold, H. 1993. *The Virtual Community: Homesteading on the Electronic Frontier*. New York: Addison-Wesley.

Riddell, R. 1997. Doom goes to war: The marines are looking for a few good games. *Wired*, April 1997, 5.04, 114ff.

Roehl. B. 1997. Shared worlds. *VR News*, 6, 9, 10–15.

Rossney, R. 1996. Metaworlds. *Wired*, June 1996, 4.06, 140ff.

Schroeder, R. 1997. Networked worlds: social aspects of multi-user virtual reality technology. *Sociological Research Online*, 2, 4, http://www.socresonline.org.uk/socresonline/2/4/5.html.

Schroeder, R., Heather, N. and Lee, R.M. 1998. The sacred and the virtual: Religion in multi-user virtual reality. *Journal of Computer Mediated Communication*, 4, 2, http://www.ascusc.org/jcmc/vol4/issue2/schroeder.html.

Sorkin, M. 1992. *Variations on a Theme Park: The New American City and the End of Public Space*. New York: Hill and Wang.

Stephenson, N. 1992. *Snow Crash*. New York: Bantam Spectra.

Stone, A.R. 1991. Will the real body please stand-up?: Boundary stories about virtual cultures. In Benedikt, M. (ed.) *Cyberspace: First Steps*. Cambridge, MA: MIT Press, pp. 81–118.

Suler, J. 1997. *The Psychology of Avatars and Graphical Space in Multimedia Chat Communities*. Unpublished paper, Rider University, July 1997. http://www.rider.edu/users/suler/psycyber/psyav.html.

Taylor, J. 1997. The emerging geographies of virtual worlds. *The Geographical Review*, April 1997, 87, 2, 172–92.

Turkle, S. 1995. *Life on the Screen: Identity in the Age of the Internet*. New York: Simon & Schuster.

Vevo. 1998. *About the AlphaWorld Map*, Greg Roelofs and Pieter van der Meulen, Advanced Technology Group at the Philips Multimedia Center, Palo Alto, California. http://awmap. vevo.com/about.html.

Vilett, R. 1998. *AlphaWorld maps*. http://www.activeworlds.com/events/satellite.html.

22 Visualizing data quality through interactive metadata browsing

Francis Harvey

Introduction

The idea of using virtual reality (VR) approaches for other non-conventional applications in architecture, planning and related fields is not unusual, but calls for some rethinking of concepts implicitly adopted in most VR work. This chapter looks at an adaptation of virtual reality techniques for facilitating the access and interpretation of geographic data quality. Although the approach I present here begins with some off-the-shelf components known as information space or the meta content framework (MCF), it is currently subject to substantial changes as it is incorporated in the next generation World Wide Web through extended markup language (XML) specifications. Concepts from MCF are becoming part of the Resource Description Framework (RDF) which is guiding Web metadata developments. Virtual reality techniques may yet have their strongest influence through these developments.

These developments are important to keep in mind, and I will return to them later. As the key issue for this chapter is showing how VR techniques can aid in making data quality more accessible, I will start there. Underlying this examination is the widely shared insight that data quality involves 'fitness for use' (Chrisman, 1984). Obviously this adage calls for going beyond metadata content standards. 'Fitness for use' can only be determined in the specific context of the activity for which geographic information is used. A content standard does not address this. Each content standard only describes the provider's evaluation of fitness to specification. If the GI community wants metadata that fulfils the 'fitness for use objective', access must be considered.

Up until now GI metadata projects have been emphasizing the fulfilling of content standards, but with limited success (Dobson, 1998). Users are expected to learn the content standard in order to understand the data and determine its quality for their uses. I will show in this chapter how information spaces can be constructed from existing websites and documents. This offers an immediate way to deal with the accessibility issue and work towards more meaningful representations of metadata.

While there have been substantial efforts in the GI community to develop and implement content standards, visualization techniques for metadata remain a shadow of these standardization efforts. Substantial efforts in this area focus on the visualization of positional aspects of spatial data quality in a statistical framework (Goodchild *et al.*, 1994) with notable results. Other important work examines the fundamental questions of accuracy, validity and uncertainty in terms of error (Buttenfield and Beard, 1994). The work I present here is much more modest and only sets out to address data quality in terms of the accessibility of metadata, a point overlooked in content standards.

Information spaces utilize one of the most basic virtual reality components: the fly-through. By use of the mouse and a few keys, a user can navigate the spatial, three-dimensional representation of metadata hierarchies. *X* and *Y* spatial proximity in the information spaces corresponds to topical correspondence, depth to hierarchy level. The base information space is constructed by the provider, and can be interactively individualized by users afterwards.

Data quality as fitness for use

The key issue this chapter addresses is the evaluation of data quality in context. 'Fitness for use' (Chrisman, 1984) needs to address metadata accessibility. This issue is overlooked in metadata content standards for geographic information. The focus of content standards on conformance to particular values is just one part of quality. Developed in the manufacturing industry, using inspection to achieve quality leaves users and customers out of the picture. For production line staff, engineers and managers, quality may be readily manageable in this form, but it can turn into a self-fulfilling system of measurements void of any relationship to user and consumer requirements. This understanding of quality as fulfilling important traits or characteristics misleads because it makes the purpose for assessing quality implicit, and cannot include user needs in determining quality. Current content standards address the needs of producers, but don't account adequately for access issues necessary to enable the assessment of 'fitness for use'.

Practical purposes demand a more mundane and pragmatic understanding of quality. The expression 'fitness for use' asserts that the quality of spatial data is only known in terms of a specific use. Isolated metadata is not sufficient for determining geographic information quality. Quality is what is important in a situation, 'Quality information is the key to putting GIS products into an understandable form' (Paradis and Beard, 1994: 26). If the data to evaluate quality is not available, metadata is not fulfilling its primary purpose from this point of view. This is perhaps the most grievous shortcoming of content standards. They contain the knowledge of a data producer, but all other information pertinent to determining quality must be added by the user. Self-contained, cloistered in a metadata catalogue or inventory, metadata cannot optimally fulfil its *raison d'être*: aiding the determination of quality.

A look at quality in a related discipline, software engineering, may offer insights for a broader understanding of geographic information quality, necessary with the growing role of GIS users as producers of information, not merely the users of data provided by national mapping agencies, etc. The ISO standard for quality management and quality assurance defines quality as the 'totality of features and characteristics of a product or service that bears on its ability to satisfy given needs' (ISO, 1989: 3.1). Clearly, this broader concept requires a paradigm shift. Approaches reflecting this shift, like Total Quality Management (TQM), contain important concepts for the producers of data.

In GIS, we also need to consider that many users are also producers, but they are users of another producer's data, and producers of data for another user too. The hybrid roles of users and consumers in GIS present an unusual conundrum for approaches like the IEEE's where collection, processing, distribution and use are all within the same organization. The following paragraphs describe an approach that opens a new way of understanding the quality of geographic information that does not merely deal with content, but addresses the question of use.

Beyond content standards

A starting point to address the broader framework of data quality than content standards permit is reconsidering the framework for evaluating data quality. As the 'fitness for use' concept underscores, any data product's quality can only be determined in relationship to use. Use is a rather vague term that only takes on meaning in distinct context. It may be impossible for producers to determine data quality without some application. A benchmark application would help, but exploring this issue lies outside this chapter. At this point, to refine the understanding of 'fitness for use', I will differentiate various categories of data quality (DQ) and their dimensions.

These categories and dimensions come from literature that is closely affiliated with Deming's work on Total Quality Management (Deming, 1989), with the central observation that quality can only be evaluated with the people choosing and using data. As more and more commercial geographic data providers offer data (for example: high-resolution satellite data), the user, as consumer, will play a much more important role, than when geographic data was largely a monopoly of national mapping agencies and surveyors.

Following Deming's tenet and the 'fitness for use' principle, this chapter applies the practical interpretation that quality is what users define to be data quality. Usefulness and usability are central here, and in this light quality consists of four categories:

- *intrinsic data quality* describes the result of production processes,
- *accessibility data quality* accounts for access to metadata and description of categories,
- *contextual data quality* refers to problems with incomplete data or inconsistent representations,
- *representational data quality* accounts more specifically for the problems users encounter interpreting and understanding data (Strong *et al.*, 1997).

These categories come from the qualitative analysis of forty-two DQ projects at an airline, a hospital and a Health Maintenance Organization (HMO). Focusing on problem-solving strategies, the analysis grouped DQ issues in the four categories.

Certainly, an important starting point for GI data providers is improving accessibility. With better accessibility a wide range of DQ problems is immediately addressed, and the foundation laid for wider-reaching dialogues with users to improve other DQ aspects. The next section of this chapter shows how information spaces can contribute to this.

Accessibility to data quality through visualization

Information spaces are a means of aiding access to information such as that relevant to data quality. Since much metadata is already widely available and hypertext a common dissemination medium, an approach that builds on existing hypertext documents and approaches has much potential.

Information space or meta content framework (MCF) may be readily described as hypertext connected to graph theory (Guha, 1997a, 1997b, 1997c).

This concept is attractive for several reasons. Recognizing that people understand metadata in a wide variety of ways, but, on the other hand, content is a relatively clear idea, the meta-content framework builds a flexible interface to content of multiple metadata catalogues or inventories. This flexibility permits the structuring of content in user-defined three-dimensional information spaces. Through the hierarchy and spatial arrangement, the user can associate contextually similar metadata. Whereas a hypertext link on a Web page is often without context, the spatial organization of MCF supplies important contextual clues that a user can enhance.

Meta Content Framework is a structure description language. Tools and algorithms are already available to manipulate it as it is based on a well-understood information organization structure known as Directed Labelled Graph (DLG). The MCF data model has nodes, arcs and labels. These are also referred to as PropertyTypes, Nodes and Properties. Nodes can represent things like Web pages, images, subject categories, channels

```
unit: "./county_gis.html"
name: "County GIS"
genls: "/Fusion.mco"

unit: "/html/administrative_boundaries.html"
name: "Administrative Boundaries"
genls: "./county_gis.html"

unit: "./html/environmental.html"
name: "Environmental"
genls: "./county_gis.html"
```

Figure 22.1 MCF HotSauce example.

and sites. For instance, in this project nodes give information about Web pages that contain metadata information. Arcs represent characteristics and relationships to other objects. Each node is a property type and property. This double role makes it possible for nodes to pass on constraints and inherit properties. This self-description also makes it possible for MCF to be its own schema definition language and be dynamically extensible. A limited number of nodes with reserved names bootstraps the objects defined in MCF to the type system.

A simple idea of MCF code shows how the data model is implemented for an application (Figure 22.1). In this case, the code was produced to conform with an older version of Apple's HotSauce, so it is much more limited than current and future versions. The full version of this code produced Figures 22.2 and 22.3.

Basically, an information space flexibly links and organizes metadata contents in a hierarchical format with a spatial dimension. The designer groups elements of an information space, proposing a framework for generic use. Each user can modify the organization of these elements to fit their specific needs. This flexibility is further enhanced by the tools provided for constructing information spaces. Based on first order logic, it uses 'context mechanisms' as arguments for predicates (Guha, 1997c). The MCF tools can be used for query processing and inferencing metadata and source data. The language can also be machine read. This opens up perspectives for future developments that are exciting, but beyond this chapter to explore.

The hierarchical 3D information space is navigated using a fly-through analogy. Holding down the cursor (mouse) button the view moves deeper into the information space, shift-click goes towards the top. By left-right and top-down movements, the user can move towards elements. When starting at the top of an information hierarchy, usually only the top-most elements are visible. Going towards any element causes lower levels of the

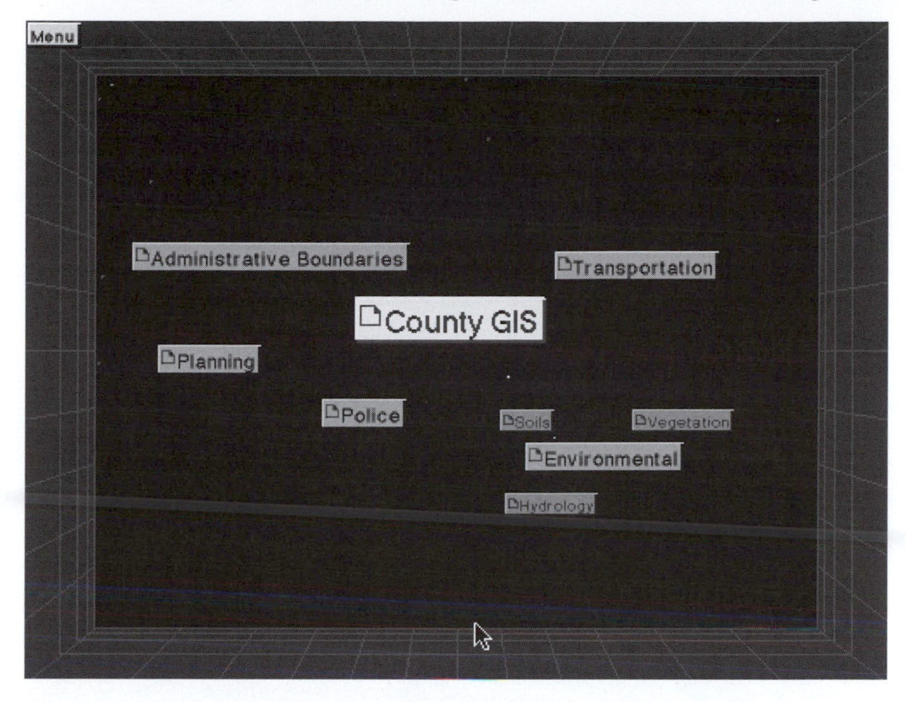

Figure 22.2 A sample information space for a county GIS showing the top three hierarchical levels.

hierarchy to become visible. A click selects an element, which can be moved to a new position as a user desires. These changes are saved in a local information space. A double-click on an element opens that element just like a hypertext link.

An information space is fundamentally a contextual ordering. Whereas metadata is well structured through standards, the myriad aspects of GIS applications are not standardizable. Context is unique and data quality as fitness for use should account for this. Information spaces facilitate the individual organization of metadata to develop a thorough understanding of quality in connection with use.

The future of information spaces

The future of information spaces is now contingent on the future of RDF and XML. At present a draft RDF/SML standard is available which does not specifically contain specifications for either information standards or MCF, but because it is built very closely on MCF principles, it does not exclude the possibility of incorporation.

Figures 22.2 and 22.3 show snapshots of what an information space could look like using Apple's version of MCF. Their version was only

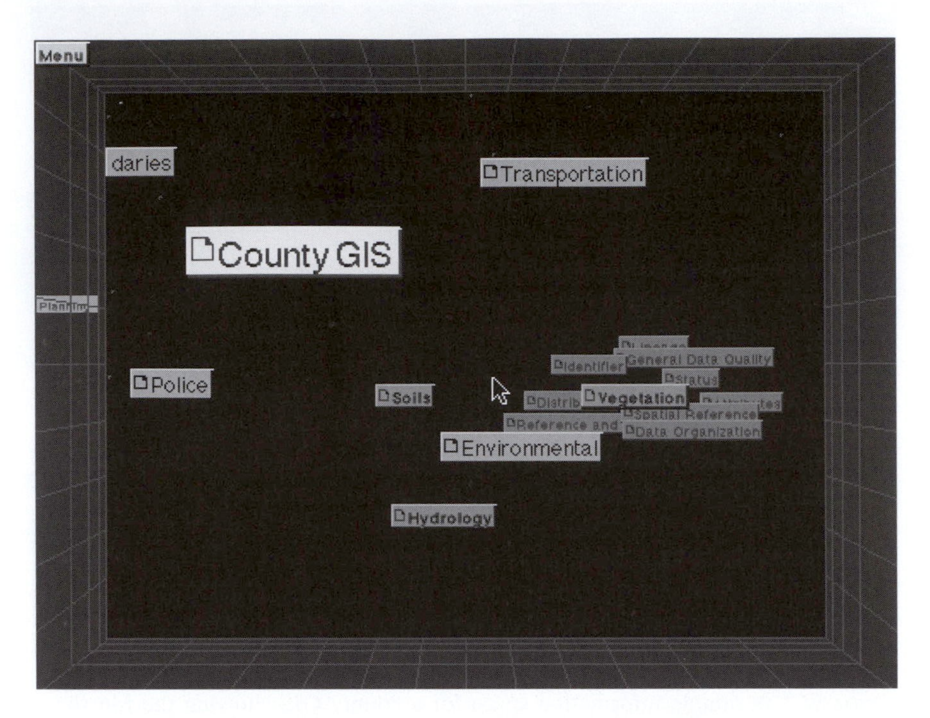

Figure 22.3 Flying in, lower levels of the hierarchy become visible.

available until 1997 after which, in a reorientation of business activities, it was dropped. The developers moved to Netscape and Textuality and Netscape proposed MCF as a W3C standard (MCF–XML) in May 1997. In August 1997, W3C formed a metadata working group to work on a new metadata standard called resource description framework (RDF). It remains 'work-in-progress' although a draft has been published and a final specification can be expected soon (December 1998). The positive side of these developments is that the version of MCF that will be supported in RDF is compliant with the Dublin Core and other Metadata standards. It will also be backwards compatible with PICS 1.1 (Platform for Internet Content Selection). Of interest to the GI community is the aim of RDF to insure metadata interoperability. It is very important to point out that RDF is primarily a further development of MCF.

The bad news in some ways is that MCF terminology is changing through the XML/RDF process and losing the spatial analogy, and the information space architecture. It appears that RDF is moving towards a tree metaphor, much like a file system is represented under Windows and Windows/NT. This, of course, also has advantages, but I believe this is not ideal. The solution is to write a graphical browser like Information Spaces since XML and RDF are open specifications, and even Mozilla, the next

```
<Categoryid="Projection">
<description>UsedasthetypeOfanyGeodata
</description>
</Category>
<Categoryid="Mercator Projection">
<superTypeunit="Projection"/>
<description>"Used as the type of any map
or geodata set using the mercator
projection"</description>
</Category>
```

Figure 22.4 An example of XML supertype.

generation Web browser from Netscape, is available as source code. It may be lots of work, but it's possible.

The change in terminology reflects a shift in emphasis that indicates the object orientation of XML. In XML, documents contain *elements*, which have *types* and are either empty or are delimited by *start-tags* and *end-tags*, and have *attributes* which have *names* and *values*. Instead of the fixed tags allowed in HTML, in XML anything can be an element which is delimited by tags. The XML representation of MCF uses an element to represent an object; properties of the object are represented by other elements contained inside it. The type of the element is the Category of the object. If the object is in more than one Category, any of them can be used for the element type. It turns out that all objects are members of a Category called 'Unit', so one way or another, you can always find an element type. Inside the element are other elements representing the properties of the object; for these, the name of the property is the element type. In Figure 22.4, Projection is defined to be a supertype. Any category whose definition includes this type, inherits its attributes, i.e. Mercator as a type of projection.

Implementation

While information spaces may not provide the ultimate solution to accessibility issues, they present the means to extend existing and metadata repositories and build on existing HTML and websites. Information space and MCF show the potential for aiding the querying and understanding of existing metadata repositories by using the spatial metaphor of proximity and VR techniques. Although MCF is no longer supported in the form it was used to prepare this chapter, it has influenced the

RDF/XML developments which probably will lead to its even wider adoption than if it remained a Web browser plug-in. Now that the RDF standardization process is nearly concluded, it only remains to wait and see how RDF is picked up. All indications are positive. Considered in the light of the open development process and public availability of the source code for next generation Web browsers, the odds are strong that another VR information browsing tool will become available. As RDF/XML move out of the development stages, it seems reasonable to assume that next generation geographic metadata will be written in RDF/XML, opening opportunities for improving access through information spaces and other means.

References

Buttenfield, B. and Beard, M.K. 1994. Graphical and geographical components of data quality. In Hearnshaw, H. and Unwin, D. (eds) *Visualization in Geographic Information Systems*. New York: Wiley, pp. 150–7.

Chrisman, N.R. 1984. The role of quality information in the long term functioning of a GIS. *Cartographica,* 21, 79–87.

Deming, E.W. 1989. *Out of the Crisis.* Cambridge, MA: MIT Center for Advanced Engineering Study.

Dobson, J. 1998. Commentary. *GIS World*, 11, 32.

Goodchild, M., Chih-Chang, L. and Leung, Y. 1994. Visualizing fuzzy maps. In Hearnshaw, H. and Unwin, D. (eds) *Visualization in Geographic Information Systems*. New York: John Wiley & Sons, pp. 158–67.

Guha, R.V. 1997a. Meta Content Framework. http://mcf.research.apple.com/hs/mcf.html.

Guha, R.V. 1997b. Meta Content Framework: A white paper. http://mcf.research.apple.com/wp. html.

Guha, R.V. 1997c. Towards a theory of meta-content. http://mcf.research.apple.com/mc.html.

ISO. 1989. *Quality Management and Quality Assurance – Vocabulary (8402)*. International Standards Organization, Zurich.

Paradis, J. and Beard, K. 1994. Visualization of spatial data quality for the decision maker: A data-quality filter. *URISA Journal*, 6, 25–34.

Strong, D.M., Lee, Y.W. and Wang, R.Y. 1997. Data quality in context. *Communications of the ACM*, 40, 103–10.

23 'There's no there there'

Virtual reality, space and geographic visualization

Rob Kitchin and Martin Dodge

Introduction

Virtual reality (VR) is providing fresh challenges for both theorisers and philosophers of space and to cartographers and those wishing to visualize the extent and form of virtual spaces. In this chapter, these challenges are identified and explored through an examination of the spatial qualities of VR, and the ways in which geographic metaphors (in particular the notion of mapping) are being employed to aid navigation in, and understanding of, VR spaces. Here, VR is defined broadly to incorporate a number of forms including (visual) virtual reality simulations, Web pages, chat-rooms, bulletin boards, MUDs (both textual and visual) and 'game' spaces. To us, VR consists of computer-generated spaces that enable the user to interact with the computer, and other people connected to the network, in ways that simulate (though not necessarily replicate) real-world interactions (e.g. movement through a landscape or chatting to somebody). As the rapidly growing website, *Atlas of Cyberspaces* (Dodge, 1999) illustrates, the mapping of cyberspace has started in earnest, and in large part has been undertaken by researchers outside the fields of cartography and geography. However, this process of mapping has largely taken place in an uncritical manner, with little thought as to the wider implications or consequences of the techniques used. In this chapter, we critically detail and assess the mapping of VR spaces, outlining some of the difficulties and issues complicit in such a project.

Why 'map' VR?

VR spaces are complex. They are, quite literally, computer-supported, informational spaces, fundamentally composed of zeros and ones and connected in a myriad of ways. Some of the information is explicitly spatial with direct geographic referents (e.g. VR models of real-world places, such as the work of Batty (1998) *et al.*'s modelling of parts of London). Other information has an inherent spatial form without a geographic referent (e.g. MUDs), or has a real-world referent but no spatial form/attributes (e.g. a list of names, Web pages), or has no geographic referent and no spatial form/attributes (e.g. computer file allocation). In addition, as discussed in Chapter 20, information with a real-world referent may have a materiality (e.g. has a mass) or be immaterial in nature (e.g. gravity, heat). Mapping in both a literal and metaphorical sense can provide a means of facilitating comprehension of, navigation within, and documenting the extent of (marking out territories) these varying forms of informational space.

It has long been recognised that geographical visualizations in all their manifestations form an integral part of how we understand the world. For example, traditional cartographic maps have been used for centuries as a method to visualize geographical distributions across a world that is too large and too complex to be seen directly (MacEachren, 1995). At the end of the twentieth century, cartography has undergone two major evolutions. One has been digitalisation, and the widespread use of computer systems such as GISs and CADs that are able to store, process, manipulate and transform spatial and attribute data. The second has been in the move away from static maps to interactive, dynamic and animated geographic visualizations that can be designed by anyone with access to software and data. These two evolutions, widespread access to map-making technology and geographic visualizations, extends the power of mapping in qualitative and quantitative ways:

1 opening up new ways to comprehend the real world,
2 providing effective ways of structuring immaterial phenomenon and material that has no geographical referent to increase comprehensibility,
3 allowing static representations to be replaced with multiple representations that can be interactive and dynamic, and
4 empowering non-cartographers to be able to access data and produce their own maps, thus breaking one of the major principles of traditional map-making theory, that is that there is a clear separation between cartographer and user (Crampton, 1999).

In the case of information that has a geographic referent and spatial attributes, constructing a map or geographic visualization provides a means to visualize and describe that form. In some cases, this means producing a virtual model of the geographic world. Here, issues such as accuracy, precision, verisimilitude (having the appearance of truth, realistic depiction) and mimesis (imitation, mimicry) come to the fore (see Chapter 3). In other cases, this means producing geographic visualizations of what Batty (1997) has termed 'cyberplace', the infrastructure of the digital world – the wires, computers and people. An example of cyberplace visualization are the 'maps' of the Internet's MBone (multicast backbone) produced in 1995 by a team of computer science and visualization researchers in California

(Munzner *et al.*, 1996). These visualizations used the powerful visual metaphor of the globe of the Earth onto which the MBone network linkages were plotted as arcs, as shown in Figure 23.1. As 3D models, in VRML format, the end-user is allowed greater freedom to interact with them – rotating and spinning them, so that they can be viewed from any position. Without these geographic visualizations, topological structure data are almost impossible for humans to interpret because they are held in large textual tables.

Another fascinating example of mapping cyberplace are the visualizations of the geography of Internet traffic computed by the National Center for Supercomputing Applications (NCSA) in 1995 (Lamm *et al.*, 1996). Figure 23.2 shows an image of one of their striking 'maps', with the traffic represented as virtual skyscrapers projecting into space from the Earth globe. These skyscrapers represent three different dimensions of the traffic data. First, the position of the base of the bar is at the approximate origin of the traffic (this is aggregated to the country level outside of North America). Second, the height of the bar represents the total volume of traffic for that time period from that region and, third, the different coloured bands on the bars indicate the type of traffic (such as images,

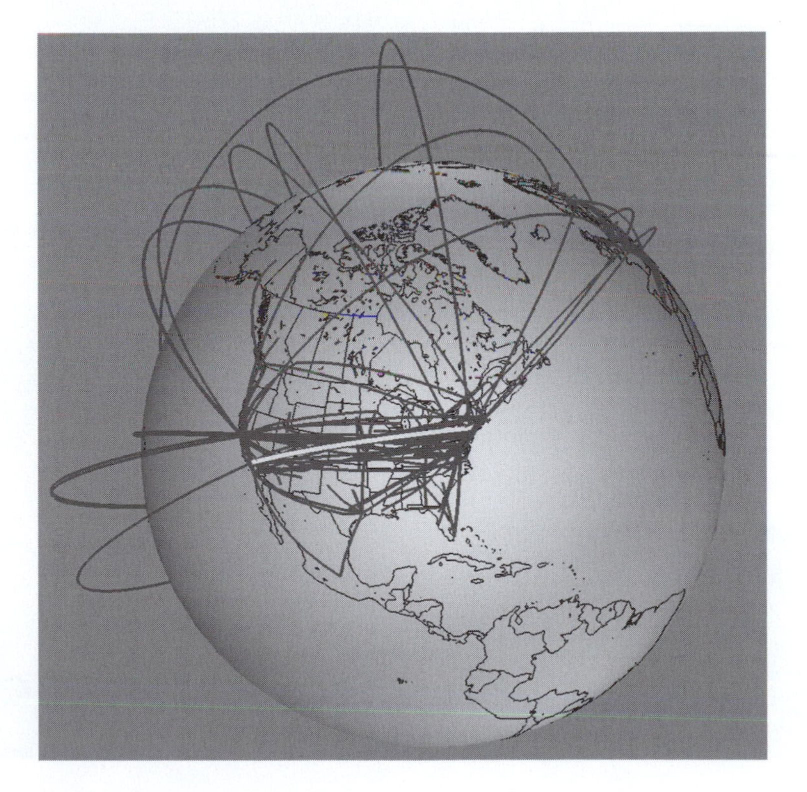

Figure 23.1 MBone linkages in the Internet.

HTML, text, video, data, etc.). Importantly, the Lamm *et al.* map could also encode another vital dimension of the traffic, that of change over time. Their 3D globe was rendered in a sophisticated VR environment that was dynamically linked to the Web server hosting the NCSA site, enabling traffic patterns to be mapped in near real-time. The VR environment enabled users to immerse themselves to a degree into the map, being able to move around the globe freely and interrogate the bars. Also provided were control panels, that can be seen in Figure 23.2 floating behind the globe, which allowed users interactively to change the display characteristics of the map.

Similarly, the spatial form attributes of VR data that have no geographic referent can be mapped in a process called spatialisation. There is an emerging body of research over the past decade into this type of information visualization (Card *et al.*, 1999; Gershon and Eick, 1995). In cases where there are no spatial attributes, we use the terms 'map' and

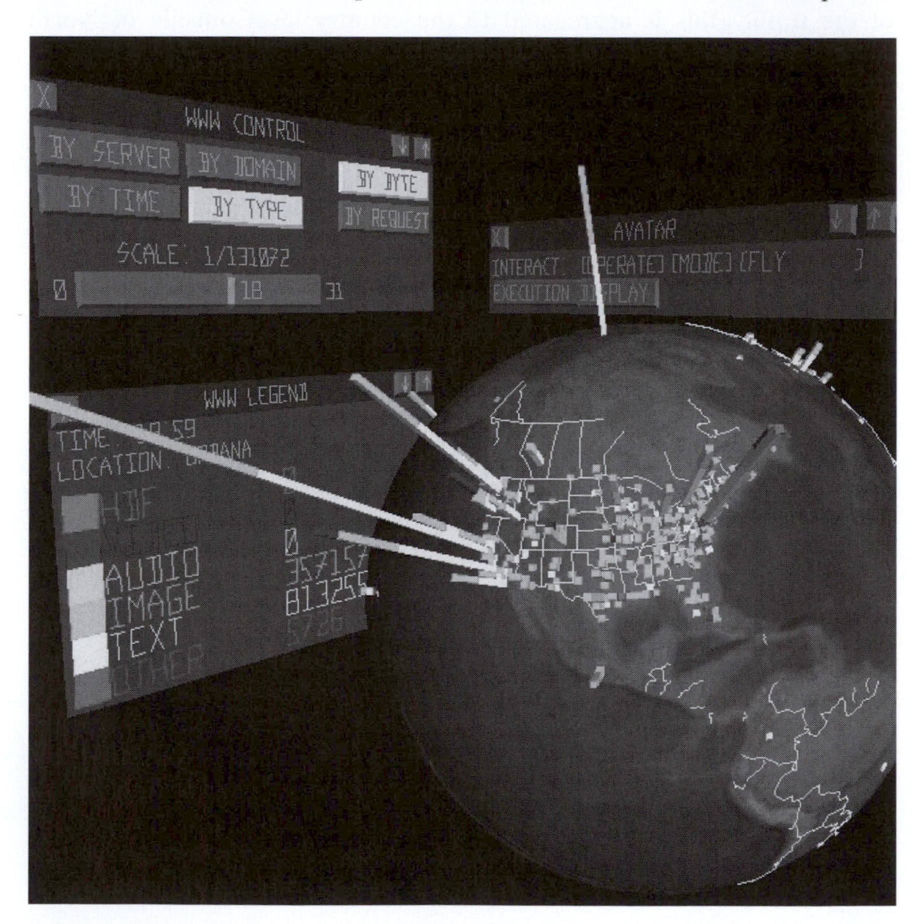

Figure 23.2 Mapping Internet traffic data.

'mapping' metaphorically. Here, a spatial structure is applied where none exists in order to provide a means of visualizing and comprehending space; to utilise the power of spatial representation to describe complex informational spaces in a new, more easily interpretable form. Here, information attributes are transformed into a spatial structure through the application of concepts such as proximity (nearness/likeness). A number of researchers are now experimenting with the application of geographic metaphors to what Batty (1997) has termed c-space (the spaces on the screen) and cyberspace (spaces of computer-mediated communication) (see p. 348)

In this chapter, we initially focus on detailing the nature of space within VR space, before examining the practicalities of mapping VR spaces. Next, we detail a topology of mapping, finishing with an exploration of wider issues.

Space and VR space

The challenge to those wishing to 'map' VR spaces with an inherent spatial form (e.g. visual virtual reality simulations; visual MUDs) is that they are qualitatively different from geographic space in a number of fundamental ways. In the following discussion the term cyberspace is often used synonymously with virtual reality. VR spaces are, to us, a sub-set of cyberspace and therefore what applies to cyberspace also applies to VR. This is particularly the case in relation to spatial qualities. As discussed, Memarzia (1997) stated:

> In cyberspace there are no physical constraints to dictate the dynamics or spatio-temporal qualities of the portrayed virtual space. Gravity or friction does not exist in cyberspace unless it has been designed and implemented. . . . cyberspace is not limited to three dimensions, since any two-dimensional plane or point may unfold to reveal another multidimensional spatial environment. . . . There are no ground rules concerning scale consistency in a virtual environment. Furthermore, the scale of the environment, relative to the user or viewer, may be altered at will. . . . Cyberspace can be non-continuous, multidimensional and self-reflexive . . . In general, all principles of real space may be violated in cyberspace and the characteristics and constraints are only determined by the specifications that define the particular digital space.

Novak (1991: 251–2) thus argued that cyberspace has a 'liquid architecture':

> Liquid architecture is an architecture that breathes, pulses, leaps as one form and lands as another. Liquid architecture is an architecture

> whose form is contingent on the interests of the beholder; it is an architecture that opens to welcome me and closes to defend me; it is an architecture without doors and hallways, where the next room is always where I need it to be and what I need it to be. Liquid architecture makes liquid cities, cities that change at the shift of a value, where visitors with different backgrounds see different landmarks, where neighbourhoods vary with ideas held in common, and evolve as the ideas mature or dissolve.

VR spaces have spatial and architectural forms that are dematerialised, dynamic and devoid of the laws of physics; spaces in which the mind can explore free of the body; spaces that are in every way socially constructed, produced and abstract. Indeed, Holtzman (1994: 210) referred to the designers of virtual worlds as 'space makers'. While some VR spaces do have an explicit spatial form (e.g. visual virtual worlds) they exist only in code; a combination of zeros and ones – objects are merely surfaces, they have no weight or mass (Holtzman, 1994). Morse (1997) describes cyberspace as an infinite, immaterial non-space, suggesting that it takes the form of a liminal space:

> Virtual landscapes are liminal spaces, like the cave or sweat lodge . . . if only through their virtuality – neither here nor there, neither imaginary nor real, animate but not living and not dead, a subjunctive realm wherein events happen in effect, but not actually.
>
> (p. 208)

Indeed, as Mitchell (1995: 8–9) explained, cyberspace is:

> profoundly *antispatial* . . . You cannot say where it is or describe its memorable shape and proportions or tell a stranger how to get there. But you can find things in it without knowing where they are. The Net is ambient – nowhere in particular but everywhere at once. You do not go *to* it; you log *in* from wherever you physically happen to be . . . the Net's despatialization of interaction destroys the geocode's key [original emphasis].

As Memarzia (1997) points out, the digital landscapes of VR spaces only possess geographic qualities because they have been explicitly designed and implemented. As such, Holtzman (1994: 197–8) professes:

> there's no there there. It only exists in some hard-to-define place somewhere inside the computer . . . [and yet in virtual reality simulations] you are completely immersed in another world. It is not a picture that is being viewed, but rather a place. This world is not being observed but experienced. You sense that you are in it.

It is a space without space, 'a nonplace' (Gibson, 1987) and yet possesses a spatiality and has virtual places. Moreover, it is a space where geographic 'rules' such as the friction of distance can be broken through the creation of what Dieberger (1996) terms 'magic' clauses, for example, teleporting. Benedikt (1991: 128) argued that virtual realities need not, and will not, be subject to the principals of ordinary space and time, which will be:

> violated with impunity. After all, the ancient worlds of magic, myth and legend to which cyberspace is heir, as well as the modern worlds of fantasy fiction, movies, and cartoons, are replete with violations of the logic of everyday space and time: disappearance, underworlds, phantoms, warp speed travel, mirrors and doors to alternate worlds, zero gravity, flattenings and wormholes, scale inversions, and so on. And after all, why have cyberspace if we cannot (apparently) bend nature's rules there?

To Benedikt, VR spaces are a 'common mental geography' (in Gibson's famous phrase – a 'consensual hallucination'), a medium in which 'ancient spaces' (mythical or imaginal spaces) become visible; the abstract spaces of the imagination freed from Euclidean geometry and Cartesian mapping; spaces where the 'axioms of topology and geometry so compellingly observed to be an integral part of nature can ... be violated or re-invented, as can many of the laws of physics' (p. 119). Indeed, many of the descriptions of VR space by novelists describe in detail its spatial qualities. In nearly all cases, Cartesian rules do not apply to the virtual spaces being envisioned. Except in relation to the body, where the mind/body distinction seemingly benefits significantly from VR space. The mind literally becomes free from the 'meat'.

Clearly these concerns relating to spatial geometries and sense of place also apply to metaphorically-constructed informational spaces – 'maps' of c-spaces and cyberspaces. They, too, are purely constructions with potentially complex spatial geometries that can bare no, or very little, resemblance to real-world geographies. Moreover, if the geographic visualizations are programmed to be interactive, the spatial representation (map) becomes the territory – map and territory become synonymous; rather than being external to a representation of data, we are navigating links within data. Here, the use of a geographic metaphor to structure the data becomes the means by which this new territory is navigated. For example, a VRML Web page is both the territory and the means in which to navigate that territory.

As discussed in Chapter 20, conceptions of space thus need to be re-analysed in light of how space is (re)formulated and used in VR spaces. VR spaces offer 'geographic-style' interaction, and yet the spaces are *not* essentialist (given) or absolutist. Instead they are purely relational (both spatially and socially). And yet, as discussed, unlike geographic space they possess a number of other qualities that set them apart. VR spaces can be Euclidean and Cartesian or multidimensional or a mixture of the two, and can be viewed from many different viewpoints in space or time. VR spaces (both those that seek to represent geographic space and those that do not) thus seem to us to offer challenges to the philosophers of space in both theorising the nature of space within VR spaces but also the consequences of that space upon geographic space.

Only a few academics have started to examine the geographies of cyberspace, the spatial geometries and forms, the intersections between different cyberspaces and the intersections between geographic space and cyberspace. For example, Batty (1997) has tried to assess the ways in which VR space and geographic spaces connect. He defined virtual geography as 'the study of place as ethereal space and its processes inside computers, and the ways in which this space inside computers is changing material place outside computers', with the space within computers (the spaces on the screen) defined as 'c-space', the space of computer-mediated communication as 'cyberspace' (of which there are different forms), and the infrastructure of the digital world (the actual hardware) as 'cyberplace'. These concepts link together and form a cyclical process of interaction and evolution, linking individual sites with real and virtual space (nodes) through distributed systems (networks). As such, c-spaces located in individual computers, and sited in real space, are linked together to form a distributed network: cyberspace. Cyberspace exists within the infrastructure of cyberplace and its use mediates the creation of new communications infrastructure and attendant services which, in turn, has material effects upon the infrastructure of traditional places. This change of infrastructure at specific sites gradually alters the geography of real space, in terms of patterns of production and consumption. To cope with,

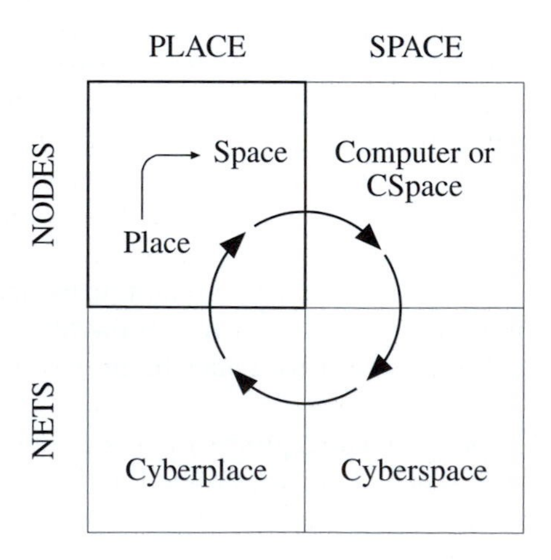

Figure 23.3 Virtual geography (Source: Batty, 1997).

and maximise competitive advantage within, these new geographies, companies, institutions and individuals are computerising their practices and processes, creating new c-spaces, and so on. Gradually, then, real geographies are being virtualised, turned into cyberplaces.

Batty suggested that the three spaces have various, differing geographies and together are key components of what Castells (1996) refers to as 'real virtuality', a reality that is entirely captured by the medium of communication and where experience is communication. Castells refers to the linkages between c-space and cyberplace, that is cyberspace, as the space of flows. He argues that the space of flows is characterised by timeless time and placeless space. Castells (1996: 464–7) explained:

> Timeless time ... the dominant temporality in our society, occurs when the characteristics of a given context, namely, the informational paradigm and the network society, induce systemic perturbation in the sequential order of phenomena performed in that context.... The space of flows ... dissolves time by disordering the sequence of events and making them simultaneous, thus installing society in an eternal ephemerality. The multiple space of places, scattered, fragmented, and disconnected, displays diverse temporalities ... selected functions and individuals transcend time.

In other words, temporality is erased (p. 375), suspended and transcended in cyberspace. Stalder (1998) extends this idea to its logical conclusion, arguing that the defining characteristic of timeless time is its binary form. Timeless time has no sequence and knows only two states: presence or absence, now or never. Anything that exists does so for the moment and new presences must be introduced from the outside, having immediacy and no history. As such, 'the space of flows has no inherent sequence, therefore it can disorder events which in the physical context are ordered by an inherent, chronological sequence' (Stalder, 1998). In a similar way, geographical distance dissolves in the space of flows so that cyberspace becomes placeless. Movements within cyberspace are immediate, presences can be multiple, and distance as we currently understand it is meaningless. There are no physical places in cyberspace, just individual digital traces that are all equally distant and accessible (traces, however, might be considered metaphorically a place such as AlphaWorld, see Dodge, Chapter 21). Every location is each other's next-door neighbour; everything is on top of everything else; everywhere is local (Staple, 1995). Stalder (1998) extends the placeless space to its logical conclusion, again using a binary metaphor, to suggest that cyberspace is a binary space where distance can only be measured in two ways: zero distance (inside the network) or infinite distance (outside the network); here or nowhere.

Adams (1998) sought to understand cyberspace by drawing parallels between their network architecture and sense of place with those of geographic spaces. He argued that an analysis of how spatial/place metaphors combined with a comprehension of how network topologies affect communications within VR spaces will lead to an understanding of social interactions within cyberspace. He thus hypothesised that the virtual geography, the topologies of the network, affects the type and nature of social interactions. In other words, a way to explore the geographies of VR is not to map it formally but to chart and quantify network topologies and the nature of social interactions within a specific topology (geography). Using combinatorial theory (a method for comparing network forms) he identifies several network typologies that mirror their geographical equivalents in terms of their structure and the social interactions performed. Adams argued that, despite VR spaces and cyberspace being incongruent, they bear significant similarity. Relationships between structure and agency are replicated online. Places within both spaces are multiple, diverse, and linked by complex paths that need to be traversed. In contrast to the discussion above, Adams postulated that the spatialities of VR, visual spaces, are very similar to the spatialities of geographic space. This is a contention that needs further empirical examination.

One source of evidence as to whether the spatialities of visual VR mirror geographic space is through the examination of how people cognise spatial relations, and spatially behave, in visual VR spaces. Tlauka and Wilson (1996) concluded from their study that navigation in computer-simulated space and real space led to similar kinds of spatial knowledge. Following learning the locations of objects within a room either through virtual navigation or viewing a map, respondents were required to point to objects that were not directly visible from both aligned and contraligned perspectives and to draw a map. No differences were noted between conditions (navigation versus map). Ruddle *et al.* (1997) tested the spatial knowledge of two groups of respondents to complete distance, direction and route-finding tasks. The first group learned a building layout (135 rooms of which 126 were empty and nine contained landmarks) by studying a floor plan and the second group of respondents learned the same layout in a non-immersive, screen-based, virtual environment. Both groups were then tested in the virtual environment. They found no significant differences in the route-finding ability of respondents who had learned a building layout within a virtual environment or through map learning. Time in the environment seems of particular importance. In their initial trials respondents were disorientated. However, there was a steep learning curve across trials with the route through the building becoming progressively more accurate with trials.

Other studies have found significant differences between spatial learning within virtual and real environments, although they conclude that the processes of learning remain the same. Turner and Turner (1997) concluded from their study of distance estimates within a small five-room virtual environment that their respondents' spatial knowledge was similar to that gained from exploring the real world but is best described as being most like that gained from a restricted exploring of a real-world environment such as a cave, or exploring with a restricted field of vision (e.g. wearing a helmet). Similarly, Richardson *et al.* (1999) compared the ability of sixty-one respondents to learn the layout of two floors in a complex building from a map, from direct experience, or by traversing through a 'desktop' virtual representation of the building. Those learning in the virtual environment performed the poorest, although similar levels of performance were displayed for learning the layout of landmarks on a single floor. They also displayed orientation-specific representations defined by their initial orientation in the environment, and were particularly susceptible to disorientation after rotation. The authors conclude that, in general, learning a virtual environment is similar to learning geographic space, using the same cognitive processes, although respondents are more likely to become disorientated and have difficulty integrating layouts of other floors.

Similarly, Satalich (1995) explored the way-finding ability of sixty-five respondents in a visual virtual environment that comprised of a U-shaped building that measured 100 feet by 100 feet. The building contained thirty-nine separate rooms and over 500 objects. Collision detection was incorporated into the environment so that respondents could not walk through the walls, but was not incorporated for objects located throughout the building. Satalich found that regardless of the measure used ((1) self-exploration (free to explore the building as they wished); (2) active guided (follow a pre-determined path using the joystick); and (3) passive guided (the respondent was moved through the environment at a constant speed with no interaction, although they could move their head to look around), the control group (who learnt the same environment by studying a map) either performed equivalently or better than the group that experienced the virtual environment. Witmer *et al.* (1996) compared the spatial knowledge of respondents that had learned an environment through a virtual medium with those who had interacted with the real environment. They reported that their respondents could successfully learn a virtual model of a real building and were able to transfer this knowledge when tested in the building, although they made significantly more route finding errors than participants who were trained and tested in the building. Respondents in the Wilson *et al.* (1996) study, who learned a three-storey building in a virtual environment, performed significantly worse at estimating direction estimates than a control group who learned the real building.

Clearly, then, VR spatialities do not currently match real spaces. Richardson *et al.* (1999) and Ruddle *et al.* (1997) explain that one might expect differences to occur between spatial learning in virtual and real environments because of the lack of proprioceptive cues during navigation causing an optic (eye movement)/vestibular (leg muscles stationary) mismatch, the need for scale translations for movement in a 'smaller' world, and, if using a desktop VE, the elimination of peripheral vision, virtual environments being less visually complex, with fewer subtle landmark cues (notices, marks on walls, etc.), and the restriction on the inclusion of sound. It might therefore be expected that, as VR spaces become indistinguishable from real spaces, that spatial understanding will become equivalent. As noted, however, VR spaces are not spatially equivalent and will only be so if explicitly programmed to be so.

'Mapping' VR space

At one level, as illustrated in Figures 23.1 and 23.2, VR space is relatively easily mapped. The physical architecture and topology of the networks (cyberplace) can be mapped into Cartesian space and the traffic through this network represented using an appropriate form of visualization. Similarly, the physical location and characteristics of hardware, software and wetware (human users) can be mapped using traditional cartographic and demographic methods.

At another level, however, VR spaces are difficult to map. As noted above, these spaces of zeros and ones provide a much greater challenge – the effective mapping of visual spatial forms and the metaphorical use of geographic visualization to provide comprehensibility for non-spatial or immaterial information that is difficult to navigate through and understand due to its complexity and mutability. Visualizers of VR spaces face a much greater challenge than their counterparts charged with mapping geographic space: to find ways to map spaces with differing spatial forms and geometries, including some with no recognisable geometrical properties; to find ways to map spaces that break two of the fundamental conventions of geographic visualizations. These conventions are that (1) space is continuous and ordered, and (2) the map is not the territory but rather a representation of it (Staple, 1995). As noted, VR spaces can be discontinuous and non-linearly organised, and in many cases the spaces are their own maps. In a deeper sense, a session in VR space is the map, with each link providing a trail to retrace (Staple, 1995); rather than being external to a representation of data, we are navigating links within data. As Novak (1991) notes, however, this is not to deny that VR spaces have an architecture (geography), contains architecture, or even are architecture, just that this architecture is their own.

The challenge for geographic visualizers, however, is only partly a matter of spatial form. As Staple (1995) notes, mapping VR space is just one part of a wider project that aims to map places that cannot be seen, such as distant galaxies, DNA, brain synapses (see Hall, 1992, for fascinating examples of these). VR spaces, though, are 'infinitely mutable' (Staple, 1995), changing daily as new computers are added, the infrastructure updated, and content refined and expanded. VR spaces are transient landscapes; spaces that are constantly changing but where the changes are often 'hidden' until encountered. As time unfolds and more and more data are uploaded, visual sophistication and detail improves, the mutability of VR spaces will increase accordingly to create spaces that are constantly evolving, disappearing and restructuring. Geographic visualizations of geographic spaces are out-of-date as soon as they are published, as the landscape portrayed is modified. The vast majority of information portrayed, however, remains stable and the shelf-life of the map can be many years. The shelf-life of a VR map, given the current and projected dynamic nature of VR spaces, particularly those accessible across the Internet, is likely to be very short. To complicate matters further, as yet, unlike geographic space, there are no agreed conventions in relation to how a space is designed or how it is traversed, providing a diverse set of spaces which differ in form, geometry and rules of interaction.

The wider challenge, then, is to construct dynamic geographies of a variety of VR spaces, some with no explicit spatial relations, some with an in-built relational (topological) geography (e.g. textual and visual virtual worlds), and to map out the intersections between virtual and geographic spaces. To produce geographies that will aid the navigation within, and comprehension of different, cyberspaces at both theoretical and practical levels.

Map topology

To our knowledge there have only been two attempts to create a topology of maps of VR spaces. The first by Dodge (1997) divided what he termed cybermaps into a number of classes: geographical metaphors, conceptual maps, topology maps, landuse maps and landscape views, virtual cities and navigation tools. The second, by Jiang and Ormeling (1997), classified maps of VR spaces using a three-fold classification centred on function: navigation; cyberspatial analysis; persuasion. Both classifications adopted a position that fails to recognise the differences between data sources, and the complexities and differences between what the maps are seeking to represent spatially. As discussed, the mappings of VR vary as a function of geographic reference, spatial form/attributes, and materiality of the information that is mapped. We therefore propose a classification that varies along three axes: (1) geographic referent (cyberplace/cyberspace); (2) spatial attributes/materiality (material, spatial form/immaterial spatial form); (3) map form (static, animated, interactive, dynamic). Each axis is discussed in turn.

It is quite clear that in mapping VR spaces there is a strong difference between maps that, on the one hand, concern aspects of the real world (Batty's cyberplace – infrastructure, hardware, people, etc.), and other immaterial aspects of the real world such as VR maps of gravity and heat, and those that on the other hand concern VR spaces (data within the computer – the spaces of ones and zeros, the social and informational landscapes of virtual worlds inside, rather than composed of, the wires). In the first case there is a geographic referent, some correspondent in the real world that the VR space is seeking to represent. In the second case no geographic referent exists and a mapping metaphor is applied to make comprehensible data that would otherwise be too complex to understand. In the first case, then, issues such as the degree of spatial equivalence are key – the extent to which the visualization corresponds with reality. In the second case, such comparisons are impossible.

Mappings can also be defined along axes of materiality and their spatial attributes. For example, maps of cyberplace can have a material geographical referent (e.g. infrastructure) or an immaterial referent (e.g. heat). In the case of a material referent, cartographic qualities most often match those of geographic space in terms of conventions and design. Essentially data are mapped onto a geographic base. For example, in Figures 23.1 and Figure 23.2 data concerning network architecture are mapped onto a globe. Another form of visualization is a virtual model of geographic location (see Figure 23.4). In the case of the immaterial referent the data are mapped metaphorically placing the values into a two-dimensional or higher display. In relation to VR spaces, mappings can portray digital data with no geographic referent but with spatial attributes (e.g. visual MUDs, see Figure 23.5), data with a geographical referent but no spatial attributes (e.g. Web pages, see Figure 23.6) or relatedness of information, (see http://www.mappingcyberspace.com/gallery/plate6_1.html), and data with no geographical referent and no spatial attributes (e.g. computer file allocation, see Figure 23.7).

Figure 23.4 A virtual model of a geographic location.

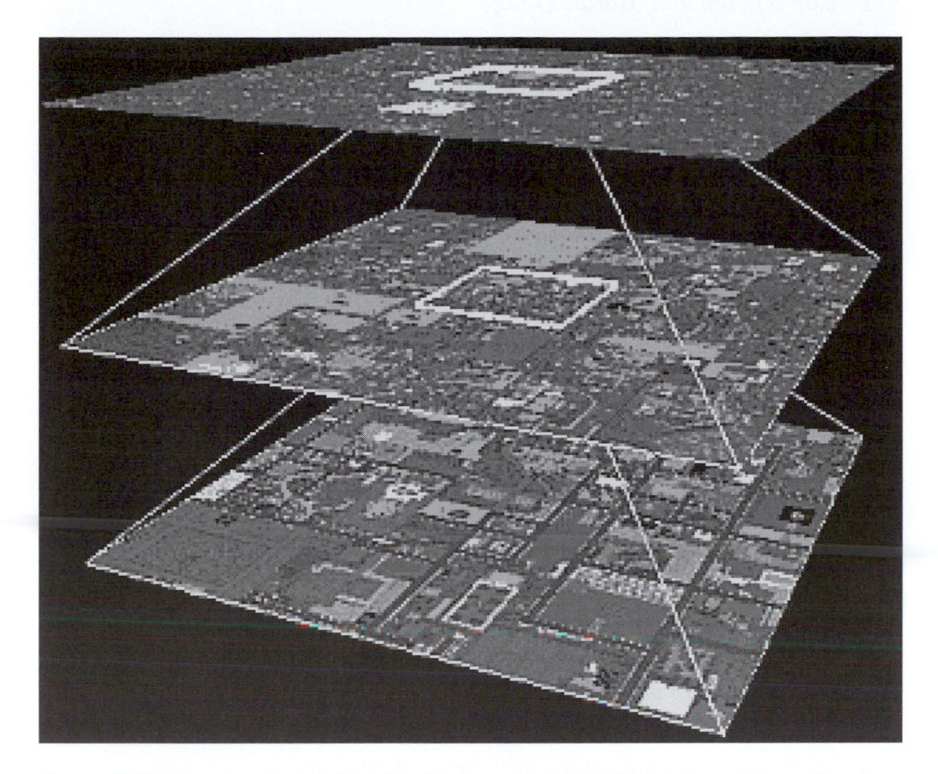

Figure 23.5 Mapping a visual MUD (Courtesy of Greg Roelofs and Peter Van Der Meulen, Philips Research, Silicon Valley.)

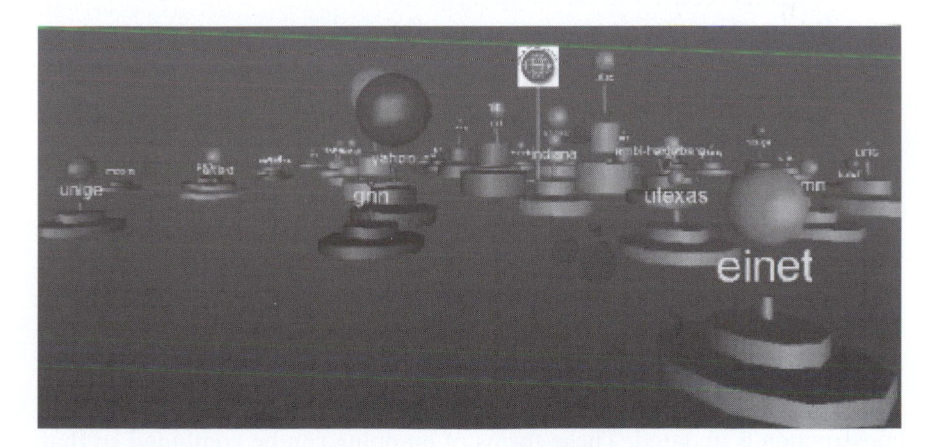

Figure 23.6 Mapping data with no spatial attributes.

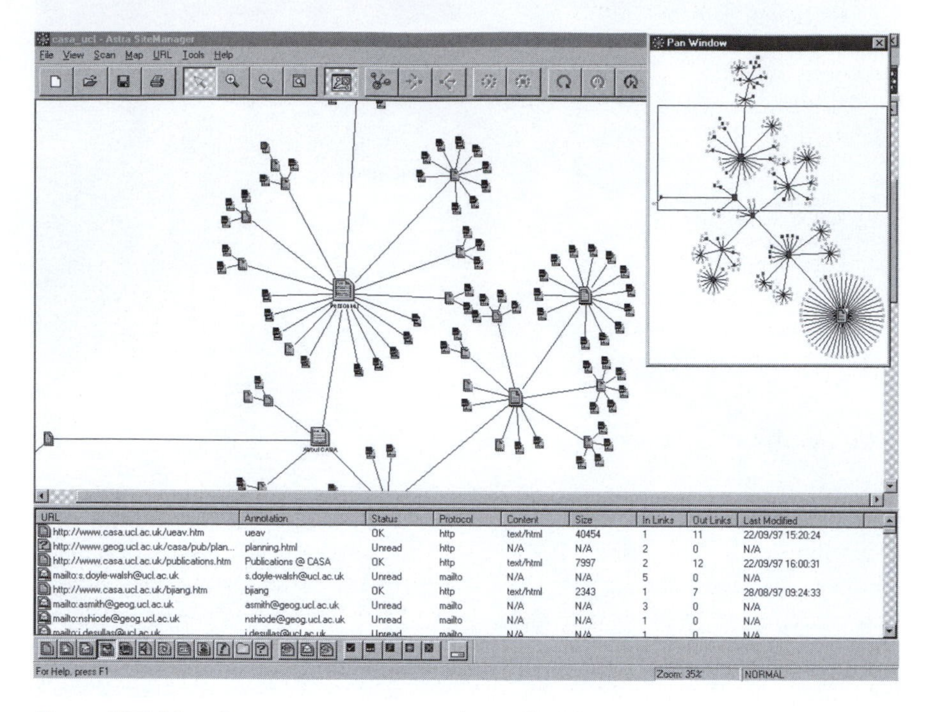

Figure 23.7 Mapping computer storage: Astra SiteManager.

At a basic level, then, mapping can be divided into one of four categories:

1 real world/material (conventional mapping),
2 real world/immaterial (metaphorical mapping),
3 VR/spatial (conventional mapping),
4 VR/non-spatial (metaphorical mapping).

Each of these four categories can be further sub-divided by mapping form, which can take one of four forms: static, animated, interactive and dynamic. Static mappings are the equivalent of traditional cartographic maps in that they are snap-shots in time. They differ, however, in that they vary in visualization technique, for example extending to three dimensions. Animated mappings portray a sequence of static maps in sequence to provide a time-series. Interactive mappings move beyond static mappings to create mappings that the user can move through and interrogate from different viewpoints (essentially 2.5D+ virtual models). Whilst the 'map' itself is static, the user becomes dynamic. Dynamic mappings are where the mapping automatically updates as the information used in its construction is updated. These forms of mapping can be combined so that a map can be static and interactive or dynamic and interactive.

This process of categorisation could continue. For example, it is possible to categorise the maps on the basis of function following Jiang and Ormeling (1997). We feel, however, that to continue our topology would be confusing, unhelpful and relatively pointless given that we think that the essential differences are captured by the topology outlined above.

Key issues

In this section we detail some of the key issues that need to be considered in relation to VR and its mapping. Some issues are omitted because they are dealt with in Chapter 20. These include: real/virtual, embodiment/disembodiment, place/placelessness, public/private.

Data quality

Geographic visualizations are only as accurate as the data used to underpin the representation. Therefore, a key question for those seeking to construct mappings of VR spaces is access to accurate information. Given the fast-growing and dynamic nature of both aspects of cyberplace and cyberspace, this issue becomes of critical importance. Online spaces such as textual and visual MUDs are in constant flux. We ourselves have encountered many dead-ends and re-routings because data and the links between them no longer existed or had changed address. Mappings will increasingly be important for understanding the connections between VR spaces and geographic spaces, and in comprehending and navigating through VR spaces, but without suitable high-quality data to underpin their construction they will be next to useless.

Naïve versus expert users

As the work of cognitive cartographers (e.g. Lloyd, 1997; MacEachren, 1995) has illustrated, maps, whilst effective at condensing and revealing complex relations, are themselves sophisticated models. For example, Liben (1991) has noted that most maps are not 'transparent' but are complex models of spatial information that require individuals to possess specific skills to process. This implies that a novice will not learn from a professionally-produced map unless they know how the map represents an area. This also applies to the mappings of VR spaces, particularly in the case of 3D interactive mappings, and metaphorical mappings. Care needs to be exercised in relation to the design of mappings so that the target audience can understand and use the information portrayed. As far as we are aware, whilst there has been some work on the legibility and design of visual virtual worlds (e.g. Darken and Sibert, 1995) and hypertext (e.g. Bachiochi *et al.*, n.d.; Kim and Hirtle, 1995; Nielsen, 1990), there has been little or no work on the legibility of VR mappings.

Representation

Geographic visualizations are spatial representations. They aim to represent, in a consistent manner, some particular phenomenon. An age-old question therefore relates to the extent to which geographic visualizations adequately represent data. Mappings necessarily depict a selective distortion of that which they seek to portray, they generalise and classify. Would the use of a different transformation, of generalising and classifying, reveal totally different relations? Harpold (1999), for example, noted that strategies of aggregation in the use of Internet demographic maps hides variation within units. Furthermore, he suggested that mappings of cyberplace reproduce particular hegemonic messages because of the ways in which they adopt traditional cartographic map units (e.g. political boundaries) in which to display data. As noted, debates concerning representation often centre around issues such as accuracy, precision, verisimilitude and mimesis. For data with no geographic referent or materiality, however, by what standard are these factors judged? When the data and mapping become synonymous, how do issues of representation apply? Here, VR spaces may become meaningless outside of their representation. The need for standards to be set and for issues of representation to be addressed, then, is of paramount importance.

Power of mapping

As Harley (1989) and others (e.g. Woods, 1993) have argued in relation to traditional cartography, maps are not objective, neutral artefacts. Mapping is a process of creating, rather than revealing, knowledge since in the process of creation decisions are made about what to include, how the map will look, what the map is seeking to communicate (MacEachren, 1995). Maps are never merely descriptive – they are heuristic devices that seek to communicate particular messages. Maps are imbued with the values and judgements of the individuals who construct them and they are undeniably a reflection of the culture in which those individuals live. As such, maps are situated within broader historical contexts and, according to Harpold (1999), they reflect hegemonic purposes through the use of historically- and politically-inflected metageographies (sign systems that organise knowledge into visual systems that we take for granted). Mappings of VR spaces are similarly the products of those that coded their constructing algorithms. They are mappings designed for particular purposes. Because they allow other users to create their own knowledge through application to their own data, these algorithms can also be empowering. In a sense, the monopoly power of the professional cartographer is severely undermined. These issues of power must be appreciated in the construction of mappings of VR spaces.

Conclusion

In this chapter we have discussed the mapping, both conventional and metaphorical, of VR spaces. Our discussion is, by our own admission, partial. The mapping of VR spaces is a recent occurrence. It has generally been conducted by information and computer scientists, with little knowledge of geographic visualization, and who construct the 'maps' for specific practical purposes. The wider implications of mapping in relation to both conceptions of space and the real world is little theorised. We have tried to give a flavour of the spatial aspects of VR spaces, how these spaces are being mapped, and some key issues that need to be considered in relation to mapping them, in order to provide context from which a more detailed analysis can occur. We would advocate two parallel and interconnected strands of research to be conducted. The first would focus on conceptualising space within VR and using these conceptions to devise effective forms of geographic visualization. The second would be a critical analysis of the process of mapping and the links between VR spaces and geographic spaces. In this way, the mapping of VR spaces can be contextualised and understood, and some of the issues such as representation, data quality and usage examined.

Acknowledgements

This chapter draws heavily on a book, *Mapping Cyberspace: Envisioning Digital Worlds*, published by Routledge (2000). We would like to thank Jeremy Crampton for comments on an earlier draft of this chapter, and the participants of the Geography and Virtual Reality meeting in Leicester (January 1999) who provided much food for thought.

References

Adams, P. 1998. Network topologies and virtual place. *Annals of the Association of American Geographers*, 88, 1, 88–106.

Bachiochi, D., Berstene, M., Chouinard, E., Conlan, N., Danchak, M., Furey, T., Neligon, C. and Way, D. 1997. *Usability Studies And Designing Navigational Aids For the World Wide Web*, in the Proceedings of the 6th WWW Conference, 511–17.

Batty, M. 1997. Virtual geography. *Futures*, 29, 4/5, 337–52.

Batty, M., Dodge, M., Doyle, S. and Smith, A. 1998. Modelling virtual environments. In Longley, P.A., Brooks, S.M., McDonnell, R. and Macmillan, B. (eds) *Geocomputation: A Primer*. Chichester, UK: Wiley.

Benedikt, M., 1991. *Cyberspace: First Steps*. Cambridge, MA: MIT Press.

Card, S.K., Mackinlay, J.D. and Shneiderman, B. (eds). 1999. *Readings in Information Visualization: Using Vision to Think*. San Francisco: Morgan Kaufmann Publishers, Inc.

Castells, M. 1996. *The Rise of the Network Society*. Oxford: Blackwell Publishers.

Crampton, J.W. 1999. *Map as Texts, Maps as Visualizations*. Department of

Geography and Earth Science, George Mason University, USA. <http://geog.gmu.edu/gess/people/jwc.html>.

Darken, R.P. and Sibert, J.L. 1995. Navigating large virtual spaces. *International Journal of Human–Computer Interaction*, 8, 1, 49–71. http://www.seas.gwu.edu/faculty/sibert/darken/publications/Navigating_IJHCI95/navigating.html.

Dieberger, A. 1996. Browsing the WWW by interacting with a textual virtual environment – a framework for experimenting with navigational metaphors. *Proceedings of ACM Hypertext '96*, Washington DC, March 1996, pp. 170–9.

Dodge, M. 1997. A cybermap atlas: envisioning the Internet. In Staple, G.C. (ed.) *TeleGeography 97/98: Global Communications Traffic Statistics & Commentary*. Washington DC: TeleGeography Inc.

Dodge, M. 1999. *An Atlas of Cyberspaces*. Centre for Advanced Spatial Analysis, University College London. <http://www.cybergeography.org/atlas/>.

Gershon, N. and Eick, S.G. 1995. Visualization's new track: making sense of information. *IEEE Spectrum*, November 1995, 32, 1, 38–56.

Gibson, W. 1987. *Mona Lisa Overdrive*. New York: Harper Collins.

Hall, S.S. 1992. *Mapping the Next Millennium*. New York: Vintage Books.

Harley, J.B. 1989. Deconstructing the map. *Cartographica*, 26, 1–20.

Harpold, T. 1999. Dark continents: critique of Internet metageographies. *Postmodern Culture*, January 1999, 9, 2. <http://www.lcc.gatech.edu/~harpold/papers/dark_continents/index.html>.

Holtzman, S.R. 1994. *Digital Mantras: The Languages of Abstract and Virtual Worlds*. Cambridge, MA: MIT Press.

Jiang, B. and Ormeling, F.J. 1997. Cybermap: the map for cyberspace. *The Cartographic Journal*, 34, 2, 111–16.

Kim, H. and Hirtle, S.C. 1995. Spatial metaphors and disorientation in hypertext browsing. *Behaviour and Information Technology*, 14, 239–50.

Lamm, S.E., Reed, D.A. and Scullin, W.H. 1996. Real-time geographic visualization of World Wide Web traffic. *Proceedings of Fifth International World Wide Web Conference*, 6–10th May 1996, Paris, France.

Liben, L.S. 1991. Environmental cognition through direct and representational experiences: a life-span perspective. In Gärling, T. and Evans, G.W. (eds) *Environment, Cognition and Action – An Integrated Approach*. New York: Plenum Press, pp. 245–76.

Lloyd, R. (1997) *Spatial Cognition: Geographic Environments*. Dordecht: Kluwer.

MacEachren, A. 1995. *How Maps Work: Representation, Visualization, and Design*. New York: Guilford.

Memarzia, K. 1997. *Towards the Definition and Applications of Digital Architecture*. School of Architectural Studies, University of Sheffield. http://www.shef.ac.uk/students/ar/ara92km/thesis/.

Mitchell, W.J. 1995. *City of Bits: Space, Place and the Infobahn*. Cambridge, MA: MIT Press.

Morse, M. 1997. Nature Morte: Landscape and narrative in virtual environments. In Moses, M. and Macleod, D. (eds) Immersed in Technology: art and virtual environments. Cambridge: MIT Press, pp. 195–232.

Munzner, T., Hoffman, E., Claffy, K. and Fenner, B. 1996. Visualizing the global topology of the MBone. In Gershon, N. and Eick, S.G. (eds) *Proceedings of the 1996 IEEE Symposium on Information Visualization*, 28–29th October 1996. San Francisco: IEEE Computer Society Press, pp. 85–92.

Nielsen, J. 1990. *Hypertext & Hypermedia*. San Diego: Academic Press.

Novak, M. 1991. Liquid architectures in cyberspace. In Benedikt, M. (ed.) *Cyberspace: First Steps*. Cambridge, MA.: MIT Press, pp. 225–54.

Richardson, A.E., Montello, D.R. and Hegarty, M. 1999. Spatial knowledge acquisition from maps, and from navigation in real and virtual environments. *Memory and Cognition*, 27, 741–50.

Ruddle, R.A., Payne, S.J. and Jones, D.M. 1997. Navigating buildings in 'desk-top' virtual environments: experimental investigations using extended navigational experience. *Journal of Experimental Psychology – Applied*, 3, 2, 143–59.

Satalich, G.A. 1995. *Navigation and Wayfinding in Virtual Reality: Finding Proper Tools and Cues to Enhance Navigation Awareness.* University of Washington, HIT Lab. http://www.hitl.washington.edu/publications/satalich/home.html.

Stalder, F. 1998. The logic of networks: social landscapes vis-à-vis the space of flows. *Ctheory, Review 46.* http://www.ctheory.com/.

Staple, G. 1995. Notes on mapping the Net: from tribal space to corporate space. *Telegeography*.

Tlauka, M. and Wilson, P.N. 1996. Orientation-free representations from navigating through a computer simulated environment. *Environment and Behavior*, 28, 647–64.

Turner, P. and Turner, S. 1997. Distance estimation in minimal virtual environments. *Proceedings of UK VRSIG 97.* http://www.brunel.ac.uk/~empgrrb/VRSIG97/proceed/030/30.htm.

Witmer, B.G., Bailey, J.H., Knerr, B.W. and Parsons, K.C. 1996. Virtual spaces and real-world places: transfer of route knowledge. *International Journal of Human–Computer Studies*, 45, 413–28.

Wood, D. 1992. *The Power of Maps*. New York: Guilford.

24 Virtual reality

An exploratory tool for investigating the cognitive mapping and navigational skills of visually-impaired people

Jo Cheesman and Chris Perkins

Introduction

The aim of this chapter is to consider virtual reality (VR) as a tool for understanding the cognitive mapping processes of visually-impaired people and how this understanding can be translated into the better design of tactile maps. Between 0.2 to 0.4 per cent of the world's population is registered as blind although some may have a certain level of residual vision (Jacobson, 1994). One of the most serious problems experienced by this group is limited independent travel. To carry this out they are dependent upon navigational aids. However, an adequate and simple to produce navigation aid has yet to be designed. This failure can be attributed to a gap in knowledge concerning the nature of spatial information, which should be conveyed via navigation aids, and the most effective way to convey it. To find answers to these questions a greater understanding is required of visually-impaired people's spatial cognition and navigation skills. The development of non-visual virtual worlds to help visually-impaired people and enhance their mobility is discussed elsewhere in this book (Jacobson *et al.*, Chapter 25). The focus of this chapter lies in the visualization capabilities of virtual reality, to aid the design of conventional navigation tools.

Mobility skills

Mobility involves the ability to travel safely, comfortably and independently. However, it is a 'complex, determined activity' (Warren and Kocon, 1973), requiring spatial cognition, navigation and motor skills. Much research related to spatial cognition and navigational processes has been carried out, a thorough review and in-depth study of which is provided by Kitchin (1994). Comprehension of how these skills are developed by visually-impaired people is imperative to the appropriate design of mobility aids.

Spatial cognition

Spatial cognition is the process of internalised reflection and construction of space in thought (Hart and Moore, 1973). It generally involves the ability to:

- locate objects accurately,
- put individual objects into a correct configurational structure with consequent comprehension of direction, orientation and intervening distances among objects,
- orient an accurate configuration with respect to a global or local frame of reference,
- understand object linkages and connectivities,
- reliably identify the names or labels of objects at particular places,
- accurately regionalise features (Golledge *et al.*, 1996).

'Cognitive mapping' is a term frequently referred to within the study of spatial cognition and is defined by Downs and Stea (1973: 7) as 'the process composed of a series of psychological transformations by which an individual acquires, stores, recalls, and decodes information about the relative locations and attributes of the phenomena in his everyday spatial environment'. In the simplest terms possible, the cognitive mapping process is a recording in the memory of the existence of an object and its known location in space, i.e. *what* is *where*. The product of cognitive mapping is a 'cognitive map', a term first introduced by Tolman (1948).

Cognitive maps are mental representations of the differentiated topography of the physical world. A cognitive map of a spatial layout codes the Euclidean relations (straight-line distances and directions) among behaviourally-relevant landmarks with a coordinate reference system centred on the environment (Sholl, 1996). They are, however, incomplete and distorted representations of the physical world. Like cartographic maps, cognitive maps can be based upon many different sources of information and encoding processes (Lloyd, 1993). Cognitive maps are created through navigating and exploring space and in turn they support navigation. When one explores and navigates an environment, landmarks are experienced sequentially in space and time (Sholl, 1996). When a cognitive map is created, a mental 'copy' of each sequentially-experienced landmark is placed into a simultaneous system that preserves information about the straight-line distance and direction of landmarks relative to one another. The spatial relations between landmarks in this simultaneous system are equally available, even though the relations between some objects may not have been directly experienced. Cognitive maps can be stored as permanent structures in long-term memory (e.g. a cognitive map of a familiar town), others may be temporary structures representing the current state of a dynamic environment (Lloyd and Cammack, 1996). In both cases the characteristics of objects are thought to be stored along with their spatial locations (Lloyd and Hooper, 1991).

The processes employed to acquire spatial knowledge appear to have a fundamental impact upon the character of a cognitive map (Lloyd and Cammack, 1996) and categories of spatial knowledge have been devised based upon different theoretical methods of acquisition. For instance, Thorndyke and Hayes-Roth (1982) argued that cognitive maps can be encoded with two different types of information.

Procedural knowledge is encoded by navigating through an environment and documents the procedures one would use to go from one location to another. Procedural knowledge is stored as verbal information and is decoded using a serial process.

Survey knowledge is encoded in our memories as mental images and can easily be obtained and stored through reading maps. Survey knowledge provides a more holistic impression of the environment and can be decoded using a parallel process.

Procedural knowledge may be transformed into survey knowledge over time. An alternative categorisation of spatial knowledge acquisition emphasises perception of the environment (Presson and Hazelrigg, 1984). In this categorisation, primary knowledge is any spatial knowledge acquired directly from the environment. This may be acquired through different perspectives – actually navigating through the environment, for instance, or from a vantage point providing a good view of an area. Secondary knowledge is acquired indirectly – when reading a map of an environment, for instance.

Navigation

Navigation is another fundamental skill required for independent mobility and this relies upon wayfinding and spatial orientation skills. Wayfinding is the cognitive and behavioural ability of a person to find their way from a specified origin to a specified destination (Golledge *et al.*, 1996). It involves the collection of information including:

- reference points for identifying the present location and destination,
- cues to signal the spatio-temporal transition from origin to destination,
- details used to maintain a heading,
- prompts signalling when to turn,
- a device for generally determining the direction in which one must continue, and
- a device for defining one's position with respect to a home point (path integration) (Golledge *et al.*, 1996).

Spatial orientation is the ability to relate personal location to environmental frames of reference. The major components of spatial orientation are:

- knowledge of spatial layout of destinations and landmarks along the way,
- the ability to keep track of where one is and in which direction one is heading, and
- comprehension of the organising structural principles embedded in a given environment (Rieser *et al.*, 1982).

People navigating through an environment orient themselves in terms of an egocentric spatial structure, i.e. ahead, behind, left or right. Human sensory modalities are tuned to integrate this information into a single coherent sensation, which allows successful navigation.

Navigating within virtual environments

Navigating in a virtual environment is not a straightforward task. If environmental information is limited to only one or two sensory modalities, then interpretations of that environment may be misleading or inaccurate, thus increasing the complexity of otherwise simple tasks. Furthermore, it is not clear how best to support navigation with design features used in interactive displays. Detailed displays take up a great deal of computer power and, as a consequence, it is often necessary to strip away much information. Often this is information describing landmarks and cues that we would normally expect to encounter in the physical world and which enable us to navigate successfully. Such an impoverished VR environment may handicap performance, it might even be analogous to the impoverished sensory environment experienced by visually-impaired people.

Research into virtual 3D environments and how they might be designed to facilitate successful navigation is increasingly based upon insights and theories of cognition, way-finding and other areas of psychology and human–computer interaction (e.g. Regian *et al.*, 1992; Tlauka and Wilson, 1994; Witmer *et al.*, 1996). This more theoretically rigorous research is accompanied by technological change that seeks to make virtual worlds more life-like. 'Smart Landscaping' is an example of a solution to this problem and involves bringing up details only when the user approaches an object. As yet, though, there is little information about what level of detail is appropriate.

Other conceptual changes are taking VR research in a different direction. Most virtual environments to date, especially fully-immersive systems, are spatial and usually 3D. They attempt to mimic the real world of buildings, landscapes and interiors. However, it is generally accepted that such virtual worlds are likely to become the exception rather than the rule as virtual reality continues to develop (Batty *et al.*, 1998; Kitchin, 1998). Representation of the environment and the users could be 'any-D', not just 2D or 3D. Virtual worlds may be visual or non-visual, or a combination of both, they may include sound- and textual-scapes, so that the spatial and non-spatial merge and multiple users (facilitated by the Net) are likely to become the norm. Navigation through such environments would not just be a matter of movement in 3D space but would involve the complete way in which users explore information in the virtual world (Batty *et al.*, 1998). It is questionable whether cognitive theories would be able to assist in the appropriate design of such environments to support navigation, since the environments would bear little resemblance to objective space as experienced in the physical world. It is the navigation of this objective space that leads to the formation of our subjective cognitive images, and most cognitive research remains grounded in the real-world experience. Batty *et al.* (1998) suggest that trial and error studies with users of new virtual environments will be needed. These would lead to representations of the environments which might help to improve navigation and lead to insights about appropriate designs. Logging and charting of navigation will be necessary so that the history of the ways in which users have interacted with their virtual world can be used to direct further interaction. So the use and application of cognitive theories for the design of virtual environments may be about to enter a new era. However, the use and application of virtual reality as an exploration tool for achieving greater understanding of cognitive processes has yet to be fully exploited. There are exciting possibilities for the use of such technology in this field, especially in relation to the understanding of spatial cognitive processes of the visually impaired, since an increased understanding can be translated into improved navigational aids required for independent travel.

Mobility of visually-impaired people

Perhaps the most serious handicap experienced by visually-impaired people is limited mobility (Foulke, 1982). Visually-impaired people are less able to travel independently because development of spatial cognition and navigation skills *largely* depends upon vision for the acquisition of spatial information. Early theories such as Von Senden's (1960) suggested that vision was a requisite condition for the acquisition of spatial concepts and abilities, but these findings have been contradicted by many subsequent studies. For instance, Jones (1975) discusses motor organisation theory and the anatomical linkages between the sensory areas of the cortex and the motor cortex, concluding that the motor cortex may therefore be a pathway between the senses. This implies that the perception of space depends upon a fusion of visual, auditory, cutaneous and proprioceptive inputs and that vision is only one element in a mutually-supportive system, rather than the primary spatial reference. More recent research has found that in groups of early-blinded, late-blinded and sighted children, visual experience facilitates the construction of spatial representations but that visual experience is not a *necessary* requirement for the ability to form integrated, global impressions of the environment (Ungar *et al.*, 1996). Different strategies of coding information are used depending on the type of experience, and these coding strategies are interchangeable. A tactile strategy could, for example, replace a visual strategy. However, it is still acknowledged that vision allows more effective comprehension, abstraction, storage, recall and use of the range of configurational layout, or spatial relational properties of large-scale complex environments.

It is now generally accepted that visually-impaired people experience the same cognitive mapping processes as sighted persons. Confident and independent mobility is therefore possible providing there is access to some of the spatial information they lack, which would allow development of the appropriate mobility skills.

What spatial information do visually-impaired people need?

The fundamental types of information necessary for safe travel by visually-impaired people are understood to be:

- micro-information about the immediate environment, which is represented in one's cognitive map and identifies obstacles and provides information about local cues, and
- macro-information about larger scale geographic space including how the terrain changes and location of buildings (Golledge *et al.*, 1996; Jacobson, 1994).

Such information is coded into a cognitive map and must be learned by the visually-impaired traveller. The information can be obtained by repetitive travel through the environment or using navigation aids. The unresolved problems are how much and which of each type of information should be presented to the visually-impaired traveller and what is the most appropriate way to present such information (Golledge *et al.,* 1996)? These are critical questions, which relate to the successful design of navigation aids.

Navigation aids

Mobility and orientation aids have been developed to be used as tools

> by visually impaired people to develop or enhance their understanding of basic spatial relationships, to facilitate a comprehension of specific travel environments, to refresh their memory of routes and areas, to further their skill in independent route planning, to enable them to travel independently in unfamiliar areas, and to add to their knowledge and enjoyment of physical space.
>
> (Bentzen, 1980: 291)

The aim of mobility aids is to offer the missing spatial data by providing information to one or more of the functioning sensory modalities of the visually-impaired traveller. Mobility aids may be categorised as obstacle avoiders or navigational aids. Obstacle avoidance aids such as canes and sensors usually have a limited spatial range and are only useful for the proximal environment. Navigation aids are designed to provide an understanding of spatial information about the environment to be traversed and the features within it. They can assist in planning routes or selecting safe path segments that encounter few obstacles. To obtain this understanding of environmental structure and layout, the blind traveller must integrate information from multiple linear segments which make up the different routes by which they cross the environment. Integration of this information is a very complex cognitive task and is often relayed by an acoustic guide or a tactile display to explain spatial relations among features and paths within the environment. From these the visually-impaired person is able to identify critically-important reference points or landmarks and 'choice' points (Golledge *et al.*, 1996).

Navigation aids providing this spatial information range from tactile maps to sophisticated portable computer systems. Current research is investigating the use of technology such as Global Positioning Systems (GPS), Geographical Information Systems (GIS), audio cues, digital spatial data and laptop computers to produce improved navigation tools. Examples include the Personal Guidance System (PGS) (Golledge *et al.*, 1998), auditory beacons (Jacobson, 1994), the Mobility of Blind and Elderly People Interacting with Computers project (MoBIC) (Petrie *et al.*, 1994) and KnowWare™ (Krueger and Gilden, 1999). Some of the systems are proving to be successful; however the expense, weight and bulk of the equipment and technological difficulties (e.g. interference of satellite signals for GPS) will limit everyday use and mass production.

In general the more sophisticated navigational aids have not been widely adopted. This failure has been attributed to social reasons and design failing to meet user needs, e.g. lack of mobility training, lack of mass production (particularly tactile maps), age of blind population, etc. Other reasons stem from design failures. Often, design is guided more by the common sense of sighted developers than by knowledge of the perceptual and cognitive abilities of the visually-impaired pedestrians who are expected to use them. For instance, the designer may have limited experience of the sequential spatial cognition process that visually-impaired people experience when gathering information from maps. In addition there are the questions of how much spatial information is required and how best to represent the spatial information. Failure to answer these questions and to overcome the design problems means that research into navigation aids continues to be a challenge. Numerous methods have been employed in attempts to further understand the cognitive processes and perception of the visually-impaired and to translate this into improving design of navigation aids, for instance, subjects drawing sketch maps, videoing and interviews. Virtual reality offers a new alternative, the potential of which has yet to be fully explored and exploited.

Virtual reality and the visually impaired

Before considering the design and production of tactile maps as navigation tools and what role VR might play in this process, it is important to draw out the aspects of virtual reality which are particularly relevant to visually-impaired people. What work has been carried out in VR using touch and sound and how this is relevant to people with visual impairments?

Sound offers the most important medium which blind people can use for perceiving virtual or real worlds. Auditory information is well established in standard desktop computer systems as an aid to functionality and to navigation through software (Brewster, 1997). The basic principals underlying the use of sound to represent spatial phenomena were defined by Krygier (1994). The most important work in this area to apply these concepts to the design of practical VR systems for visually-impaired people is being carried out by Jacobson and his colleagues, in the haptic soundscapes project (Jacobson, 1999). Other work (reviewed by Wilson *et al.*, 1997) has shown that people can recognise geometric shapes and alphanumeric characters presented through a 'virtual speaker array' as sequential arrays of sound.

However, touch appears to be more useful in our application, because it is the sense most employed by visually-impaired people when reading conventional mobility aids such as tactile maps and a tactile rather than auditory interface does not obscure other important audio information to the user (Wilson *et al.*, 1997). A substantial amount of research has established how touch works as a sense (Klatzky *et al.*, 1993; Lederman *et al.*, 1988). VR research in this area has so far concentrated on the design of haptic feedback devices to simulate sensations of contact, rather than true multi-modal systems integrating different sensory modalities (Srinivasan, 1997). This work has emphasised the development of devices to simulate vibration (Minsky and Lederman, 1996), small-scale shape and pressure distribution information (Howe *et al.*, 1995), thermal feedback (Ino *et al.*, 1992), and other tactile display modalities such as ultrasonic friction displays. Howe (1998) argues that this emphasis has much in common with previous work on sensory substitution for disabled people, in which attempts were made to map one sense onto another. In contrast, a multi-modal VR interface will seek to replicate the stimuli of the original sense instead of translating sensory inputs.

A number of researchers have tested haptic devices with disabled people, in the fields of education, training, communication, rehabilitation and access to information technology (Wilson *et al.*, 1997). Specific and relevant recent examples include: Bowman (1997) on stroke patient rehabilitation and Hardwick *et al.* (1997) on tactile access to Web interfaces. Little research has yet been carried out into how blind people might benefit from these studies. A notable exception is Colwel *et al.* (1998) who describe an experiment to investigate the perception of textures and objects by blind and sighted subjects using a force feedback device, in the real world and in a virtual environment. Their results suggested that blind people were more discriminating than sighted people in their assessment of roughness, and that the device was able to simulate the complexity of real-world textures with sufficient accuracy to allow virtual haptic perception by blind users. The visually-impaired subjects were able to assess an object's size and angle of presentation. However, their study concluded that major problems remain in the perception of complex objects.

Tactile maps

Tactile maps are often criticised for design failings and the difficulty of mass production. Despite this, they are well-established and proven navigational tools. Their use depends upon haptic perception that collects information sequentially. In the absence of vision, tactile perception is considered to provide the most information about the composition of space (Foulke, 1982). Despite the criticisms, there is still support for the use of tactile maps (Perkins and Gardiner, 1997). Ungar *et al.* (1996) carried out tactile map research with visually-impaired adults and children. With both groups they found that tactile maps can facilitate the construction of accurate and integrated cognitive maps and suggested that they are a more effective means of familiarising visually-impaired people with an environment than direct locomotor experience. They believe that tactile maps provide a more integrated and global impression (i.e. survey knowledge) of the geographic space to be navigated. This is particularly significant with visually-impaired children since they generally exhibit a relatively poor ability to form coherent cognitive maps from direct experience of environments. Ungar *et al.* (1996) believe that if visually-impaired children are trained to use tactile maps effectively, they might form the basis for improving their general spatial skills and particularly help the development of their cognitive maps. Tactile maps are also relatively cheap to produce in terms of the actual hard output. Attempting to overcome the design and production problems associated with tactile maps is therefore a task worth pursuing.

The production of tactile maps can be divided into two processes: (i) the creation or design of the map layout and (ii) the physical production of the medium (Michel, 1996). Design involves selecting features to be mapped and choosing how these should be represented as symbols. The second stage involves output of a hard copy of the map and a conversion into a medium that may be read by touch.

Design

The problems associated with tactile map design relate to layout, scale and symbol choice, and which spatial information to incorporate. Symbols have to be labelled and placed relative to one another so that they may be read. The map scale is also a crucial influence on successful design. Design improvements can only be achieved through a greater comprehension of the cognitive processes of visually-impaired people, i.e. how they perceive, collect and assimilate tactile information, and of the task performed when they travel independently, i.e. what critical information is required.

One of the chief difficulties in designing tactile maps for the visually-impaired person is that the sighted designer experiences a different form of spatial perception. When reading a tactile map the blind user acquires information in a sequential or serial process, whereas the sighted designer is able to scan the whole map and acquire spatial information synoptically (Tatham, 1991). Many published tactile maps fail to meet user needs and the differences in spatial perception may account to some extent for designers' failure to design maps which are more appropriate for the customer. Greater insight into visually-impaired people's map-reading processes is therefore needed if the design of tactile maps is to improve.

Effective design also requires a better understanding of the task performed by visually-impaired people when they travel independently. Knowledge of what information is critical to mobility, where to find it in the environment, and how to convey these data in a mapped form is needed.

Production

Two main production technologies dominate, a raised thermoformed image vacuum-moulded in plastic, and the microcapsule medium where the raised tactile image is created from expanded alcohol-impregnated paper. The computer has been used to assist in the creation of hard copy tactile output from both systems. Digital graphics data can now be used to automatically to control a drilling machine to produce a master for vacuum-forming. In the microcapsule process black and white computer printouts from graphics packages can be used to assist the standardising symbol size and placement. These are photocopied onto microcapsule paper containing a layer of alcohol-filled cells, which is exposed to heat, the alcohol boils in the hotter, black-printed areas which then rise to produce a tactile map. Such automatic creation of tactile maps from digital graphics data offers many advantages over traditional manual methods. In theory it requires less experience and effort than a completely manual process. A single data set could be used, along with appropriate and standardised routines, to generate customised products designed for different users. Potentially, automation could increase the availability and use of tactile maps. However, there are production problems associated with the creation of appropriate digital graphics data.

Digital spatial data used and produced using Geographical Information Systems (GIS) offer significant potential in this area. Appropriate GIS software is now readily available and, in many countries, regularly updated spatial data sets are produced and maintained by a national mapping agency. For instance, in the UK, the Ordnance Survey provides urban LANDLINE™ data at a resolution of 1:1250, which is sufficient to map such areas in the UK where visually-impaired people might want tactile mobility aids. However, utilisation of such data within a fully-automated tactile map production process is limited, and these limitations relate back to the design requirements. Commercially-available data sets with pedestrian-specific information such as post-boxes, bicycle stands, or other obstacles do not exist. Standard available digital data can be used but these must be augmented with special information for blind pedestrians. Digital data sets with the most appropriate information content have polygon data structures that provide information on the shape of areas. However, transformation of these data is required for tactile applications, which results in displacement of relative positions of objects and in topological errors. Line junctions merge, buildings disappear, and features overlap. Alternative data sources that have structures based upon connectivity and geometric extensions (such as the Ordnance Survey OSCAR™ data) lack the information content and neither data type is immediately suitable for translation into a tactile map. Also different map scales and times of data capture limit combination of data sets.

These data problems mean that some form of manual correction is necessary and additional capture of data by sighted persons has been used. Software has been developed to allow interactive manipulation of data in mapping (Michel, 1996), but many potentially misleading representations are likely to go unnoticed and no method of automatically detecting these errors has yet been designed. In many instances visualization of the data would aid in identifying potential structural problems.

While the physical production of tactile maps has moved forward considerably from the days of spaghetti and string, the employment of computer technology for the design of the map layout has yet to be fully exploited. It is possible that VR could assist within this area in three related applications. The first possibility is in the visualization of the sequential spatial cognition processes of visually-impaired people and in determining the critical geographical information required for independent travel. The second application relates to the actual training process for effective use and understanding of tactile maps. The third lies in the visualization of spatial digital data to allow exploration of data structures and thus aid in the automation of the design procedures.

Using virtual reality for the design of tactile maps

Virtual reality provides an intuitive means of interacting and visualizing complex three-dimensional data representing a virtual or synthetic environment. It is the powerful visualization capabilities of virtual reality that make it an attractive tool for design purposes. Visualization enhances the human visual system by employing a computer to produce images that were previously impossible or very difficult to generate. Such images can provide the map designer with previously unimagined displays from previously unattainable perspectives of previously non-visual phenomena (Friedhoff and Benzon, 1989). MacEachren *et al.* (1992: 101) believe that 'visualization ... is definitely not restricted to a method of computing ... it is first and foremost an act of cognition, a human ability to develop mental representations that allow us to identify patterns and create or impose order'. The externalisation of visualization (i.e. the visual depiction of data and/or mental images) facilitates the distinction between process and product. Particular characteristics of visual representations can be exploited to examine data in more depth. These include dimensionality (i.e. the number of data dimensions), abstraction (as opposed to realism) and the number of variables displayed (i.e. multivariate information). Visualization with VR allows exploration of all these characteristics and could promote improved design of tactile maps by allowing a greater insight into the cognitive mapping processes of the visually impaired and a clearer image of data structures. Furthermore, the dynamic and active perceptual simulations possible within VR can provide more useful means for investigating aspects of spatial cognition than conventional media (Wilson *et al.*, 1997).

Virtually exploring sequential cognition

VR provides the opportunity to re-create a street or route that a visually-impaired pedestrian may wish to navigate. For a VR environment, a designer would normally incorporate landmarks, buildings and other features that are recognisable to a sighted person, thus allowing successful travel. However, as previously stated, a visually-impaired person depends upon a much smaller and different subset of information and cues to navigate an environment than a sighted person. It is difficult for the sighted designer fully to appreciate the topological structure of the visually-impaired person's cognitive image of an environment or the importance of certain buildings and landmarks, and how these alter during a journey, since it is impossible for the designer directly to experience it.

In the past a wide variety of methods have been used by researchers in order to understand spatial cognitive processes and images of the visually impaired. Examples include graphic tests, sketching on a tactile raised-line drawing pad, reconstruction tests, constructing models using a variety of materials such as magnets or velcro, estimating distances between locations, recognition tasks, selection of correct orientation, quantitative scaling methods such as non-metric multidimensional scaling, interviews and language tests (Jacobson, 1999). Recent studies looking at orientation biases and perspectives when acquiring spatial information have used computers to generate maps and provide viewers with secondary knowledge, incorporating three-dimensional, multiple orientations and viewer perspectives with indirect navigation experiences through map animation (Lloyd and Cammack, 1996; MacEachren, 1992). The following suggestion is that VR could be employed to form a re-creation of the mental image or cognitive map in order to help the map designer understand cognitive processes of the visually impaired.

Cognitive information could be collected through a more traditional method, for instance interviews with visually-impaired people about well-known routes. Such interviews can be conducted before, during and after a journey. During a journey, the interviewee must constantly provide details of which are the important wayfinding features being utilised. Descriptions of the mental impressions of the features, i.e. size, shape, locations, distributions, densities, patterns, connections, distances and hierarchies of importance, should also be recorded. In addition, the times should be noted for when new features start being used and others becoming redundant in the navigation process. From this a simulation of the journey can be re-created using VR and be experienced by the tactile map designer as a 'walk-through', the idea being that the designer can get an impression of the key geographic features and how the importance of these grow and diminish during the journey. The before and after VR-simulated mental images would allow examination of the incremental improvement in a traveller's cognitive map. While the VR impression would certainly be far removed from the cognitive image, it would provide some indication of the importance of particular buildings and locational references. In a sense, it provides the designer with an insight to the mind images of the visually-impaired person. If many such environments were constructed from interviews with many visually-impaired people, patterns of the most important features, linkages, etc., could be found – such patterns could then be used as templates for tactile maps of unvisited areas. In travelling through the impoverished environments it should provide the tactile map designer with an education of the critical geographic building blocks which should be used on a tactile map. Clearly the interpretation of the interview materials by a VR environment designer is subjective. Tests would be necessary to examine the range of interpretations by a number of different designers. The aim of the above exercise is for the tactile map designer to understand the critical geographic information that visually-impaired people use to form their cognitive impressions of an environment and then what they use to travel through it. However, it is also important that the map designer understands the process and limitations of tactile perception and sequential information retrieval, in order to translate the critical information successfully onto a tactile map. Virtual reality also has potential in such an area.

Virtual environments would allow the examination of how the sequential information-collecting behaviour of visually-impaired people develops, and investigation of how feedback links develop between perceptions of environmental cues and behavioural responses. This could be carried out by the simulation of a route so that it resembles the actual scene as closely as possible. Using a tactile map as the navigation aid, the objective is for the visually-impaired traveller to navigate through the environment. The aim here is to assess how well the user is able to understand and translate the information provided by the tactile map into successful navigation – within a 'safe' situation. Close observation of feedback links between map information and behavioural responses can be carried out and be fed back into the design of the maps. Such an approach clearly has another application, that of training visually-impaired people, especially children, in the use of tactile maps and the improvement of their spatial skills and cognitive perception.

Training in a virtual environment

For some time now VR has been used as a training tool for able-bodied individuals, for instance flight and combat simulations. Increasingly, however, VR applications have been suggested or used for the treatment, training and diagnosis of patients with motor disturbances, neurological rehabilitation, paralyses and other physical disabilities, Parkinson's disease, impaired cognitive function, visual defects, speech defects and for occupational retraining (e.g. Brooks *et al.,* 1999b; Rose *et al.*, 1996, 1997). Many of these studies have shown that skills and knowledge attained in VR training environments are transferable to the real world (Brooks *et al.*, 1999a, b; Rose *et al.*, 2000; Wilson *et al.*, 1996). In particular it has been suggested that VR can help to train people to navigate spatially-complex environments and these spatial skills are transferable (Brooks *et al.,* 1999a, b; Peruch *et al.*, 1995; Regian *et al.*, 1992; Ring, 1998; Wilson *et al.*, 1996, 1997).

It has been shown that people can orient themselves in virtual environments and that simulations of three-dimensional space can be used as tools for investigating aspects of spatial cognition (Regian *et al.*, 1992). Studies have also considered whether passive or active participation in the navigation process allows better spatial acquisition (Brooks *et al.*, 1999a, b; Peruch *et al.*, 1995; Wilson, 1993). In general it has been shown that active participation is better. It is suggested that this is because active participation encodes information in a more automatic, elaborate form, with multi-modal components that include conceptual, visual, sensory and motor components. It is considered that the motor information provided by the actual doing is a critical component. Brooks *et al.* (1999b) have also shown that this motoric information does not have to be gathered overtly but a symbolic method also works. Brooks *et al.* (1999b) carried out a study in which active participants were required to navigate a VE using a joystick. Their results showed that active participants had enhanced spatial memory compared to the passive participants. This suggests that critical motor information may be encoded in a more symbolic form, i.e. the muscular movements required to perform a task do not necessarily have to be the same as those that would be required to carry out the task in the real world. Brooks' study also indicates that full immersive VR is not necessary for the training of spatial skills. It has also been shown that VEs require more cognitive capacity to carry out a task than is required to carry out the same task in the real world (Rose *et al.*, 2000). Rose *et al.* (2000) hypothesise that the mismatch of visual feedback and vestibular and proprioceptive feedback, which is characteristic of virtual environments, makes the virtual test more difficult than the real-world counterpart. Virtually-trained participants, when moving to the simpler real-world task, may have surplus cognitive capacity to cope with an interference, which would obviously be an advantage. There has not, however, been any examples of using VR for training visually-impaired people.

The suggested method for using virtual environments for training visually-impaired individuals to use tactile maps would be similar to the method explained in the previous section for understanding sequential information collection behaviour; this is that the virtual environment should be designed to resemble the actual environment as closely as possible. The participant is then required to navigate the route using information from a tactile map. Feedback to the participants about their progress and mistakes can be in the form of haptic or audio information. One of the major advantages of using VR for such training is that it provides individuals with a safe environment in which practically unlimited exploration can be carried out without the adverse effects of pain or danger (Brooks *et al.*, 1999a; Latash, 1998; Wilson *et al.*, 1997). Participants can make mistakes without suffering the real consequences. Furthermore, individuals can be trained without distractions that would usually occur in a real environment (Brooks *et al.*, 1999a). A learning task simulation can be designed to emphasise the key aspects and important features of the environment while distracting or irrelevant information can be left out. VR training allows consistent analysis because the VE conditions can be repeated precisely and consistently. Analysis of the individual's performance during a navigation exercise can help participants correct iterative errors that interfere with performance (Ring, 1998). Furthermore, virtual environments allow a trainer total control of both the stimulus situation and the nature and pattern of feedback, and also allow comprehensive monitoring of performance and progress (Rose *et al.*, 2000).

Many authors express concern about the safety implications of using VR for training (Jones, 1998; Korpela, 1998; Latash, 1998). The main concern is whether the removal of potentially dangerous objects or situations, both physical and social, will actually desensitise individuals, thus putting them in more danger once the real-world task is to be carried out. Feedback mechanisms can be incorporated into VR, but by including the sensations of discomfort or pain, this effectively eliminates a major advantage of VR, i.e. the lack of pain or discomfort that allows the patient to explore a wider range of motor strategies (Latash, 1998). Wilson *et al.* (1997), however, point out that feedback can be incorporated by a number of other entertaining methods – for instance, images. Obviously images would be of little help for the visually-impaired; alternatives are sound and haptic feedback.

Visualization of spatial digital data

If readily-available spatial digital data are to be employed in the production of tactile maps, then the impact of changes in scale and width of roads, removal of non-essential data, and integration with other data sets must be determined. Such changes have important implications for the presentation of the tactile information (for instance, topological structure of a road junction) to the user and its transfer to real-world navigation situations. Realistic three-dimensional visualization of the data using VR provides an effective method for identifying such changes, evaluating how the information will be interpreted, assessing the impact upon the tactile output and determining effective solutions to problem areas. The effective VR visualization of digital data representing real-world environments, in particular urban, is covered by other chapters within this volume (Batty *et al.*, Chapter 15).

Conclusion

Early developments of virtual environments experienced problems with navigation and many of these have been addressed by using the theories and insights of cognitive research. However, virtual reality environments are beginning to evolve into multidimensional, multi-sensory, multimedia and multi-user worlds in which such theories may be of limited assistance. This chapter suggests the role of virtual reality within cognitive research has yet to be fully explored. The ability to construct artificial worlds, which may or may not mimic our physical, 'real-world' geographies, offers exciting opportunities for further exploration of human cognitive processes. Such exploration is essential for the further understanding of the cognitive processes experienced by the visually impaired in order that effective navigation tools can be designed that allow independent travel.

Traditional mobility tools are limited in their success and, while new navigational methods are being developed using the latest sophisticated technology, they are restricted for everyday use. The traditional tactile map still offers one of the simplest, most effective methods of providing information to the visually-impaired traveller. However, the design of tactile maps are often criticised basically because they fail to meet users' needs. Design failings are related to lack of knowledge about what spatial information is needed by visually-impaired people and how this should be presented to them.

This chapter put forward three application areas which could help to improve our knowledge in this area and thus produce improved design of navigation tools, in particular the tactile map. The application areas include: (i) visualization of the sequential spatial cognitive processes of visually-impaired people and determining the critical spatial information required for safe, independent travel; (ii) training visually-impaired people to use effectively and understand the information presented by tactile maps, and (iii) visualization of spatial digital data, which could be employed for automated production of tactile maps, to explore misleading data structures that misrepresent the real world. Suggestions and ideas for carrying out such work are put forward.

The possibility that visualization using virtual reality may lend some insight into the process of spatial cognition through an investigation of the learning of spatial map information using tactile perception, suggests that it may also be of assistance within the larger discipline of environment–behaviour interaction. Many problems remain to be solved and many questions will be raised. For instance, how can we measure whether this technique captures the subjective knowledge, or 'what's in the head' more effectively than the traditional techniques? Also cognitive knowledge often consists of fragmentary information that contains contradictory information when integrated. How can this be visualized in a virtual reality environment? On the other hand, the ability to build synthetic worlds using virtual reality may provide the first opportunity to effectively visualize the complex spatial cognitive understanding of our surrounding environment.

References

Batty, M., Dodge, M., Doyle, S. and Smith, A. 1998. Modelling virtual environments. In Longley, P.A., Brooks, S.M., McDonnell, R and Macmillan, B. (eds) *Geocomputation: A Primer*. Chichester: John Wiley, pp. 139–61.

Bentzen, B.L. 1980. Orientation aids. In Welsh, R. and Blasch, B. (eds) *Foundations of Orientation and Mobility*. New York: American Foundation of the Blind, pp. 291–345.

Bowman, T. 1997. VR meets physical therapy. *Communications of the ACM*, August, 40 , 8, 59–60.

Brewster, S.A. 1997. Using non-speech sound to overcome information overload. *Displays (Theme issue on multimedia displays)*, 17, 179–89.

Brooks, B.M., Attree, E.A., Rose, F.D., Clifford, B.R. and Leadbetter, A.G. 1999a. The specificity of memory enhancement during interaction with a virtual environment. *Memory*, 7, 1, 65–78.

Brooks, B.M., McNeil, J.E., Rose, F.D., Greenwood, R.J., Attree, E.A. and Leadbetter, A.G. 1999b. Route learning in a case of amnesia: the efficacy of training in a virtual environment. *Neuropsychological Rehabilitation*, 9, 1, 63–76.

Colwell, C., Petrie, H., Kornbrot, D., Hardwick, A. and Furner, S. 1998. Haptic virtual reality for blind computer users. In *Proceedings of ASSETS '98, April, Los Angeles.* http://phoenix.herts.ac.uk/SDRU/pubs/VE/colwell.html.

Downs, R.M. and Stea, D. 1973. *Maps in Minds: Reflections on Cognitive Mapping.* New York: Harper & Row.

Foulke, E. 1982. Perception, cognition and the mobility of blind pedestrians. In Portugali, J. (ed.) *Spatial Abilities: Development and Physiological Foundations.* New York: Academic Press, pp. 55–76.

Friedhoff, R.M. and Benzon, W. 1989. *Visualization: The Second Computer Revolution.* New York: Harry N. Abrams.

Golledge, R.G., Klatzky, R.L. and Loomis, J.M. 1996. Cognitive mapping and wayfinding by adults without vision. In Portugali, J. (ed.) *The Construction of Cognitive Maps*. Dordrecht: Kluwer Academic, pp. 215–46.

Golledge, R.G., Klatzky, R.L., Loomis, J.M., Speigle, J. and Tietz, J. 1998. A geographical information system for a GPS based personal guidance system. *International Journal of Geographical Information Science,* 12, 7, 727–49.

Hardwick, A., Furner, S. and Rush, J. 1997. Tactile access for blind people to virtual reality on the World Wide Web. *IEE Digest No. 96/012 (Colloquium on Developments in Tactile Displays)*, 1997, 9/1–9/3.

Hart, R.A. and Moore, G.T. 1973. The development of spatial cognition: A review. In Downs, R.M. and Stea, D. (eds) *Image and Environment: Cognitive Mapping and Spatial Behaviour*. Chicago: Aldine, pp. 246–88.

Howe, R. 1998. *Introduction to Haptic Display: Tactile Display.* Available online at: http://haptic.mech.nwu.edu/intro/tactile.

Howe, R.D., Peine, W.J., Kontarinis, D.A. and Son, J.S. 1995. Remote palpation technology for surgical applications. *IEEE Engineering in Medicine and Biology Magazine*, 14, 3, 318.

Ino, S., Shimizu, S., Hosoe, H., Izumi, T., Takahashi, M. and Ifukube, T. 1992. A basic study on the tactile display for tele-presence. In *Proceedings of the IEEE Workshop on Robot and Human Communication.* IEEE: Tokyo, pp. 58–62.

Jacobson, D.R. 1994. GIS and the visually disabled – the spatial contribution to mobility. *Mapping Awareness*, July, 34–6.

Jacobson, D.R. 1997. Talking tactile maps and environmental audio beacons: an orientation and mobility development tool for visually impaired people. In *Maps and Diagrams for Blind and Visually Impaired People: Need, Solutions, Developments, Proceedings of the Conference*. Ljubljana: ICA Commission on Maps and Graphics for Blind and Visually Impaired People, 1–22.

Jacobson, R.D. 1999. *Haptic Soundscapes.* http://pollux.geog.ucsb.edu/~djacobson/haptic1.html.

Jones, B. 1975. Spatial perception in the blind. *British Journal of Psychology*, 66, 4, 461–72.

Jones, L.E. 1998. Does virtual reality have a place in the rehabilitation world? *Disability and Rehabilitation*, 20, 3, 102–3.

Kitchin, R.M. 1994. Cognitive maps: What are they and why study them? *Journal of Environmental Psychology*, 14, 1, 1–19.

Kitchin, R. 1998. *Cyberspace: the World in Wires*. Chichester: John Wiley.

Klatzky, R.L. and Lederman, S.J. 1993. Spatial and nonspatial avenues to object recognition by the human haptic system. In Eilan, N., McCarthy, R.A. and Brewer, W. (eds) *Spatial Representation: Problems in Philosophy and Psychology*. Oxford: Blackwell, pp. 191–205.

Korpela, R. 1998. Virtual reality opening the way. *Disability and Rehabilitation*, 20, 3, 106–7.

Krueger, M.W. and Gilden, D. 1999. KnowWare™: virtual reality maps for blind people. In Westwood, J.D. (ed.) *Medicine Meets Virtual Reality*. Amsterdam: IOS Press, pp. 191–7.

Krygier, J.B. 1994. Sound and geographic visualization. In MacEachren, A.M. and Fraser-Taylor, D.R. (eds) *Visualization in Modern Cartography*. Oxford: Pergamon, pp. 149–66.

Latash, M.L. 1998. Virtual reality: a fascinating tool for motor rehabilitation (to be used with caution). *Disability and Rehabilitation*, 20, 3, 104–5.

Lederman, S.J., Browse, R.A. and Klatzky, R.L. 1988. Haptic processing of spatially distributed information. *Perception & Psychophysics*, 44, 3, 222–32.

Lloyd, R. 1993. Cognitive processes and cartographic maps. In Garling, T. and Golledge, R. (eds) *Behaviour and Environment: Psychological and Geographical Approaches*. Amsterdam: Elsevier Science Publishers, pp. 141–69.

Lloyd, R. and Cammack, R. 1996. Constructing cognitive maps with orientation biases. In Portugali, J. (ed.) *The Construction of Cognitive Maps*. Dordrecht: Kluwer Academic, pp. 187–213.

Lloyd, R. and Hooper, H. 1991. Urban cognitive maps: computation and structure. *The Professional Geographer*, 43, 15–27.

MacEachren, A. 1992. Learning spatial information from maps: can orientation-specificity be overcome? *Professional Geographer*, 44, 431–43.

MacEachren, A., Buttenfield, B., Campbell, J., Dibiase, D. and Monmonier, M. 1992. Visualization. In *Geography's Inner Worlds: Pervasive Themes in Contemporary American Geography*. New Jersey: Rutgers University Press, pp. 101–37.

Michel, R. 1996. Creation of tactile maps using digital map data. In *Maps and Diagrams for Blind and Visually Impaired People: Need, Solutions, Developments, Proceedings of the Conference.* Ljubljana: ICA Commission on Maps and Graphics for Blind and Visually Impaired People, pp. 1–5.

Minsky, M. and Lederman, S.J. 1996. Simulated haptic textures: roughness. In Danai, K. (ed.) *Proceedings of the ASME Dynamic Systems and Control Division*, DSC-58, pp. 451–8.

Perkins, C. and Gardiner, A. 1997. What I really, really want ... how visually impaired people can improve tactile map design. In *Technical Proceedings of the 18th International Cartographic Association Conference*, Stockholm, June 1997. Stockholm: ICA, pp. 1159–66.

Peruch, P., Versher, J.L. and Gauthier, G.M. 1995. Acquisition of spatial knowledge through visual exploration of simulated environments. *Ecological Psychology*, 7, 1–20.

Petrie, H., Johnson, V., Holmes, E., Jansson, G., Stothotte, T., Michel, R. and Raab, A. 1997. Access to digital maps for blind people: results from the MoBIC project. In *Maps and Diagrams for Blind and Visually Impaired People: Need,*

Solutions, Developments, Proceedings of the Conference. Ljubljana: ICA Commission on Maps and Graphics for Blind and Visually Impaired People.

Presson, C. and Hazelrigg, M. 1984. Building spatial representations through primary and secondary learning. *Journal of Experimental Psychology: Learning, Memory and Cognition*, 10, 716–22.

Regian, J.W., Shebilske, W.L. and Monk, J.M. 1992. Virtual reality: an instructional medium for visual–spatial tasks. *Journal of Communications,* 42, 4, 136–49.

Rieser, J.J., Guth, D.A. and Hill, E.W. 1982. Mental processes mediating independent travel: implications for orientation and mobility. *Journal of Visual Impairment and Blindness*, 76, 6, 213–18.

Ring, H. 1998. Is neurological rehabilitation ready for immersion in the world of virtual reality? *Disability and Rehabilitation*, 20, 3, 98–101.

Rose, F.D., Attree, E. and Johnson, D.A. 1996. Virtual reality: an assistive technology in neurological rehabilitation. *Current Opinion in Neurology*, 9, 461–7.

Rose, F.D., Johnson, D.A. and Attree, E.A. 1997. Rehabilitation in the head injured child: from basic principals to new technology. *Journal of Pediatric Rehabilitation*, 9, 3–7.

Rose, F.D., Attree, E.A., Brooks, B.M. Parslow, D.M., Penn, P.R. and Ambihaipahan, N. 2000. Training in virtual environments: transfer to real world tasks and equivalence to real task training. *Ergonomics*, 43, 494–511.

Sholl, J.M. 1996. From visual information to cognitive maps. In Portugali, J. (ed.) *The Construction of Cognitive Maps*. Dordrecht: Kluwer Academic.

Srinivasan, M.A. 1997. Multimodal virtual environments: paper presented to the *Sixth Annual Symposium on Haptic Interfaces for Virtual Environment and Teleoperator Systems ASME International Mechanical Engineering Congress and Exposition, November 17, 1997 – Dallas, Texas.*

Tatham, A. 1991. The design of tactile maps: theoretical and practical considerations. In Blakemore, M. and Rybaczuk, K. (eds) *Mapping the Nations: Proceedings of the 15th International Cartographic Association Conference, Bournemouth, 1991.* London: ICA, pp. 157–66.

Thorndyke, P. and Hayes-Roth, B. 1982. Differences in spatial knowledge acquired from maps and navigation. *Cognitive Psychology*, 14, 560–81.

Tlauka, M. and Wilson, P.N. 1994. The effect of landmarks on route-learning in a computer-simulated environment. *Journal of Experimental Psychology*, 14, 305–13.

Tolman, E.C. 1948. Cognitive maps in rats and men. *Psychological Review*, 55, 189–208.

Ungar, S., Blades, M. and Spencer, C. 1996. The construction of cognitive maps by children with visual impairments. In Portugali, J. (ed.) *The Construction of Cognitive Maps*. Dordrecht: Kluwer Academic, pp. 247–73.

Von Senden, M. 1960. *Space and Sight.* (Translated from the 1932 edition by P. Heath.) Glencoe, IL: Free Press.

Warren, D.H. and Kocon, J. 1973. Factors in the successful mobility of the blink: a review. *American Foundation for the Blind Research Bulletin,* 28, 191–218.

Wilson, P.N., Foreman, M. and Tlauka, M. 1996. Transfer of spatial information from a virtual to a real environment in physically disabled children. *Disability and Rehabilitation*, 18, 633–7.

Wilson, P.N., Foreman, M. and Standon, D. 1997. Virtual reality disability and rehabilitation. *Disability and Rehabilitation*, 19, 213–20.

Witmer, R.G., Bailey, J.H., Knerr, B.W. and Parsons, K.C. 1996. Virtual spaces and real world places: transfer of route knowledge. *International Journal of Human–Computer Studies*, 45, 4, 413–28.

25 Multi-modal virtual reality for presenting geographic information

R. Daniel Jacobson, Rob Kitchin and Reginald Golledge

Introduction

Humanity's desire to represent the world spatially is strong. From the earliest times we have tried to communicate and represent the spatial aspects of the world to each other through a variety of media: cave paintings, drawings in the sand, models, maps, works of art, photographs and, in present times, with satellite images, computer-generated worlds and virtual environments. Spatial representations also extend beyond the visual. For example, we also communicate spatial relations through the spoken and written word. In this chapter, we examine the multi-modal representation and communication of spatial information using VR technologies.

In common with most methods intended to convey spatial information, VR relies heavily on visual display. This is not surprising given that it is well noted that vision is the most useful of the senses for understanding space (Foulke, 1983). For visually-impaired people, the over-reliance on visual display for conveying spatial relations denies them access to these media. In recent years a number of studies have shown that people with visual impairments can understand spatial relations when communicated in a manner that is accessible to them. This understanding relates both to learning a new environment (Jacobson *et al.*, 1998), learning from secondary sources (Ungar, 1994), and using information gleaned from secondary sources to interact with environments (Espinosa *et al.*, 1998). Visually-impaired people are able to gain a comprehension of spatial relations because walking through an environment and interacting with a secondary source are multi-modal experiences – information is gathered through visual, auditory, haptic, kinaesthetic, proprioceptive, vestibular, olfactory and gustatory means.

Given this ability to comprehend real-world spatial relations due to their multiple modalities, it is our contention that multi-modal VR can be used successfully to augment interaction in real-world environments and also to provide a media for secondary learning for people with severe visual impairments. Multi-modal VR offers an opportunity for a better quality of life through enhanced education and mobility, opening up arenas of work and leisure. Such developments will also augment VR usage for sighted individuals by enhancing human–computer interaction (HCI).

Geographic representation, visual impairment and VR

As noted, geographic representations are mainly visual in nature and reliant on people to be sighted in order to access them. People with visual impairments are thus denied access to them and, in part, to navigating the environments they represent. Over the past few decades several different means for conveying spatial information to people with severe visual impairment have been explored. As detailed in Jacobson and Kitchin (1997), these means can be divided into those that seek to convey spatial representations using modified geographic representations (e.g. tactile or talking maps) and those that seek to act as navigation/orientation aids (e.g. talking signs and personal guidance systems).

As our case examples will illustrate, in our opinion VR systems offer qualitatively improved means of both representing spatial data and providing navigation aids. This is because of the nature of VR as a medium. Virtual reality is a wide-ranging term, with many differing meanings to different researchers. Wickens and Baker (1995) list the core 'reality giving' features in a virtual system as: dimensionality, motion, interaction, frame of reference and multi-modal interaction. These five aspects interact to make VR spaces qualitatively different to other forms of geographic visualization (e.g. maps), with a wider range of applications.

Dimensionality encompasses a continuum from two to three dimensions, including perspective and stereoscopic viewing. A 3D VR system is considered more 'real' as this replicates our perceptual experiences in the natural environment. Moreover it offers a greater potential for visualization. For example, a 2D contour map is difficult to interpret, requiring training and experience, and is more cognitively demanding than viewing a three-dimensional representation of the same terrain.

Motion refers to the degree of dynamism within the representation. VR is generally considered to be dynamic with the view changing as the user moves position. Maps, on the other hand, are static representations and the view remains the same despite the viewer moving position. However, as a static map is rotated, the ability to relate the map to the real world changes.

Interaction refers to the degree to which the user can alter or 'play' with the representation. Interaction is either closed-loop or open-loop in nature. In the open-loop form the display runs in its entirety, from beginning to end, and viewing is passive. In the closed-loop form the viewer can interact with the data or display and actively direct their navigation through the information. VR allows interaction through the adoption of a closed-loop approach, with a desktop VR 'walking' the user through a building, for example.

Frame of reference refers to the viewing position of the VR user and varies as a function of a spectrum between an ego-reference (inside-out) and a world-reference (outside-in). In an ego-referenced presentation, the display is presented from the viewer's perspective (e.g. as if looking from the user's eyes). In a world-referenced display, an exocentric view is taken, where the scene containing the user is viewed from an external location (e.g. from the top of a lamppost). Generally speaking, the egocentric perspective is easier to interpret and it is this view which is most often utilised with VR.

Multi-modal interaction consists of interaction through several different, complementary sources. For example, a multi-modal VR system might employ a number of display and input devices including conventional computer input peripherals such as keyboards, mice, joysticks, computer screens as well as novel input peripherals such as speech recognition, eye gaze tracking, gesture tracking, tactile feedback from surface textures, data gloves, force feedback, head-mounted display, screen display and so on.

To this list we add *mimetic representation* and *scalar changes*. VR systems provide representations that detail in mimetic (imitation, mimicry) visual form the multifaceted, dynamic and complex nature of geographic environments. Here, VR users see a VR landscape that has a high degree of verisimilitude (having the appearance of truth) to the real environment rather than an abstracted representation (such as a map). As with maps, mimetic representations can provide representations of both material (physical objects) and immaterial (such as heat) information. However, mimetic representation allows users to make the link between representation and reality more clearly as it partly removes the degree to which the abstraction needs to be processed in order to make connections between the abstraction and reality.

VR representations also differ qualitatively from other forms of representation because they allow users to explore the same data through a set of seamless *scalar changes*. These scalar changes can only be achieved in map-based representations by viewing a set of maps, and then always within the frame of the abstraction (flat, symbolic map containing degrees of inaccuracy and imprecision).

Although VR provides a qualitatively-enriched form of geographic representation, their combination of dimensionality, motion, interaction, frame of reference, mimetic representation, scalar changes and multi-modal peripheral input is designed to provide an augmented visual display, and therefore does not seek to cater for people without sight. We contend that these seven defining aspects can, however, be reconfigured to allow people with severe visual impairments to benefit from using these qualitatively different representations. This is achieved by providing additional multi-modal input devices. Traditional forms of spatial representation are static, unintelligent and inflexible. For example, tactile maps can only be read by one person at any time, they cannot be 'questioned', they cannot be manipulated to change scale or perspective. Like any cartographic product, tactile maps are subject to certain constraints: scale reduction, 'bird's eye view' perspective, classification, symbolisation, generalisation, etc. These constraints are exacerbated by the need for the information to be read tactually. To illustrate our contention we briefly detail two on-going projects which utilise VR. Before moving to these studies, we feel it instructive to first redefine the scope of VR in light of the discussion so far, and to detail by what means we can compensate for lack of vision.

Redefining VR

In conventional definitions of VR, reality is *re-created* as a virtual reality. Here there is a high degree of mimesis – the VR seeks to mimic reality. However, for our purposes, this definition is limiting. In a system where another dimension such as sound or touch is being used symbolically to represent visual components, the process of re-creation becomes one of abstract representation. Here, the VR system is not seeking graphically to mimic the geographic environment but provide a representation which augments interaction with either a spatial representation or the real-world environment. Moreover, geographic information within an accessible VR can be a representation of a representation. For example, a map in its creation has already undergone a degree of abstraction via scale change, selection, symbolisation and so on. When this is conveyed either through sound or touch through a VR, a new level of abstraction is inserted. As such, a more complicated process than re-creation is occurring. In an extended definition of VR, the VR system then seeks to provide either re-creation or augmentation. These categories are not mutually exclusive.

Using multi-modal VR to present geographic information

At present, it seems that touch (haptics) and sound (auditory) provide the best options for constructing augmented VR systems. Indeed, most work to date has concentrated on exploring these two alternative means of communicating spatial information. We will give a brief summary of the merits and limitations of each in turn.

Haptics

Haptic perception involves the sensing of the movement and position of joints, limbs and fingers (kinaesthesia and proprioception) and also the sensing of information through the skin (tactile sense) (Loomis and

Lederman, 1986). A virtual reality expressed through a haptic display allows a user to *feel* a virtual environment. The added dimensionality that haptic, kinaesthetic and tactile interfaces provide are important for three reasons. First, the display of information through the haptic senses can lead to greater immersion and realism in a virtual environment. For example, users are able to sense, feel and interact with objects represented in the virtual world. Second, the haptic channel may augment visual information, but also offers an independent channel for interaction for the non-visual or multi-modal presentation of geographic space. Third, haptic interfaces have enormous potential for enhancing virtual environments, and by the provision of information through a modality other than vision, they extend the range of applications to a wider section of the population (Brewster and Pengelly, 1998; Colwell *et al.*, 1998; Hannaford and Venema, 1995). Haptics provide a very natural interaction within body-sized spaces, for example, grasping objects.

Tactile, haptic or force feedback output can be achieved through several techniques including pneumatic (driven by air pressure), vibrotactile (vibrating tactile stimulation), electrotactile (electrical charge driven), and electromechanical stimulation. Systems adopting these outputs can be finger-based, hand-based or exoskeletal (body). Commonplace zero technology systems for haptic communication include Braille, sign language and Tadoma (non-verbal reading of speech via a hand placed on a speaker's face and neck). There are good reasons to include haptic interfaces in VR systems, particularly when an object needs to be manipulated. This is because once vision is occluded, such as an object passing out of sight behind another object, there is no perceptual feedback and it becomes difficult or almost impossible to control the occluded object. In addition, haptic interfaces are able to provide accessibility to representations of geographic space without the need for vision (Porter and Treviranus, 1998).

Auditory

Along with haptics, sound is a useful addition to VR systems providing a number of augmentations. The sound we perceive in the environment contains information about the source of that sound, and its distance and possible direction (Kramer, 1994). Vision gathers a large amount of information but only from one direction at a time. In contrast, auditory perception is omnidirectional. Moreover, as a species we are good at ascribing causal meanings to sound in a very natural way, such as identifying the ring of a doorbell or the backfire of a car exhaust (Ballas, 1993) and picking up repetitions and correlations of various kinds which give rise to a certain rhythm (Bregman, 1994).

Loomis and Soule (1996) describe the advantages of auditory displays over traditional visual or tactile displays. Auditory displays have high temporal bandwidth, specialised mechanisms for temporal pattern processing, allow simultaneous monitoring of all spatial directions, and have low sensory thresholds, which permit the use of displays that minimally consume power. Kramer (1994) provides a comprehensive summary of the benefits of auditory display.

- An auditory display's presence is generally as an augmentation and non-interfering enhancement to a visual display.
- An auditory display increases the available dimensionality of the representational space.
- Auditory displays have superior temporal resolution. Shorter duration events can be detected with auditory displays.
- Auditory displays create user interest and engagement by decreasing learning time, reducing fatigue and increasing enthusiasm.
- Auditory displays have complementary pattern recognition capabilities by bringing new and different capacities to the detection of relationships in data.
- Auditory displays provide strong inter-modal correlations, reinforcing experiences gained through visual or other senses.
- Auditory displays enhance realism by adding immersive qualities and making virtual reality situations more realistic.
- Auditory displays are synesthetic, i.e. they replace insufficient or inappropriate cues from other sensory channels.
- Auditory displays enhance learning by providing a presentation modality suited to many students' learning style.

Clearly, the auditory sense has several advantages and benefits as a method for displaying information. It is complimentary in how and what information it can communicate to a computer user, by broadening the human–computer communication channel and taking advantage of unused 'bandwidth'.

Auditory data of use in a VR system can be divided into two categories: speech and non-speech. The non-speech category can be further subdivided. Sonification is the use of data to control a sound generator for monitoring and analysis of the data. This would include mapping data to pitch, brightness, loudness or spatial position. This is highly relevant to auditory cartography (Krygier, 1994). Within an auditory computer interface, an auditory icon is an everyday sound that is used to convey information about events in the computer (Gaver, 1994). This is done by an analogous process with an everyday sound-producing event. For example, a file being deleted may be indicated by the sound of a trashcan, and a computer process running represented by an engine running. An earcon is a tone-based symbol set where the 'lilt' of verbal language is replaced with combinations of pitch and rhythmic structures (Blattner *et al.*, 1994; Brewster *et al.*, 1994). Audiation is the direct translation of data waveform to the auditory domain for monitoring and comprehension. Examples of audiation would include listening to the waveforms of an electroencephalogram, seismogram or radio telescope data (Kramer, 1994). At present, this appears to have minimal application in the presentation of geographic data, but might be useful for representing continuous data such as rainfall or temperature that has clearly-defined spectral properties.

A set of key auditory variables that can be used in the presentation of geographic data were presented by Krygier (1994) and include:

Location – the location of a sound in a two- or three-dimensional space such as a spatially-referenced verbal landmark on a touch pad.
Loudness – the magnitude of a sound.
Pitch – the highness or lowness (frequency) of a sound.
Register – the relative location of a pitch in a given range of pitches.
Timbre – the general prevailing quality or characteristic of a sound.
Duration – the length of time a sound is (or is not) heard.
Rate of change – the relation between the durations of sound and silence over time.
Order – the sequence of sounds over time.
Attack/decay – the time it takes for a sound to reach its maximum or minimum.

Historically, sound has been applied to maps by verbal commentary and voice-over effects (Thrower, 1961). More recently multi- and hypermedia have added new dimensions to spatially-referenced data in encyclopaedias and digital atlases. Within the context of human–computer interaction, auditory icons and earcons are now regularly used. Fisher (1994) used such sounds for conveying uncertainty in a remotely-sensed image, and Jacobson (1996) used spatial auditory icons (the sound of traffic, and the bleep of a pedestrian crossing) on an audio–tactile map. Researchers have investigated navigating the World Wide Web through audio (Albers, 1996; Metois and Back, 1996) and as a tool to accessing the structure of a document (Portigal and Carey, 1994). Data sonification has been used to investigate the structure of multivariate and geometric data (Axen and Choi, 1994, 1996; Flowers *et al.*, 1996). Auditory interfaces have successfully been used in aircraft cockpits and to aid satellite ground control stations (Albers, 1994; Ballas and Kieras, 1996; Begault and Wenzel, 1996).

Two projects

Haptic–soundscapes

A current project is to construct a haptic–soundscape prototype VR system for conveying spatial representations. Work on this project has just started and builds on an earlier project which implemented a sound map system. The design and effectiveness of the sound map system is reported elsewhere (Jacobson, 1998, 1999; Jacobson and Kitchin, 1997) but we briefly outline the systems and findings here for context, before detailing the present haptic–soundscape work.

The original soundscapes project was initially conceived as a way to produce interactive maps accessible to people with visual impairment, which avoided some of the difficulties of tactile maps. Many tactile maps are relatively expensive, difficult to construct, largely unportable because they cannot be folded, and inaccessible to many visually-impaired people because they demand an ability tactilely to identify features using the fingertip or palm. Labelling is particularly problematic; when enough labels are applied to facilitate suitable understanding, the map often becomes cluttered and illegible (Tatham, 1991). Using labels in a separate legend or key reduces the immediacy of the graphic and introduces interpretative problems as referencing is disrupted (Hinton, 1993). Computer systems that link sound with touch, enhancing the tactile map with the addition of audio, have increased the utility of tactile maps. For example, when a raised area on a tactile map is touched, a corresponding sound label is triggered. Two such systems include NOMAD (Parkes, 1988) and 'talking tactile maps' (Blenkhorn and Evans, 1994). Fanstone (1995) has exploited the GIS capabilities of NOMAD to build a hierarchical audio–tactile GIS of Nottingham University campus. Access within a map is very efficient with users reporting enjoyment and ease of use (Jacobson, 1996). However, access from one map to the next remains problematic. For example, to 'zoom in' or to move to an adjacent area the user has to locate the speech label indicating that it is possible to zoom in, then remove the tactile map, search for another tactile map, register this on the touchpad and then continue map exploration. This break in the continuum of map reading is disrupting and confusing.

The sound map project sought to overcome some of these difficulties by utilising hypermedia software. Within a hypermedia environment a user can navigate between textual and cartographic information nodes in order to get a well-documented, multi-faceted representation of space, from varied sources and differing viewpoints (Milleret-Raffort, 1995). Conventional hypermedia systems are predominantly visual in nature. They can, however, also offer people with visual impairments a 'virtual' way of exploring the world (Jacobson and Kitchin, 1997). In this case, hypermedia was used to construct a bold visual (for sighted people and those with residual vision) and audio environment consisting of a set of hierarchically-distributed sound maps which could be accessed across the Internet.

In this system, bold visual media (text and image) were augmented by spoken audio and metaphorical sound (e.g. sound of waves crashing on a beach). The work sought to build upon other projects such as Webspeak and the Graphical User Interface for Blind People Project (Petrie *et al.*, 1996) which aim to make traditional computer interfaces accessible.

The sound map system comprises a conventional personal computer, running a World Wide Web browser such as Netscape Navigator. The only specialist addition is a 'touch window', a peripheral device consisting of an electro-resistive glass screen. Touching or dragging a finger or stylus on the glass screen provides an alternative to mouse-controlled input. The touch window can be attached to a monitor so a user with limited vision is able to view the screen through the touch window, or used at table-top level where a totally blind individual is able to scan the pad with their fingertips. Spatial information is presented as an auditory map. Areas of the touch pad are overlain with sound and when the map user's finger enters the designated area the sound is played. By touching adjacent areas of the pad, users are able to determine the size and shape of a map feature by the change in sound. A collection of sounds are used to represent map information that is usually conveyed by visual symbols (text, colour, line style, shape, etc.).

For the purposes of development, an off-line World Wide Website was built which utilised inter-linking auditory maps that could be traversed solely by sound and touch. As the user's finger is dragged across the touch pad, the system 'talks', playing audio files which are triggered by the position of the user's finger. This audio–tactile hypermedia conveys cartographic information through the use of spoken audio, verbal landmarks and auditory icons. Audio consists of environmental sounds (such as traffic noise for a road) and icons which denote specific events like the edge of a map, links to further maps, or allows the user to press for more information.

Rather than direct manipulation of a tactile surface, such as pressing on the tactile maps in NOMAD, this system uses a touch window. Therefore the user has no direct cutaneous stimulus from tactile relief. The encoding from the audio–tactile stimulus meant that map information is built up from kinaesthetic sensing of movement across the pad, sensing of the distance traversed across the pad, proprioceptive sensing of the location of the fingers and location information obtained by referencing with the hands to the outside frame of the touch pad. Linking enables a blind user to traverse from one auditory map to another. As each map loads, a verbal overview describing the map is played. From all maps there is direct access to a help screen that explains the system and the modes of interaction.

Figure 25.1 displays the simple user-interface for the auditory hypermap system. As the map-reader's finger moves across the touch pad and over the 'SOUTH' bar, the audio message 'Press to go south' is played. Once this part of the touchpad is pressed the central area is filled with an auditory map to the south of the previous one. If no maps are available, this is relayed to the user verbally. NORTH, WEST and EAST all work in a similar manner. HOME returns the user to the main auditory map. The HELP button explains how to use the system. When exiting from help the

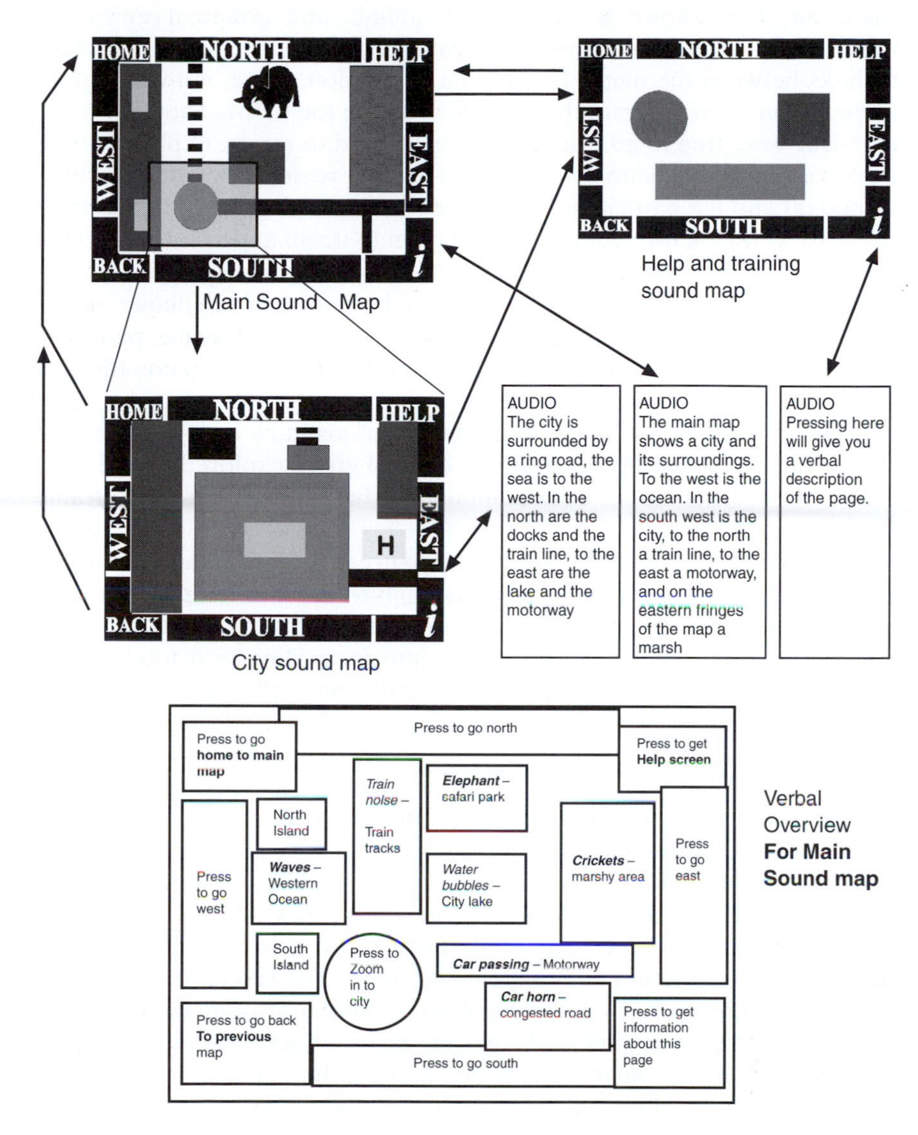

Figure 25.1 Interface and functionality for sound map prototype.

user is returned to the correct map. The '*i*' button plays information about the map in view (e.g. 'the map shows a city and its surroundings, in the south-west is the city, etc.'). The BACK and FORWARD buttons allow the user to traverse through the 'history' of their links.

The system was evaluated by five visually-impaired people and five blind people. Initial training took place for fifteen minutes using the help screen of the sound map system. Users were familiarised with the touchpad, were shown how to follow a link, and obtain more verbal information. They were given no information about the content, structure or links between the maps. During the evaluation phase, individuals had fifteen minutes to navigate through and explore the maps. They were told that they were free to go where they wished and to return to places previously visited. At the end of this fifteen-minute period, the computer was turned off and the participant was assessed using techniques adapted from cognitive mapping for people without vision (Kitchin and Jacobson, 1997). These included a verbal description of the maps and map layout, imagining they had to explain the maps to somebody over a telephone, and a graphical reconstruction of the maps using a tactile drawing pad. The whole process was videotaped and a log made of people's paths through the audio–tactile maps. Semi-structured interviews were used to get impressions of the system, feedback on how it could be improved and for ideas of where it may be beneficial. A control group explored a hard-copy tactile map version of the map obtaining verbal information from Braille labels.

All of the sound map users were able successfully to interact with the system. This included people who had never used a computer before. Interview responses suggest that the system aroused great interest and that map access was 'simple, satisfying and fun' (according to a totally blind participant). Users were able both graphically and verbally to reconstruct the maps with varying degrees of accuracy (Jacobson, 1998). The level of integration and information recall was greater among sound map users than the control group. However, the tactile map users had greater accuracy of shape (map object) reconstruction, due to the raised line tracing nature of their representations.

The current project, haptic–soundscapes, seeks to build upon the developmental sound map work in several ways. It will carry out experiments into the nature of non-visual audio–tactile perception, looking at such issues as resolution and shape recognition. The results from the perceptual experiments will be implemented to build a working demonstrator to display a map, a graphic and a graph, where experimentation and user feedback will lead to further refinements of the system. The most notable feature will be development work with a haptic mouse. A haptic mouse works in a manner similar to a conventional mouse with two key differences. First, it is an absolute pointing device rather than a relative one, so a certain mouse position will always correspond to a certain cursor position. Second, the device is able to apply force-feedback, to offer varying degrees of resistance in a two-dimensional plane. However, by the use of acceleration techniques, the mouse user perceives a three-dimensional movement, such as rolling their hand over a small hill. As such, a third dimension of haptic feedback will be added to bold visuals and auditory information to provide an interactive, multi-modal VR system. Table 25.1 and Figure 25.2 show the haptic effects.

Table 25.1 Virtual effects available with VRM haptic mouse (numbers refer to Figure 25.2)

- Virtual Wall: [1]
  When a mouse cursor contacts a virtual wall, extra force will be required to pass through. This allows the user to detect different regions of the screen surrounded by walls or borders.
- Gravity Well: [2]
  When the cursor enters the active region of a gravity well, the VRM is physically drawn towards the centre of the virtual object. Thus, objects with gravity (e.g. a map object) can be found by feel alone.
- Damping: [3]
  Sections of the screen with damping provide a resistive force that is proportional to the speed the mouse is moving. This allows the user to detect different regions of the screen and to stabilise unsteady hand movements.
- Friction: [4]
  Screen areas with friction exhibit a uniform resistive force to the movement of the mouse offering the user an additional effect to differentiate various screen objects.
- Rubber Bands: [5]
  Objects with the rubber band effect have a spring-like feel so that it is easy to remain on the object while performing mouse clicks or moves (e.g. map objects, or buttons, slider bars in a desktop environment).

Source: Adapted from Control Advantage, http://www.controladv.com.

Personal Guidance System

The sound maps/haptic soundscapes projects seek to make geographic representations accessible to visually-impaired people. VR is also being used in another way to provide navigation aids for traversing environments. The Personal Guidance System developed at Santa Barbara by a team of geographers and psychologists has been extensively reported elsewhere (Golledge *et al.*, 1991, 1998; Loomis *et al.*, 1998). Therefore we will only give a brief overview. The PGS seeks to create a virtual auditory layer that is draped across the real world to augment interaction with that world. The PGS comprises three modules: a locator unit (GPS), a detailed spatial database (GIS), containing an algorithm for path selection, and a user-interface (VR). The interface of the PGS is a virtual auditory display (Loomis *et al.*, 1990). Here, the labels of the objects within 'real space' such as 'tree', 'path', 'library' and so on are spoken through stereo headphones and appear as virtual sounds at their correct (real-world) locations within the auditory space of the traveller. As such, objects appear to 'announce' themselves with the sound emanating from the geographic location of the landmark. In this system, the PGS circumnavigates some of the more troublesome aspects of spatial language in conveying spatial relations by adopting a system whereby virtual space is overlain on real space. The PGS has evolved into a fully-functional system, adapted so a naïve untrained novice can use the system, and is an example of a VR-based 'naïve' GIS

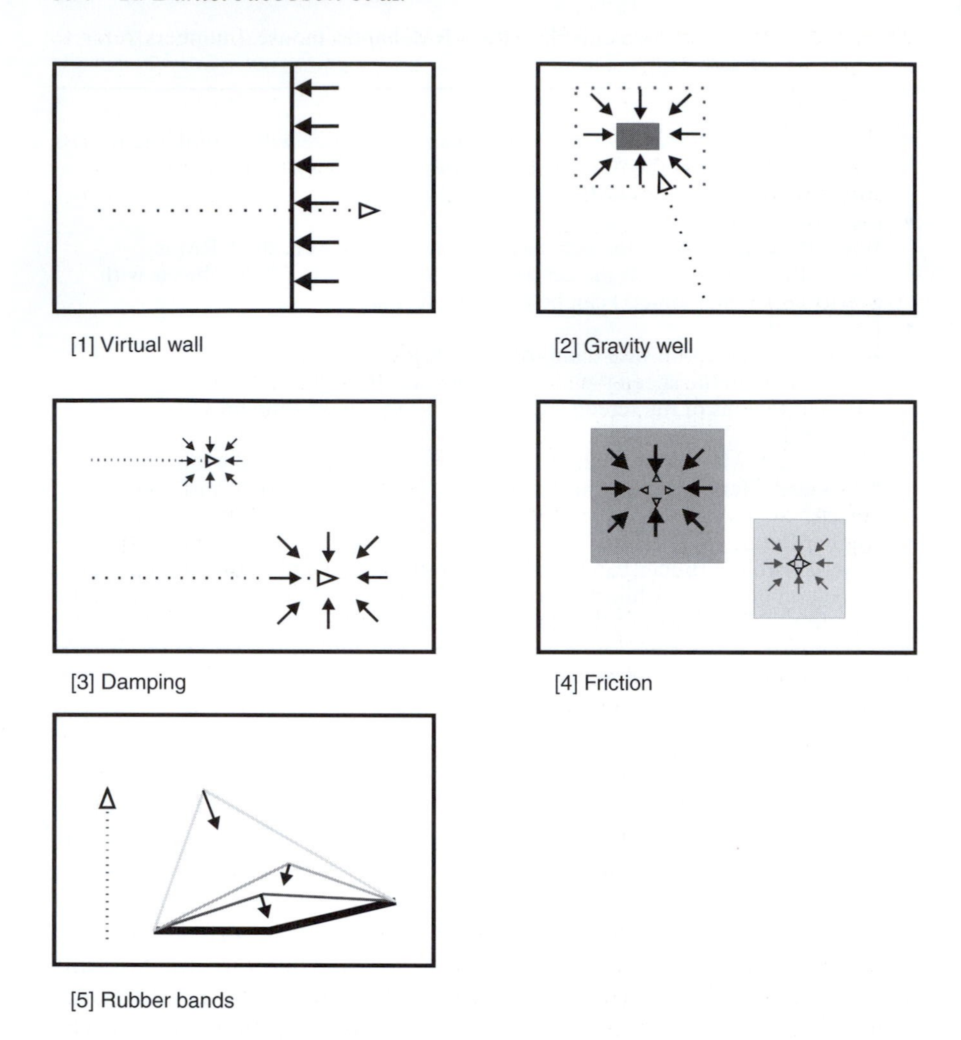

Figure 25.2 Diagrammatic illustration of haptic effects.

Note
Numbers refer to description in Table 25.1.
Hollow arrows represent the direction of mouse movement.
Black arrows indicate direction of force applied.

(Egenhofer and Mark, 1995). The problem of deciding which information to present to a blind user has been overcome by using common GIS techniques: hierarchies, order, buffering and corridoring. For example, a buffer of a predetermined size is created around the user. Any features which fall within the buffer, have been allocated weights that transcend a chosen value, and 'call' the user as if sited in their real location. This feature allows visually-impaired users access to the macro environment normally experienced by vision. Features can be given a salient value within the database, so those which pose the greatest danger, or are of particular interest, are highlighted first. In addition to buffering, whole routes can be corridored. If the traveller veers from their desired route by leaving the corridor, an error is signalled and directions for their return provided. At present, the user interacts with the system by using a small keypad. In the future it is hoped that interaction will be speech controlled.

Other studies

It should be noted that other teams are exploring the use of VR for communication, education, training and rehabilitation in relation to disabled people. Many of these approaches are multi-modal and some non-visual, demonstrating the salience of presenting information non-visually. Hardwick *et al.* (1996) used a force feedback device to present the structure of the World Wide Web to blind people through a haptic interface. Data was presented through a Virtual Reality Modelling Language (VRML), to build tangible surfaces and, by combining objects into scenes, primitive haptic shapes were constructed. Porter and Treviranus (1998) have used VRML and a haptic interface to provide haptic sensing of the relief 'maps' of the continental US for children and blind individuals. Work by Fritz *et al.* (1996) was successful in presenting scientific data through 'non-visual data rendering' (auditory and haptic interfaces) to blind people. Kurze (1997) has rendered drawings for interactive haptic perception by blind individuals who were able to obtain more accurate representation of spatial relations via the haptic method than conventional tactile maps tagged with audio. Bowman (1997) used joysticks and other physical interaction devices to provide muscle re-education for stroke patients. Virtual reality technology can allow communication between non-speaking deaf people and non-signing hearing people. Speech is signed while wearing a data glove, and this is translated to synthesised speech for the hearing individual (Kalawsky, 1993). Other vision-based systems have included training cognitively-impaired students to travel independently (Mowafy and Pollack, 1995) and using virtual replications of science laboratories to allow students with physical disabilities to carry out experiments (Nemire, 1995). Wilson *et al.* (1997) provide an overview of VR in disability and rehabilitation.

Future research

There is a need for future research to address the further development and use of new interface technologies such as voice recognition, touch screens and tactile display, force feedback devices and gestural interaction. Probably the most pressing need is to improve the user-interface, as this is the largest barrier to successful and meaningful interactions with representations of spatial information in a virtual environment. There have been several novel and interesting approaches that require further investigation. A vibro-tactile mouse which registers the mouse's position over a desired spatial object on a map (Nissen, 1997), tonal interfaces for computer interaction (Alty, 1996), and 'The Voice' which can convert a two-dimensional picture, map or representation into a 'tonal soundscape' (Meijer, 1992). Further research is also needed on sonification issues in VR, haptic interaction, and their combination.

One particular problem that is frequently overlooked is that of scale, not just the mapping from one physical scale to another representational scale (i.e. from the real world to a virtual environment) but intra-virtual geography scale changes (i.e. the equivalent to a zoom in a digital cartographic database or geographic information system). This scale transformation is one that is particularly difficult to appreciate without vision. However, by seeking techniques to present this information non-visually, it offers the potential methods to present this information in situations where it is critically important, such as where visual attention may be diverted (e.g. teleoperation, medicine, driving automobiles, flying planes and in military situations).

Three preliminary means of communicating scale without vision seem worthy of further investigation. Natural language could be used to explicitly or implicitly state a scale, such as 'the map is 1 mile across' or 'the map is now a neighbourhood/town/city/region size'. With the addition of a haptic mouse, which generates force feedback from designated regions on the screen, giving rise to such sensations as 'gravity wells', 'rubberbanding', 'stickiness' and impressions of relief such as bumps and hollows, it would be possible to generate a haptic scale referent. Options may include a haptic rendering of a visual scale bar, or equating resistance to distance. The third option is the presentation of relative map distance, and hence to imply scale through the auditory domain, such as using sound decay to equate with map distance or a pulse of sound travelling at a set speed across the map. The applied reasons for presenting information in modalities other than vision are compelling, namely in situations of a visual disability, where data is visually occluded, and for the augmentation of visual displays. Multi-modal information offers another independent channel for providing information to the user.

Conclusion

In this chapter we have outlined the need for multi-modal VR systems. VR systems provide qualitatively different forms of spatial representation from traditional media such as maps. However, they remain predominantly visual in nature. A move to a multi-modal configuration opens up VR systems to people with severe visual impairment, providing media that could improve quality of life, and qualitatively improves human–computer interaction by augmenting visual presentation. The two case studies illustrated that multi-modal VR systems can be successfully developed, that they have a number of advantages over other media, and that they can be used successfully by visually-impaired people to both learn spatial relations and to navigate through a geographic environment. These systems are, however, prototypes and more experimental work on both their development and use needs to be undertaken.

Acknowledgements

The authors wish to express their gratitude to the participants who took part in the pilot study. Funding for on-going research is supported by the University of California at Santa Barbara, Research Across the Disciplines Program. Partial support was provided by the National Center for Geographic Information Analysis, Project Varenius.

References

Albers, M.C. 1994. The Varese System, hybrid auditory interfaces and satellite-ground control: using auditory icons and sonification in a complex, supervisory control system. In Kramer, G. and Smith, S. (eds) *Proceedings of the Second International Conference on Auditory Display*. Sante Fe, New Mexico: ICAD, pp. 3–13.

Albers, M.C. 1996. Auditory cues for browsing, surfing, and navigating the WWW: the audible web. In Frysinger, S.P. and Kramer, G. (eds) *Proceedings of the Third International Conference on Auditory Display*. Palo Alto, California: ICAD, pp. 85–90.

Alty, J.L. 1996. Tonal interfaces for the blind and visually handicapped. Paper read at Conference on Technology for the Blind and Visually Impaired. Grassmere, UK: August.

Axen, U. and Choi, I. 1994. Using additive sound synthesis to analyze simple complexes. In Kramer, G. and Smith, S. (eds) *Proceedings of the Second International Conference on Auditory Display*. Sante Fe, New Mexico: ICAD, pp. 31–43.

Axen, U. and Choi, I. 1996. Investigating geometric data with sound. In Frysinger, S.P. and Kramer, G. (eds) *Proceedings of the Third International Conference on Auditory Display*. Palo Alto, California: ICAD, pp. 25–34.

Ballas, J.A. 1993. Common factors in the identification of an assortment of brief everyday sounds. *Journal of Experimental Psychology: Human Perception and Performance*, 19, 250–67.

Ballas, J. and Kieras, D.E. 1996. Computational modelling of multimodal I/O in simulated cockpits. In Frysinger, S.P. and Kramer, G. (eds) *Proceedings of the Third International Conference on Auditory Display*. Palo Alto, California: ICAD, pp. 135–6.

Begault, D.R. and Wenzel, E.M. 1996. A virtual audio guidance and alert system for commercial aircraft operations. In Frysinger, S.P. and Kramer, G. (eds) *Proceedings of the Third International Conference on Auditory Display*. Palo Alto, California: ICAD, pp. 117–21.

Blattner, M.M., Papp, A.L. and Gilnert, E.P. 1994. Sonic enhancement of two-dimensional graphic displays. In Kramer, G. (ed.) *Auditory Display: Sonifica-*

tion, Audification, and Auditory Interfaces. Reading, MA: Addison-Wesley, pp. 447–70.

Blenkhorn, P. and Evans, D.G. 1994. A system for reading and producing talking tactile maps and diagrams. Paper read at the 9th International Conference on Technology and Persons with Disabilities. Los Angeles, California: California State University, Northridge.

Bowman, T. 1997. VR meets physical therapy. *Communications of the ACM*, 40, 8, 59–60.

Bregman, A. 1994. Forward. In Kramer, G. (ed.) *Auditory Display: Sonification, Audification, and Auditory Interfaces*. Reading, MA: Addison-Wesley, pp. x–xi.

Brewster, S.A. and Pengelly, H. 1998. Visual impairment, virtual reality and visualization. In *Proceedings of the First International Workshop on Useability Evaluation for Virtual Environments*. Leicester, UK: British Computer Society, pp. 24–8.

Brewster, S.A., Wright, P.C. and Edwards, A.D.N. 1994. A detailed investigation into the effectiveness of earcons. In Kramer, G. (ed.) *Auditory Display: Sonification, Audification, and Auditory Interfaces*. Reading, MA: Addison-Wesley, pp. 471–98.

Colwell, C., Petrie, H., Kornbrot, D., Hardwick, A. and Furner, S. 1998. Haptic virtual reality for blind computer users. Paper read at ASSETS '98, at Monterey, CA.

Egenhofer, M.J. and Mark, D.M. 1995. Naive geography. *Technical Report 95–8. National Center for Geographic Information and Analysis (NCGIA)*. Santa Barbara, California: University of California at Santa Barbara.

Espinosa, M.A., Ungar, S., Ochaita, E., Blades, M. and Spencer, C. 1998. Comparing methods for introducing blind and visually impaired people to unfamiliar urban environments. *Journal of Environmental Psychology*, 18, 277–87.

Fanstone, J. 1995. Sound and touch: a campus GIS for the visually impaired. *GIS Europe*, April, 44–5.

Fisher, P. 1994. Hearing the reliability in classified remotely sensed images. *Cartography and Geographical Information Systems*, 21, 1, 31–6.

Flowers, J.H., Buhman, D.C. and Turnage, K.D. 1996. Data sonification from the desktop: should sound be a part of standard data analysis software? In Frysinger, S.P. and Kramer, G. (eds) *Proceedings of the Third International Conference on Auditory Display*. Palo Alto, California: ICAD, pp. 1–7.

Foulke, E. 1983. Spatial ability and the limitations of perceptual systems. In Pick, H.L. and Acredolo, A.J. (eds) *Spatial Orientation: Theory, Research and Application*. New York: Plenum Press, pp. 125–41.

Fritz, J.P., Way, T.P. and Barner, K.E. 1996. Haptic representation of scientific data for visually impaired or blind persons. Paper read at CSUN – Technology and Persons with Disabilities Conference '96, at Los Angeles, CA.

Gaver, W.W. 1994. Using and creating auditory icons. In Kramer, G. (ed.) *Auditory Display: Sonification, Audification, and Auditory Interfaces*. Reading, MA: Addison-Wesley, pp. 417–46.

Golledge, R.G., Klatzky, R.L., Loomis, J.M., Speigle, J. and Tietz, J. 1998. A Geographical Information System for a GPS-based personal guidance system. *International Journal of Geographical Information Science*, 12, 727–49.

Golledge, R.G., Loomis, J.M., Klatzky, R.L., Flury, A. and Yang, X.-L. 1991. Designing a personal guidance system to aid navigation without sight: progress on the GIS component. *International Journal of Geographical Information Systems*, 5, 373–96.

Hannaford, B. and Venema, S. 1995. Kinesthetic displays for remote and virtual environments. In Furness, T.A. and Woodrow Barfield, E. (eds) *Virtual Environments and Advanced Interface Design*. New York: Oxford University Press, pp. 415–70.

Hardwick, A., Furner, S. and Rush, J. 1996. Tactile access for blind people to virtual reality on the World Wide Web. *Proceedings IEE Colloquium on Developments in Tactile Displays (Digest No.1997/012)*. London, UK: IEEE, pp. 91–3.

Hinton, R.A.L. 1993. Tactile and audio–tactile images as vehicles for learning. *Non-visual Human Computer Interaction*, 228, 169–79.

Jacobson, R.D. 1996. Auditory beacons in environment and model – an orientation and mobility development tool for visually impaired people. *Swansea Geographer*, 33, 49–66.

Jacobson, R.D. 1998. Navigating maps with little or no sight: a novel audio–tactile approach. In *Proceedings, Content, Visualization and Intermedia Representation*. Montreal, Canada: ACL, August 15.

Jacobson, R.D. 1999. Geographic visualization with little or no sight: an interactive GIS for visually impaired people. Paper read at 95th Annual Conference of the Association of American Geographers, March 23–7, at Honolulu, Hawaii.

Jacobson, R.D. and Kitchin, R.M. 1997. Geographical information systems and people with visual impairments or blindness: exploring the potential for education, orientation and navigation. *Transactions in Geographical Information Systems*, 2, 4, 315–32.

Jacobson, R.D., Kitchin, R.M., Garling, T., Golledge, R.G. and Blades, M. 1998. Learning a complex urban route without sight: comparing naturalistic versus laboratory measures. Paper read at Mind III: The Annual Conference of the Cognitive Science Society of Ireland, at Dublin, Ireland.

Kalawsky, R.S. 1993. *The Science of Virtual Reality and Virtual Environments: A Technical, Scientific and Engineering Reference on Virtual Environments*. Reading, MA: Addison-Wesley.

Kitchin, R.M. and Jacobson, R.D. 1997. Techniques to collect and analyze the cognitive map knowledge of persons with visual impairment or blindness: issues of validity. *Journal of Visual Impairment & Blindness* 91, 4, 360–76.

Kramer, G. 1994. An introduction to auditory display. In Kramer, G. (ed.) *Auditory Display: Sonification, Audification, and Auditory Interfaces*. Reading, MA: Addison-Wesley, pp. 1–75.

Krygier, J.B. 1994. Sound and geographic visualization. In MacEachren, A.M. and Fraser-Taylor, D.R. (eds) *Visualization in Modern Cartography*. Oxford: Pergamon, pp. 149–66.

Kurze, M. 1997. Rendering drawings for interactive haptic perception. In Ware, C. and Dixon, O. (eds) *Proceedings CHI '97*. Electronic publication accessed from http://www.acm.org/sigchi/chi97/.

Loomis, J.M., Golledge, R.G. and Klatzky, R.L. 1998. Navigation system for the blind: auditory display modes and guidance. *Presence-Teleoperators and Virtual Environments*, 7, 2, 193–203.

Loomis, J.M., Hebert, C. and Cicinelli, J.G. 1990. Active localization of virtual sounds. *Journal of the Acoustical Society of America*, 88, 4, 1757–64.

Loomis, J.M. and Lederman, S.J. 1986. *Tactual Perception. Handbook of Perception and Human Performance.*

Loomis, J.M. and Soule, J.I. 1996. Virtual acoustic displays for real and synthetic environments. In *Society for Information Display International Symposium Digest of Technical Papers, Volume XXVII*, San Diego, California, May 12–17, 1996. Santa Ana, California: Society for Information Display, 43, 2, pp. 965–8.

Meijer, P.B.L. 1992. An experimental system for auditory image representation. *IEEE Transactions on Biomedical Engineering*, 39, 2, 112–21.

Metois, E. and Back, M. 1996. BROWeb: an interactive collaborative auditory environment on the World Wide Web. In Frysinger, S.P. and Kramer, G. (eds) *Proceedings of the Third International Conference on Auditory Display*. Palo Alto, California: ICAD, pp. 105–10.

Milleret-Raffort, F. 1995. Some cognitive and technical aspects of hypermaps. In Nyerges, T.L., Mark, D.M., Laurini, R. and Egenhofer, M.J. (eds) *Cognitive Aspects of Human–Computer Interaction for Geographic Information Systems.* Dordrecht: Kluwer Academic, pp. 197–212.

Mowafy, L. and Pollack, J. 1995. Train to travel. *Ability (Journal of the British Computer Society Disability Group)*, 15, 18–20.

Nemire, K. 1995. Virtual environment science laboratory for students with physical disabilities. *Ability (Journal of the British Computer Society Disability Group)*, 15, 22–3.

Nissen, J. 1997. Pers. Comm.

Parkes, D. 1988. NOMAD – an audio–tactile tool for the acquisition, use and management of spatially distributed information by partially sighted and blind people. In *Proceedings of the 2nd International Conference on Maps and Graphics for Visually Disabled People*. Nottingham, UK: Nottingham University Press, pp. 24–9.

Petrie, H., Morley, S., McNally, P., Graziani, P. and Emiliani, P.L. 1996. Access to hypermedia systems for blind students. In Burger, D. (ed.) *New Technologies in the Education of the Visually Handicapped.* Montrogue: John Libbey, pp. 210–36.

Porter, L. and Treviranus, J. 1998. Haptic to virtual worlds. Paper read at CSUN 98 – Conference on Technology and Persons with Disabilities, at Los Angeles.

Portigal, S. and Carey, T. 1994. Auralization of document structure. In Kramer, G. and Smith, S. (eds) *Proceedings of the Second International Conference on Auditory Display*. Sante Fe, New Mexico: ICAD, pp. 45–54.

Tatham, A.F. 1991. The design of tactile maps: theoretical and practical considerations. In *Proceedings of the International Cartographic Association: Mapping the Nations.* London, UK: ICA, pp. 157–66.

Thrower, N.J.W. 1961. Animated cartography in the United States. *International Yearbook of Cartography*, 1, 20–30.

Ungar, S.J. 1994. *The Ability of Young Visually Impaired Children To Use Tactile Maps.* University of Sheffield: Unpublished Ph.D.

Wickens, C.D. and Baker, P. 1995. Cognitive issues in virtual reality. In Furness, T.A. and Woodrow Barfield, E. (eds) *Virtual Environments and Advanced Interface Design.* New York: Oxford University Press, pp. 514–41.

Wilson, P.N., Foreman, N. and Stanton, D. 1997. Virtual reality, disability and rehabilitation. *Disability and Rehabilitation*, 19, 6, 213–20.

Index